MANAGING
TECHNOLOGY
IN THE
DECENTRALIZED FIRM

WILEY SERIES IN ENGINEERING & TECHNOLOGY MANAGEMENT

Series Editor: Dundar F. Kocaoglu

MANAGING TECHNOLOGY IN THE DECENTRALIZED FIRM

ALBERT H. RUBENSTEIN
Walter P. Murphy Professor
Northwestern University
and
President of International Applied Science and Technology
 Associates, Inc.

WILEY

A WILEY-INTERSCIENCE PUBLICATION
JOHN WILEY & SONS
NEW YORK   CHICHESTER   BRISBANE   TORONTO   SINGAPORE

Library of Congress Cataloging in Publication Data:
Rubenstein, Albert Harold, 1923–
  Managing technology in the decentralized firm.

  "A Wiley-Interscience publication."
  Bibliography: p.
  Includes index.
  1. Technological innovations—Management.
2. Technology—Management.   3. Research, Industrial—
Management.   4. Decentralization in management.
I. Title.

HD45.R77   1988   658.5'14   88-20875
ISBN: 0-471-61024-0

To My Wife, Hildette

CONTENTS

PREFACE

A lot has happened in the art of managing technology in the firm since the initial research for this book was undertaken and since I began consulting with several dozen large, medium, and small companies on improving the management of their technology. During those three decades many changes have occurred: in the organizational structuring of U.S. (and other) corporations; in the structure of markets and even the definition of many traditional markets; in the knowledge base in most fields of science and technology and the ways of adding to and taking from those fields; in the number and skills of university graduates seeking employment in industry; in the role of governments at all levels in providing incentives and restrictions on technological development; in instrumentation, facilities, and equipment; and in the patterns of funding for technology projects and whole programs.

The managers of technology at all levels, from corporate top management to project managements, have had to learn to cope with such changes in the firm's environment and to many other changes within the firm itself.

The art of managing technology has changed, as a consequence, but not always in the direction of improving the effectiveness of the R&D/innovation process and not always to the long-term benefit of the firm and to the national economy. Unlike progress in many fields of science and technology, progress in *management* of science and technology (R&D/innovation) does not occur in steadily positive increments. Some technology programs in some firms do not appear to be as well managed today as they were years or decades ago, and some are managed independent of the many lessons learned by leading companies over the past three decades.

All this suggests that we are still far from a testable, cumulative, comprehensive "science" of managing technology (or most other management activities). Effective management is still heavily dependent on the skills and insights of individual managers and the shared skills and experience of the managers in a particular firm. Where successful trial and error, good decisions, and good luck have been combined, technology programs in that firm may be highly effective.

In view of this state of affairs, the most that this book can accomplish is to assist those managers who already understand the process of R&D/innovation to improve that understanding and increase the effectiveness of their technology programs by trying out some of the ideas presented herein. For those managers having little experience with R&D/innovation specifically and technology in general, this book may help provide a framework for viewing this complex set of activities in the firm and for making decisions and formulating policies that will contribute to the effectiveness of their firms' technology programs.

Except for the first chapter, which provides an introduction, some cases, and a preview of some of the major trends in technology management, the format for each chapter includes (1) some questions and issues; (2) discussion of these questions and issues in detail, based on my consulting experience with technology management in many companies and the research of my colleagues and myself on this subject through the Program of Research on the Management of Research, Development, and Innovation (POMRAD) at M.I.T. and Northwestern University over a period of more than 30 years; (3) implications for managers of technology and the managers who manage *them*; and (4) a watch list for monitoring technology in the firm to see where problems are occurring and where improvements may be made.

Rather than organizing the book the way we do our academic research, in a scientific format,[*] I have organized it around major questions, ideas, and issues that managers involved in and with technology have indicated are important to improvements in the art of managing it. The chapter and section headings are addressed to what I believe are "real" and significant questions in the minds of practicing managers.

The book is directed at the problems of organizing and managing *technology*[†] in decentralized companies. However, I expect that a great deal will spin off from it of relevance to the general problems of designing and operating decentralized organizations. In this sense, the book is intended as a companion study to several previous works on particular functions in decentralized organizations.[‡]

In addition, early work on this book stimulated us to undertake a major study of staff relations in large companies, most of them decentralized. This is the study we refer to as the "birth and death" or "life cycle" of OR/MS (short for Operations Research/Management Science) Groups. Several other research programs related to the management of technology have also been generated in the course of doing the research for and writing this book. They are referred to in relevant chapters.

---

[*] Some sections in Chapter 6 on sources of ideas and technology and Chapter 8 on commercialization are structured around our research format.

[†] See my definition of Technology in the first section of Chapter One.

[‡] See Appendix B for references to such studies.

SOME COMMENTS ABOUT EVIDENCE AND CREDIBILITY
OF THE MATERIAL IN THIS BOOK

Reporting on a stretched out, meandering program of theoretical–empirical aca-demic research on a subject of high immediate interest to practitioners of the man-agement art presented a major problem. Portions of the results from the research study, which started in 1955 and is still continuing (since the phenomenon contin-ues to evolve), are laced with statistics, tests of hypotheses, and rigorously defined propositions and variables. That material is included in Ph.D. dissertations, M.S. theses, and journal articles and reports.* Much more of the data, however, are in the form of anecdotes, case histories, fragmentary observations, and reports of events and circumstances that do not lend themselves to rigorous treatment. The reader will encounter, and probably be annoyed by, the constant use of terms such as "in some of the sample companies," "many of the division managers," "fre-quently the corporate labs," and so on. A student of organization or technology management may want to see the sample sizes on which we base these vague "de-scriptive statistics." We would have liked to be able to provide them. However, most of the raw data from the study are not in clean enough form to provide such quantitative evidence. One reason is that the organizations studied and the people interviewed were seldom able to devote enough time to the study (although some sessions lasted 3–4 hours) to give clean, unambiguous answers on all of the liter-ally hundreds of issues in the study. In addition, when it was not a matter of time, such as in some of our in-depth, "real-time" studies (where we spent 1–2 years in and out of a number of companies) or studies where we spent the equivalent of several man-days or man-weeks in a score of companies, stretched over several years, much of the data were still very "soft" and qualitative. In particular, there are many variables in the study where the ultimate classification of the company or the situation had to be made on an arbitrary basis, weakly supported by a variety of evidence. For example, such issues included the degree of centralization of orga-nizations and technology activities, the relative status of the corporate technology managers, the success of a particular lab or set of labs, and the collective attitudes of division managers toward R&D.

But perhaps the dominant factor in some of the assertions and generalizations in the book is that I have been practicing, studying, and soaking in the technology management environment for over three decades and have gathered many impres-sions and formulated many implicit propositions and generalizations that were not even obvious to me until the writing of this book actually started. Thus, when I describe an event or situation, such as what happens when a division manager diverts his R&D people to work on a new crash product (in the section on the division manager—Chapter 3) or the problems in getting a new laboratory started up (in the section on organization of technology—Chapter 2), the reader can be sure that it is based on systematic observation of actual cases; that is, it is not hypothetical. The problem is this: How general or representative is the event/situ-

---

* See Appendix B.

ation and how many cases does it actually represent in industry? Faced with the self-imposed dilemma of responding to that anticipated question, I recognize that this use of generalized instances and case studies represents a risk to the credibility of conclusions or recommendations in this book, where statistical or other direct evidence is not presented to back them up. I decided then to take the risk and to rest on a combination of modes of supporting evidence such as "face validity" of specific statements, the general credibility of our[*] descriptions and interpretations, and the reader's sense of critical judgment about how far he can or should generalize or particularize to his own organization from *our* generalizations and particular cases.

A SENSE OF FRUSTRATION

One of the side effects of performing research and consulting in the field of technology management and writing up the results and their implications is a deep sense of frustration at the slow pace of learning by organizations in the area of organizational design. In the more than three decades during which we have systematically studied, casually observed, and participated in the field of technology management, there appears to have been very limited advance in the state of the art in organizing for technology. The same questions about organization are being asked in the late 1980s that were asked in the early 1950s in the same ways and, in some cases, by the same people! In addition, many of the same attempts at resolution of key issues are being made today and are leading to the same failures.

This is not to say that managers have not learned a lot about the organization and operation of technology in the past 30 years. Certainly, methods of accounting for expenses, methods of budgeting and project selection, methods of project planning and control have advanced in general. Many of these advances have come from the example of military R&D in the era of systems analysis, program budgeting, and cost effectiveness. However, the effects of these management advances on the organization and control of technology on the primarily commercial companies in our sample are hard to detect. Particularly with respect to major policy and organizational design questions such as internal charging for technical services and the microstructure of technical groups and functions, there seems to have been very limited cumulative learning and improvement.

One of our many "longitudinal" studies involved periodic discussions with the CTO[†] of a diversified medium-to-high technology company who later became president of the company. He reported, in the mid-1960s (and *his* middle sixties), that he had discovered "the" solution for funding and controlling corporate R&D, by a method of charging direct fees to the divisions for all work done by the Corporate Research Lab (CRL). Unfortunately, for definitive improvement of

---

[*]My colleagues in our research group (POMRAD) and my consulting firm (IASTA, Inc.).
[†] Chief Technical Officer.

the art of managing R&D, this "solution" had been tried at one time or another by many of our sample companies and, ultimately, rejected by the majority of them as leading to less, rather than more, effective corporate R&D.

Our conclusion is that learning about organization design of technology and methods of operating it effectively is not yet systematized enough to lead to a cumulative, credible body of knowledge. This is in spite of the already voluminous and still increasing number of seminars, conferences, and workshops in which technology managers swap experience and presumably learn from each other's mistakes.

I deeply believe that technology, of all the functional areas in the firm, can and should develop a more scientific and rigorous approach to resolving its organizational problems. There is currently enough literature on technology and R&D management representing the results of empirical and theoretical research and practice to provide a reasonable base for improving the art of technology management. The minimum requirement, however, is for an internal capability to take in this information, evaluate it in the context of the specific organization's circumstances, and to adapt it to the task of improving organizational design for technology.

The current tendency is to attribute most problems and failures to personality problems and to individual weaknesses, rather than to a combination of these and general organization phenomena which are shared by many organizations. Certainly, the uniqueness of individual organizations is closer to the truth than the paper stereotypes of R&D organizations that one finds in much of the literature. But perhaps the "useful" truth is somewhere in-between: individual differences and uniqueness operate on a base of patterned behavior to produce what appears to be overall uniqueness. In our work in the field, we have been struck by the overwhelming convergence and similarity of responses, rather than their wide variations. We have frequently been disappointed with the very low level of searching for alternatives to organizational problems and the very narrow range of alternatives that are considered and put into operation.

Of course, merely knowing what the "correct" or a "better" solution is does not necessarily lead to improved operations, let alone to "optimal" solutions. Perhaps the behavior of management specialists in the university struggling through problems of organizing themselves for curriculum design are as sufficient evidence as any that there is a long road and a lot of work between knowing or suspecting a good solution to an organization problem and putting it into practice effectively.

It would be very satisfying to my colleagues and me if we could claim that our work on technology management in decentralized companies over the past 30 years has led to a set of rigorous "principles" for the organization and conduct of such activities. Unfortunately, as the discussion of the various topics in the book will readily indicate, the situation is too complex and subject to too many individual differences among organizations to produce such a result. In the search for underlying regularities and the tentative generalizations that might be credibly derived from them, we *have* been able to come up with a set of less rigorous results that may be of use to practicing managers involved with technology. These results are presented in two forms in most of the chapters: (1) *implications for management*, which can serve as guides in organizing technology activities and reorganizing those that are either not meeting performance or output expectations, or are to be given changed objectives and missions; and (2) *watch lists,* which can help management monitor the behavior and performance of the technology network in the firm.

SOME CONCLUSIONS

Rather than keep the reader in suspense all through the book (assuming he or she will go through it systematically from front to back), I would like to state some conclusions at the outset. None of them should be surprising to the experienced manager of technology or the general manager, within whose organization he has a technology activity. The principal conclusion is that no general, definitive "solution" to the problem of how to organize technology in a decentralized company emerged from our study and other involvement with over 200 of the largest decentralized companies in the United States, but that most such firms are still seeking this elusive solution. The underlying reason for this lack of success is discussed in some detail in succeeding parts of the book such as those on "organizational form" and "organizational change." It can be summarized briefly here. Due to the complexity of life in a large industrial organization, the influences of individual personalities, the pressures for change from both within and outside the organization, the "organizational aging" process, and many other factors, any organizational form for technology encounters pressures toward change. If it is too tightly controlled (centralized), forces are set in motion which attempt to loosen up the constraints on it. If it is too loosely controlled, forces are set in motion which constrain it toward more centralization. Some organizations appear to be somewhat stable over relatively long periods, apparently resisting the external and internal pressures to move too far or too fast in either direction along the centralization · · · decentralization continuum. When change does come to this kind of organization, it is likely to be explosive or at least highly disruptive. This appears to be the result of holding in check the various forces for change which might have permitted the organization to adapt smoothly to changed environmental or internal pressures. Examples of *apparent* stability are some cases of corporate research laboratories (CRLs), which led or dominated R&D in their companies for many years, despite a buildup of dissatisfaction among its clients. Inevitably, a major opportunity for explosive change occurred, such as a new management group taking over, a merger or major acquisition, a spectacular failure that could be attributed to R&D, or the breaching of an action threshold of dissatisfaction. Of the increasing number of dissolutions of a CRL we have encountered, most of them occurred under such traumatic circumstances, rather than as a result of systematic, planned organizational design. I do not mean to imply that the corporate research laboratories in most of our large decentralized companies are in danger of being dissolved as soon as the opportunity is right for their critics and enemies to strike. However, I do conclude that those CRLs that resist change and ignore frantic signaling from the environment that something is wrong with the way they are organized or operating are open to such unplanned, sometimes quite "irrational" reorganizations and redirections.

Another, nonstartling conclusion is that the post-WWII period of grace for corporate R&D is long over. Again, it is not in mortal danger in some leading firms, since it has become indispensable to the modern well-managed corporation. But the period of laissez-faire and watch-in-wonder is over in most firms. There are a number of pieces of evidence for this, including the first period of widespread reduced R&D spending in post-war history which occurred in the late 1960s, the appointment of an increasing number of non-R&D executives to the top technology management positions, and the trend toward more detailed and critical scrutiny of all technology proposals and budgets. Each of these trends or situations is discussed in the book, with attempts at explanations and, where appropriate, some suggested remedies for the underlying causes of the trend or situation. I have tried, where feasible, to report significant regularities or anomalies we have encountered, which may be instructive to the practitioner and/or student of technology management.

ALBERT H. RUBENSTEIN

Evanston, Illinois
December 1988

ACKNOWLEDGMENTS

Selecting those of my associates—colleagues, students, and industry advisors—whose research, ideas, and stimulation contributed most to sections of this book was very difficult. Many of the research studies that are summarized or referred to under various headings were conducted by teams of faculty, graduate students, and consultants—some teams as large as a dozen people on a given project (Idea Flow, Decentralization, R&D/Production Interface). Some of the consulting engagements referred to were individual assignments by myself, while others involved teams of two, three, or half a dozen. Since some of my associates were involved in more than one project, as well as in discussions of others in which they were not formally engaged, my original plan of providing a matrix of "people versus projects" for acknowledgment seemed awkward when I began constructing it. Therefore, I decided to extend my appreciation for their help, without identifying them with specific projects, in this section of the book. In the appendixes, however, publications, reports, dissertations, and working papers of POMRAD and IASTA[*] are listed along with citations to the open literature. Individual names are associated with written outputs—intellectual property—connected with the many projects referred to throughout the book. Here goes. My thanks to Robert W. Avery, Norman R. Baker, Frank Baker, Richard T. Barth, Alden S. Bean, Michael S. Bergman, Victor Berlin, David M. Birr, Frank Bolen, John W. Bonge, Dawson Brewer, Susie Y. Brown, Charlene Brown, Bruce S. Buchowicz, Alok K. Chakrabarti, John Chambers, Donald W. Collier, Donald B. Cotton, William A. Davig, Carlos Davila, Burton Dean, Charles F. Douds, Bruno Drews, Israel Dror, William Emerald, John

[*]Program of Research on the Management of Research, Development and Innovation (POMRAD) and International Applied Science and Technology Associates (IASTA).

E. Ettlie, Bruce Fischer, R. Patrick Forster, Warren Fredericks, James E. Freeman, Donald N. Frey, Judith Fuss, Jean Gagen, Eliezer Geisler, Horst Geschka, Martin E. Ginn, Milton A. Glaser, Edward M. Glass, Bela Gold, Joel Goldhar, Barbara Grabowski, Sue Growney, Chadwick Haberstroh, Walter A. Hahn, Richard Hannenberg, David Heiman, David B. Hertz, William A. Hetzner, Stephen C. Hill, Gerald Hoffman, Charles Hollocker, Ira Horowitz, Cynthia Huffman, Ronald Hughes, Arthur P. Hurter, Jamal Husain, Allen D. Jedlicka, Paul Jervis, Ronald O. Johnston, Isabel Juan, Nickolas S. Kaskovitch, Abe Katz, Takeshi Kawase, George Kazan, Daniel L. Kegan, John A. Kernaghan, Kenneth Kirsch, Connie Knapp, Lynda Kuykendall, Peter Lai, Robert E. Large, Jon Larson, C.C. Lee, Jinjoo Lee, Howard R. Lipson, P. Michael Maher, Jacques Marcovitch, Robert B. Martin, John B. McColly, John P. Miller, William C. Moor, Evelyn A. Moore, Horst Morgenbrod, Robert D. O'Keefe, Enrique Ogliastri, Alan W. Pearson, James Pelissier, Barbara Peters-Koehler, Ken Porrello, William H. Pound, Lakshmanan Prasad, Michael Radnor, Gustave J. Rath, Ed Rifkin, Robert Rifkin, Shlomo Rozen, Roberto Sbragia, Theodore W. Schlie, Hans Horst Schroeder, Heinz Schwaertzel, George Seeger, Falguni Sen, Herbert A. Shepard, C. Richard Shumway, Jack Siegman, Jaime Silva, Jon Soderstrom, Ruth Sohngen, William E. Souder, Raymond St. Paul, Edward Sullivan, Kirstin Synnestvedt, David A. Tansik, Charles W. N. Thompson, Richard W. Trueswell, James M. Utterback, Eduardo Vasconcellos, Atul Wad, David Watkins, David Werner, Ann B. Woelflein, Earl C. Young, H. Clifton Young, Gerald Zaltman, and Liang-Yi Zhou. In connection with the main theme of this book—managing technology in decentralized firms—I would like especially to acknowledge Elie Geisler, George Kazan, and Mike Radnor, with whom I have worked on a number of projects in this area, and David B. Hertz, who first introduced me to the field of R&D/technology management. Dundar Koacaglu, editor of the Engineering Management series for John Wiley & Sons, Burton Dean, and Wiley's editor Frank Cerra should also be mentioned. Susan Arden, my long-term assistant, did much to protect my time for the writing and proofing of the manuscript as did Kirsten Synnestvedt in previous years. Janet Goranson was the first person to persuade me to use word processing, although vicariously, for a whole manuscript and did a fine job of preparing it. Finally, to my wife, Hildette, without whose encouragement the manuscript might still be "on the shelf," thank you.

MANAGING TECHNOLOGY IN THE DECENTRALIZED FIRM

_____1
INTRODUCTION

SOME TERMINOLOGY USED IN THIS BOOK

Definitions and terminology differ widely in the literature of technology management. Rather than add to the confusion by a set of new terms and definitions, I have attempted to adapt some of the more common ones in current management usage. Here are some of the ones that are used frequently and their abbreviations or acronyms, as well as some other terms I use in the book.

Technology. This covers a wide range of activities and functions in the firm which are devoted to producing new and improved products and processes, materials, and know-how for the firm. All these activities are aimed at maintaining the firm's competitive position. Some of the activities and functions included in this broad area of "technology" are research, development, advanced engineering, product development or engineering, process design and engineering, market research (related to new products and markets), manufacturing engineering (also related to new process start-up), and other activities tied to getting new and improved products and processes into the plant and/or out into the market. Both hardware and software are included in the broad term technology, as well as manufacturing and service technology for products and processes.

R&D/Innovation. This is another term for the process whereby new and improved products, processes, materials, and services are developed and transferred to plant and/or market. Typically, this process is represented in the firm by a number of formally organized laboratories, departments, groups, teams, and functions. The most easily recognized are the "upstream" functions, which involve scientists and engineers in laboratories. Although the names of these laboratories vary widely

1

(research, research and development, product development, research and technology, science and technology, etc.) we refer to them generally as R&D labs.

Corporate Research Lab (CRL). Again, names for this corporate department or function vary widely. We use this term for all such activities, which are supported at the corporate level and whose assigned functions are described in Chapter 2.

Divisional Labs. These are laboratories, departments, functions, and so on maintained at the operating division level to look after its product lines, manufacturing processes, and general technology base.

Operating Divisions. These are semiautonomous units in the decentralized corporation which have responsibility for a product, product line, market, or area of technology and which operate on a "profit center" basis. There are some exceptions to this double definition in some decentralized companies. Some divisions do not have control over their own manufacturing (they may support a common manufacturing facility with one or more other divisions) and some do not control their own marketing. Typically, however, operating divisions, even in small decentralized companies, have their own production, marketing, and technology, in some form.

Chief Technical Officer (CTO). Since this book is about managing technology in the firm, the CTO will play a prominent role throughout. Again, terminology for this position varies. In earlier days, the top R&D person in the firm was generally called the Director or Vice President of R&D (VP R&D) or some permutation of the words science, technology, development, or engineering (in many production-oriented firms the top technology officer is still called the VP Engineering).

I/We. Since many of the observations in this book are based on my direct experience with technology in the firm and research on this subject over many years by myself and my colleagues, the terms I and we may appear more frequently than some readers prefer. However, since the field is still more heavily based on experience, insight, and judgment than scientific principles and facts, the use of such pronouns reminds the reader of the sources of the observations, conclusions, and advice contained herein.

He/She. The facts of life in the large corporation are that almost all the key positions referred to continually in this book—CEO, CTO, division manager, group executive—are occupied by men. There are exceptions, including some outstanding top management and top technology management women. However, with apologies, I have opted to avoid the sometimes awkward he/she terminology in favor of the generic he.

The Sample. The observations in this book are based on direct experience, research, consulting, and close observation of more than 250 decentralized firms over a period of 30 years. Most of them are large—among the Fortune 500. Others are small and medium size. A list of the firms in this "sample" is given in Appendix

A. Some of them were included in the formal research on R&D in decentralized companies conducted by my associates and myself at M.I.T. and Northwestern over a 15-year period. The others include consulting clients and firms where we have conducted training programs and research on other aspects of the overall R&D/innovation process. Many of the firms included in the original study have changed form, ownership, and focus during the past three decades. A few have even disappeared from the U.S. industrial scene. The patterns of organization and management and the issues they have faced in attempting to manage technology effectively seem to persist. That is, for example, observations made in the early 1960s about difficulties between corporate research lab (CRL) people and operating division people still appear across all sizes and kinds of firms in the late 1980s.

1.1 THE NOTION OF THE PROMINENT ISSUE OR PROBLEM OF TECHNOLOGY MANAGEMENT IN THE DECENTRALIZED FIRM

Each technology laboratory or activity in a decentralized company, or anywhere for that matter, will have to cope with a set of problems and situations that are common to many (or most) such activities, plus some that are (or appear to be) unique to that laboratory at that time. These "problems of technology management" have been catalogued in many places. They deal with problems of personnel, budgets, decisionmaking on projects and programs, scheduling and control of work, assessment of results, relations with other activities in the firm, fighting technological obsolescence, and so on. In our interviews and interactions with top managers and managers of technical activities we have been particularly alert to those specific problems that were salient at the time of our visits and that appeared to be closely related to the fact that the setting was a decentralized company. In reviewing the data from these interviews and the in-depth consulting and field studies as well as other, informal observations and experiences, I selected the specific issues discussed below which were prominent in the minds of the technology management and/or the general management of the company. Some of these issues have served as foci for extended discussion in later chapters. Initial statements of the issues in some brief case studies are given in this chapter, with more general discussion left to the other chapters.

The illustrations given below are certainly not exhaustive of all the issues we have encountered relative to organizing and conducting technology in decentralized companies. But they are representative of the issues on the minds of managers (technical and nontechnical) in a broad cross section of large- and medium-size decentralized companies.

1.1.1 A Small Chemical Company That Had Been Acquired by a Large Chemical Company

When acquired it operated as a fully integrated subsidiary with its own decentralized structure and labs. This company faced several decentralization-related issues, some of which had been gnawing at the management for several years and some of which had become salient only recently. A dominant issue was the problem of "competing" in technology with other laboratories in the parent company. There was little overt duplication of specific projects, but it was clear that some of the subdivisional laboratories of the subsidiary were in the same fields as existing laboratories of the parent company. The latter were, in some cases, much larger and better equipped. In addition, the subsidiary had a nascent "corporate research laboratory" (CRL), which was clearly not yet viable in terms of size and maturity. This was viewed as unnecessary overhead by some members of the parent corporation management. Although the corporation had not yet set up a CRL of its own at that time (it since has), it seemed clear that the subsidiary's small CRL was not the proper nucleus on which to build. This unstable situation had persisted for several years subsequent to the acquisition, and aspects of it were coming to a head at the time of our visits (we held several dozen interviews with all levels of general and technology management over a 1-year period). Compounding the situation were the facts that several of the subsidiary's divisional labs were clearly below critical size and that one, possibly two, of the subsidiary's divisions were in trouble in the marketplace. In some measure, the marketplace trouble could be partially attributed to shortcomings of their technology. A further set of compounding factors, quite common to similar situations, included (1) the geographical distance of the subsidiary and all its operations to the center of gravity of the corporation, where the major technology actions and decisionmaking for the corporation took place; and (2) the historical domination of all its technology programs by the founder of the company, who remained in control until several years after the acquisition.

Issues

1. Integrating the technology activities of acquired companies into the corporation.
2. Location of one or more CRLs in the corporation.
3. Competition, overlap, and lack of cooperation between labs of different divisions.
4. Minimum effective size of divisional labs and subdivisional labs.
5. Deployment of technology at the subdivisional level.

1.1.2 The Government-Oriented Half of a Merged Commercial–Government Contract Company

This major government contractor had been party to a large-scale merger with a predominantly commercial-oriented company a few years before. (*Note*: It might be fun for the reader to try to guess the identity of this or other companies described in the book. Good luck! Although I am reporting the situations as accurately as possible, I have carefully tried to omit any major clues that might give away the identity of the firms. The above general description of the firm in this illustration, for example, could apply to at least seven or eight companies in our sample. The purpose of these specific illustrations and the caselets in other sections of the book is to illustrate general situations and issues that were encountered and not to disclose the adventures or problems of particular firms. The major mission in developing these brief case studies was to find and attempt to explain patterns and regularities in organizational behavior which might lead to improved organizational practice and organizational design. As such, our interest in developing specific individual cases was limited to the possibility of drawing common threads from them.)

In this company, one of the objectives of the merger appeared to be a mutual strengthening of the R&D capabilities of the overall corporation. Although each half had a network of laboratories of substantial size, it was evident that each had developed major specialties that were essentially lacking in the other. At the time of our interviews (which lasted over a period of about 6 months), a major reorganization of the government-oriented half of the corporation was under way due to several factors. Among the major ones were large cutbacks in government spending in their areas of specialty, leading to the layoff of a large number of technical people; shifts in the state of the art away from some of their main historical fields of technology; and change in top management (cause or effect?). During this reorientation period, the question of diversification out of primarily government work was being considered and also the question of getting closer to and more integrated technically with its commerically oriented half. Although we talked to division and top general management during our visits, we concentrated on the technology managers and found great reluctance or lack of interest in getting closer technically to the "other half." Geographical distance was not a major issue, but major differences in outlook, habit patterns, specific interests, and perceived threats did appear to make closer integration an unlikely event. As far as we know, following the situation from a distance since our intensive field study, such integration has yet to occur, more than two decades after the merger.

Issues

1. Integrating the R&D activities of two equal parties to a merger.
2. Taking advantage of technical strengths and specialties in independently operated parts of the corporation.

1.1.3 A Medium-Size Manufacturer of Engineered Products Which Had Grown by Acquisition

This company had half a dozen divisions that were based on certain common technology, plus a relatively newly formed division that was in an entirely different field of technology. The managers of the older divisions were very powerful and continued to oppose the formation of a corporate laboratory for support of the new division's work, more long-term and more basic work in support of existing divisional product lines, and attempts at diversification from within. The manager of the new and "different" division was the major proponent of the establishment of a CRL, but his influence was very low, due to the relatively small contribution to profits of this division and due to one major product development debacle in which his division had engaged. Virtually none of the other division managers had any interest in research or even advanced development, since their product lines had traditionally been supported by fairly competent but routine engineering. They were cautious about any added costs to them that might accompany increased corporate spending on R&D, as well as skeptical of the potential benefits of more advanced R&D to their individual divisions. This situation remained in deadlock for many years, since the top manager of the corporation was on record as being "fully committed" to decentralized operations and control in the company. His central office staff included little more than a few lawyers and bookkeeping and accounting personnel.

Issues

1. The need for a corporate research lab.
2. Imbalance between the recognized technical needs of different divisions.

1.1.4   A Medium-Size Chemical Company Whose R&D Was Centralized

This company was highly marketing oriented, and marketing people dominated most operating decisions. It had established a modest-size CRL several years before our visits, which occurred over a period of about $1\frac{1}{2}$ years. In recent years there had been numerous complaints by division managers that the CRL was not helping them and was not paying its way. The CRL was staffed predominantly with advanced degree people. Since the operating divisions had no R&D of their own, except for some process control and engineering, those divisions that needed an advanced technology input had to rely on the CRL. Divisional managers complained that the CRL was not responsive enough and that there had been no major product developments that could lead to new product lines. Part of the problem, as in other companies with "pure" centralized R&D (see Chapter 2), was that there was no dependable or routine transition path from the CRL to the production or marketing functions in the divisions. That is, the Ph.D. chemists in CRL spoke a different language and operated at a different level of abstraction from the operating people in the divisions.

Issues

1. The feasibility of centralized R&D in the decentralized firm.
2. Lack of transition paths and mechanisms from the CRL to divisions.
3. Feelings by division managers that they were not being adequately supported in the R&D area.

1.1.5   A Medium-Size Maker of Engineered Products Which Had Not Grown Much Recently

Over a 5–6 year period, we followed the attempts by members of the planning and corporate technology staff to set up a small, long-range corporate research laboratory (CRL). The project appeared to have moderate support from top management, although there was some question of how effective it could really be in contributing to company growth and profit. The "champion" of the proposed CRL was a working scientist who had been serving in a consulting capacity to some of the operating divisions. He proposed a series of plans for a small CRL and seemed on the verge of pulling it off several times. However, he had no real leverage to use with the division managers, whose support was deemed essential for the venture to succeed. His boss, the corporate vice president for technology, served in strictly a staff advisory capacity to the divisions and did not have enough organizational or personal power to get the CRL launched. One problem that was evident to an outside observer was that the plans and proposed programs of the CRL were not phrased in language that would "grab" an operating division manager. There appeared to be little or nothing in the way of specific divisional benefits to be expected from a CRL for a long time after its initiation, if it *were* initiated.

Issues

1. Criteria for establishing a CRL.
2. Lack of support for attempts to establish a CRL.

1.1.6 A Very Large Diversified Company With Several Dozen Divisions

This company had, during the 5 years over which we conducted interviews, and during the preceding decade, gone through three or four top corporate technology managers and several reorganizations of its corporate research laboratory. Most of the divisions had originated by acquisition and retained a high degree of autonomy. Some of them had substantial engineering development laboratories and felt that they were self-sufficient in technology. Over the entire 15-year period, the major preoccupation of the top corporate technology person and his staff appeared to be how to get closer to the operating divisions. In some of the divisions, it was well known that CRL personnel were not welcome. In others, there might have been some interest, but there was no dovetailing group that could smoothly accomplish the transition of things developed in CRL into the divisions. Many techniques of establishing better relations were tried by the CRL, none of them being generally successful. Its relations with a few divisions were cordial and there was some cooperative work, but by and large the CRL existed for many years in extreme isolation from the operating divisions.

The continued existence of the CRL and its fairly substantial budget was assured by unwavering support of the top manager of the corporation. His enthusiasm was not shared, however, by the many group and divisional executives to whom we talked.[*]

Issues

1. Long-term absence of payoff from a CRL.
2. Lack of interest among operating divisions in using a CRL, despite top management pressure.

[*]The CRL was finally shut down during the proof reading of this book!

3. Inadequate mechanisms for transition from corporate research to divisional development.

1.1.7 A Two-Division Company Split Along Technology Lines

One division of this company had a historically dominant position in a major commercial field, backed up by a highly competent traditional engineering development department. The other half of the company was heavily engaged in research and advanced development at the state of the art in government contracting. The first division's market position was being threatened by rapid changes in the technical base, which required skills beyond the capabilities of the division's engineers. It was known in this division that the kind of talent and capability needed to meet the technological challenge in its field was available in abundant supply in the other, government-oriented division. Despite this, virtually no moves had been or were being made to take advantage of the technology capability in the second division. There was, in fact, virtually no contact between the two R&D groups. Since the two divisions were run quite independently of each other and there was only a token corporate staff, there was no initiative taken by corporate management to bridge the gap. This, together with the possibility of "looking bad" by comparison, kept the first division's technical people in isolation from a group that could have been of great help to them and to the corporation.

Issues

1. Inability or unwillingness of a division in technical difficulty to use the help of another division with the needed skills.
2. Extreme isolation over many years between the commercial and government-oriented halves of the company.
3. Lack of any corporate-level coordinating mechanism for R&D.

1.1.8 A Giant Corporation Formed by Merger of Two Very Large Companies in Technically Related Fields

This was one of the almost-a-dozen cases we encountered of mergers between two very large companies where each had a substantial network of laboratories. In this case, the process of assessing them, sorting them out, combining or integrating them had been going on for over a dozen years and was still not completed to everyone's satisfaction. Geographical dispersion of the laboratories was one problem; the existence of two major central laboratories (one in each half) was another. Overlap in fields of technology was a third factor and perhaps one of the most difficult to handle, when combined with the first two. For several years during our visits to the company, the questions of (1) whether there would be only one corporate central lab and (2) if so, where it would be located, remained matters of deep concern and anxiety for large numbers of the R&D people. Although our formal interviews were made in the period just before and just after the merger, we have continued to have close contact with the technology activities in both halves of the company until the present time. Every 2–3 years major moves are made which are intended to change the deployment of technology in the corporation and the situation still has not settled down.

Issues

1. Reorganization of technology after a major merger.
2. Frequent organizational change.
3. Overlap and duplication of activities in company laboratories.

1.1.9 A Multidivisional Company With a Technical Center

This manufacturer of consumer products had gathered most of its divisional laboratories into a technical center with the corporate research laboratory. As in most decentralized companies, the corporate vice president of R&D had line responsibility for the CRL, but only staff advisory relations with the divisional labs and technology programs. As a result of the physical proximity, communication and interpersonal relations were good between the CRL and the divisional labs, and some informal influence over divisional project portfolios was exercised by people in corporate R&D. However, on major questions of funding, programming, and project selection, the vice president of R&D had very little influence where it mattered—with the division general managers. In addition, some of the division managers were unhappy about the location of their laboratories in the technical center, which in most cases was geographically separate from their divisional manufacturing and other operations.

Issues

1. Influence of corporate vice president of R&D on divisional technology.
2. Location dilemma for divisional R&D—close to divisional operations or close to other laboratories.

1.1.10 A Company With a New Corporate Vice President of Technology

At the time of our interview, the new VP had only been with the company for about 1 year. His predecessor had been fired because of dissatisfaction with the performance of R&D. The new VP figured that he had less than 3 years to turn the situation around and show that corporate R&D could pay off for the company. The dilemma he faced was that he knew it would be difficult to set up projects or programs that were guaranteed to pay off within the 3-year time frame and still meet the expectations of the company and the corporate research staff that this was a true longish-term, basic-type CRL. He ultimately was also fired and the company has since had a succession of top technology managers.

Issues

1. Strategy for demonstrating effectiveness of a CRL.
2. High turnover of top corporate technology managers (cause or effect?).

1.1.11   A Company That Completely Centralized Its Technology

This was one of the few large companies we encountered over a 20-year period which abolished all its divisional laboratories and completely centralized R&D. Soon after this move was completed, there was evidence that some of the division managers were allowing or encouraging their remaining technical people in plant engineering, process control, and quality control to begin undertaking what appeared to be bona fide development projects. Part of this tendency was attributed by the people in corporate R&D to the feeling by some division managers that they were not being adequately serviced by the CRL and/or that they wanted closer control over activities that they were paying for (they were taxed for general support of the CRL as well as for sponsored projects in CRL) and that they considered vital for the health of their divisions.

Issues

1. Can "pure" centralized R&D work in a large diversified decentralized company?
2. How can a centralized R&D effort adequately serve a large number of clients or sponsors?

1.1.12   A Company With Several Dozen Divisions, One of Which Was Developing Its Own CRL

The company had firmly maintained a "combination" deployment of technology for many years (see Chapter 2), dividing work along the technology spectrum in the conventional way between the operating (product) divisions and the CRL. One of the operating divisions had, a few years before, embarked on a major new field of technology, leading most of industry. The size of technical effort and the numbers of technical people were of the same order of magnitude as the rest of the company's total R&D effort combined and began to dwarf the CRL. The division manager was running rapidly in the direction of establishing a fully self-sufficient technology activity within his division and was duplicating some of the specialties, equipment, and services of the CRL at a rapid rate. There was concern among corporate technology people that building such an organization based on a single project was unwise in the long run and would serve to make the overall company technology effort less effective.

Another division of the same corporation was geographically and technically isolated from the main center of corporate activities and from the CRL. They, in contrast to the first division, were trying to squeeze minor product improvements out of a tired technology and an almost obsolete product line. They had no contact at all with other technical groups in the company who were working in fields related to the division's, but at a much higher technical level. They neither sought nor received technical assistance from the CRL.

Issues

1. Imbalance in the technical level of divisions.
2. Failure to use the CRL in either advanced areas or traditional areas by some of the divisions.
3. Funding of multiple CRLs.

1.1.13 Top 300[*] Machinery Maker

Events and Conditions

1. Historically, the corporate research laboratory (CRL) has been relatively independent due to income from outside contracts.
2. Recently, corporate management had tried to get operating divisions to use and support the CRL.
3. The vice president of technology was put in the dual position of staff support group and auditor or monitor of divisional programs.
4. The CRL was caught in a backlash against corporate staffs by division managers.
5. The CRL staff and budget were cut.

Key Issues

1. Lip service to long-range R&D by top management, but no support when chips are down in conflict between corporate staff and divisions.
2. Who pays for longer-range research for the benefit of the operating divisions?
3. How can the CRL penetrate a brick wall of resistance to "outsiders" built by the operating divisions?
4. Frustration of CRL personnel in seeing mistakes and wastage of resources by divisions due to outdated or poor technology.
5. Lack of understanding of research and technology by top management personnel.

1.1.14 Top 100 Chemical Company

Events and Conditions

1. Historically, the company had been organized into half a dozen product divisions.

---

[*]In list of Fortune 500.

2. Corporate technology and the corporate research laboratory (CRL) have not been considered major contributors to growth and profits.
3. A new application-oriented chief technical officer (CTO) is appointed to get new products developed and into the market.
4. The divisional structure is being broken down further into subdivisional product groups with P&L responsibilities.
5. Divisional R&D labs are broken up and divided among new business groupings.
6. The CTO sets up a new ventures group for several new technology-based businesses. It takes a year or more for them to be organized, staffed, and started.
7. Four out of five new ventures do not succeed/survive.

Key Issues

1. Fragmentation of divisional technology makes interface with CRL difficult to impossible.
2. Managers of business groups shorten their time horizons and scope of interest.
3. No divisional receptivity for accepting front-end risk for new ventures.
4. New (nonchemical) technology ventures resisted by board and corporate management.

1.1.15 Top 50 Supplier to OEMs

Events and Conditions

1. Chairman unhappy about general innovation climate in company.
2. Major penetration of market by a competitor's new product line.
3. Poor coupling between CRL and divisional technology groups.
4. Domination of divisional technology by production.
5. Retirement of VP R&D and imminent retirement of next level in CRL.
6. Several key new products stalled between R&D and production.
7. Lack of ideas for new products that excite top management.
8. Sharp compartmentalization of divisional development groups.
9. Personnel and funding for new technology below minimum effective level.

Key Issues

1. Top management desire for new products without providing adequate resources.
2. Dominance of production.
3. Weak technology leadership.
4. Inappropriate accounting and controls for R&D (production/engineering criteria).

1.1.16 Top 50 Electronics Company

Events and Conditions

1. Dominance of market for many years.
2. Entry of new, strong competitors.
3. Corporate commitment to long-range strategic business planning.
4. VP R&D included in planning activity.
5. Diversification of company into new markets.
6. Power struggles between segments of corporate staff.

Key Issues

1. Dominant role of business planners with little technology experience.
2. Lack of mechanisms for technology inputs to planning.
3. Lack of resources for corporate staff to influence technology in operating divisions.

1.1.17 Another Top 50 Electronics Company

Events and Conditions

1. Further decentralization and increased autonomy of division managements.
2. Long tradition of technological strength in company—with engineering, development, and research highly respected.
3. Changes at top management level.
4. Questions raised (gently) about the value of corporate research and costs of CRL.

Key Issues

1. No CRL is immune from questions about its cost effectiveness.
2. Corporate technology management initiative in self-examination before questions are raised by outsiders.

1.1.18 Top 50 Equipment Maker

Events and Conditions

1. A history of stop and go, cut and grow in R&D.
2. Increasing competitive pressures.
3. Obsolete plants leading to high cost.
4. Slow progress of new product development.
5. Imminent retirement of chief technical officer.
6. Advent of new top management team.

7. Commitment to increasing decentralization and delegation of power to operating divisions.
8. Adoption of the strategic business unit (SBU) concept.
9. Fragmentation of divisional technology efforts.

Key Issues

1. Impact of SBU concept on technology.
2. Autonomy of division managers.
3. Lack of stable base for CRL.
4. No mechanisms for new products other than operating divisions.
5. Lack of a strong CTO.

1.1.19 A Third Top 50 Electronics Company

Events and Conditions

1. New CRL established by corporate management.
2. Division managers preoccupied with current, narrowly focused technological issues—see no role for CRL.
3. CRL making major thrust in a field new to the company.

Key Issues

1. Who will fund CRL?
2. How can CRL establish credibility with division managers for long-term support?
3. Lack of transfer mechanisms for new products/technology to divisions.

1.1.20 Top 75 Chemical Company

Events and Conditions

1. Establishment of CRL.
2. Diversification of company.
3. Lack of strong corporate technology leadership.
4. Lack of strong interest by operating divisions in CRL.
5. Role of CRL not clearly spelled out to divisions.

Key Issues

1. CRL role in company.
2. Leadership of corporate technology not strong.
3. Lack of support for CRL.

1.2 TRENDS IN MANAGEMENT OF TECHNOLOGY[*]

1.2.1 Some Questions

- What has been happening in technology management in the decentralized company?
- What is liable to happen in the future?
- How can the management deal with trends in technology management?

1.2.2 Main Body of This Section

About a dozen trends relating to management of technology in the firm are discussed briefly here (see list in Figure 1-1). Some of these are expanded on as individual sections or chapters of this book.

1.2.3 Implications of These Trends for Management

Some of these trends are industry-wide and reflect a reaction to situations or events outside the firm—in the economy, in the work force, in the advance (or stagnation) of specific or general fields of technology, and in the marketplace whether it be domestic and/or foreign. It may be difficult for the management of an individual firm to counter such "macrotrends" or to do more than be aware of them and to tailor the firm's policy and operating decisions to cope with them. Other trends may reflect general tendencies in industrial R&D/technology, but it may be within the control of the individual firm to adapt or to counter any adverse effects.

[*]©1985 IEEE. Reprinted with permission, from Rubenstein, Albert H., *IEEE Transactions on Engineering Management*, **EM-32**(4), 141–145, November 1985. Adapted from my farewell editorial as editor of the *Transactions* for 25 years.

Technical entrepreneurship

Corporate research laboratories (CRLs)

Networking of divisional R&D/technology

Long-range technology planning

R&D/production interface

Evaluation of R&D projects and programs

Expert systems

Make or buy

Technology policy and imbedded technology capability

Software development process

Sources of chief technical officers (CTOs)

Figure 1-1. Current trends in management of technology.

Few of the trends are "good" or "bad" per se. If not observed and dealt with in planning for and operating technology, however, they can catch the firm unawares and unprepared for potentially damaging consequences. For example, failure to note and act on the trend toward technical entrepreneurship in competing firms that have not been successful at it in the past may allow such competitors to "creep up" on or leapfrog the firm's new product development efforts and get to the market first. Or failure to note and respond to the trends in the increasing role of software versus hardware in production systems can result in longer lead times for process development, higher costs, and less effective production processes.

1.2.4   A Checklist/Watch List

- Set up a system and group to monitor and evaluate trends.
- Provide an action mechanism to respond to both favorable and unfavorable trends or events.

Technical Entrepreneurship

Although there has always been some level of entrepreneurship associated with R&D/Innovation, top management interest in it is increasing in the face of tough competition—domestic and foreign—and the need to get new products and processes out the door and into the market and factory more quickly. The large corporation has many drawbacks as a setting for "true" entrepreneurship, and many failures have been and are being experienced in this new wave of internal entrepreneurship. However, as long as top managers are willing to keep trying and to back up their internal entrepreneurs, there will also continue to be some successes. What is not clear yet is what set of conditions are necessary and sufficient in a given firm or in industry in general to foster such nonroutine and failure-prone activities. We have some clues, but general theories and rules for management are in the future.[*]

This trend provides great potential opportunities for some technology managers, but not for most. It is a high-risk game with large potential rewards for success and penalties for failure. It is not an arena for the faint of heart or the technical manager who wants to be comfortable and safe. (See Chapter 8 for more on this trend.)

Corporate Research Laboratories (CRLs)

Although it was not clear until the past 7–10 years, there is now a definite trend toward the elimination and reduction of size and scope of corporate research labs in large companies. This trend is also occurring in medium-size companies but is not so evident, because their CRLs have not been as large, visible, or numerous. Major factors relate to the theory and practice of decentralization and the struggle between operating units (generally product divisions) and corporate staff over turf, costs, and differentiation of roles. This trend is not a great threat or

---

[*]My colleagues and I, with support from industry, are currently attempting to develop an "expert system" or knowledge-based system of propositions on the factors influencing success and failure of entrepreneurship in the firm.

disadvantage to technical professionals who are more interested in highly applied work and associating directly with operating units and specific product lines or manufacturing processes. It does seriously diminish the opportunities for technical people who want to push the technical base of their organizations further toward the state of the art and beyond and further afield from current products and processes. (See Chapter 2.)

Networking of Divisional R&D Technology

Like the trend toward a reduced role for CRLs, this one is also driven by the extreme decentralization that characterizes a majority of U.S. (and many European and Japanese) firms. In a sense, it is an attempt to overcome some of the gap created by lack of a corporate-level technology activity—for example, a CRL—which (ideally) has an overview of all the company's technology and which can help fill in the gaps, push the technical frontiers, and help counteract duplication. However, the forces against cross-divisional cooperation in technology—sharing of ideas, facilities, equipment, people, and experience—are formidable. Many organizations are still moving away from, instead of toward, effective networking and technology sharing. Corporate managers are in a paradoxical situation on this issue: they may espouse extreme divisionalization and independence and yet would like to "encourage" cooperation to avoid gaps and duplication and to prevent lost opportunities. Unless this encouragement takes material form in the way of enforced incentives for cooperation—positive or negative—such cooperation is not likely to go very far. (See Chapter 4.)

Long-Range Technology Planning

Finally, after many years of lagging seriously behind the advance of strategic planning in other parts of the firm, planning of technology and R&D are beginning to show signs of life. In a recent study we did, about half of the very large firms we contacted have made attempts at a systematic procedure for strategic technology/R&D planning. Many of them are still following-type plans—driven by general business, marketing, or financial planning. Others are done as afterthoughts, after the "real" business planning has been essentially completed. Here is a frontier area for technical managers—especially those in R&D—to lead the company and to be the source of ideas for future directions rather than to be in the position of trying to make things happen consistent with the plans already made by others in the firm. (See Chapter 5.)

R&D/Production Interface

Despite the lack of experience and interest of most top managements in production as such, increased serious attention is being paid to barriers and opportunities in the flow of products and processes from R&D into and through production. Manufacturing executives still have to fight for recognition in the "front office," but issues of cost, quality, product liability, time to market, and introduction of new products and manufacturing technologies are finally rising toward the top of management's priority list—but perhaps not fast enough to save many firms from loss of competitive position or even demise. (See Chapter 8.)

Evaluation of R&D Projects and Programs

Evaluation is a difficult area in which to improve management practice significantly, due to major time lags, politics, uncertainties, and many other factors that can influence the outcome of specific projects and whole technology programs. However, technical managers must do a better job of evaluating their own activities or this will be done for or to them. Although corporate managements have not yet found satisfactory techniques or methodologies for effective evaluation of technology, they are tending to use what can best be described as nonappropriate techniques, derived from other fields, with often serious results for technology projects and programs (e.g., the issue of eliminating or cutting back on CRLs).

In view of the quantitative training of most technology managers and the high stakes, this area provides an opportunity for them to contribute to the art of evaluation in their own areas and the organization as a whole. The many articles published over the past two to three decades on this subject should provide a base for such efforts—including some of the very old articles (20 years or more) that still have technical merit and application potential. (See Chapter 7.)

Expert Systems

Managers are currently drowning in articles and books and exhortations about the potential impacts of artificial intelligence (AI) on their products, manufacturing processes, and even their own jobs. A lot of the rhetoric in this field is futuristic stuff that may or may not come about. However, specific applications of AI and its related subfields have already been made in many areas related to the whole technology spectrum, from R&D (knowledge-based systems for integrating what is "known" in a given field of science or technology) to technical service (expert systems for diagnosing maintenance problems in machinery).

There are many opportunities in the technology area of the firm for potential uses of expert systems, including the core area of R&D itself. As specialists in this field and R&D personnel themselves continue to look for feasible and cost-effective applications, radical changes can occur in the way in which technology projects are initiated and conducted in the firm. One area—the concept of the "super designer"—alone has the potential for revolutionizing product and process design and providing tremendous leverage for capitalizing on huge bodies of knowledge about design features and their impacts. (See Section 9.2.2 for more on this.)

Make or Buy

Related to some of the issues already discussed, there is a clear trend toward considering or actually doing more "buying" of technology outside the firm rather than depending on internal groups to develop it. Most firms have done some licensing, joint venturing, purchasing of technical consulting, and so on. However, only a few as yet have an overt policy of *buy* rather than *make* (with the exception of software development) and it is not clear how far the dependence on outside sources is likely to progress over the next decade or so. From a national point of view, this trend, if it accelerates, may be an economic and technological disaster. From the viewpoint of the individual firm, it may rob the firm of its internal ability to react to or anticipate events that may provide technological threats or opportunities. We have seen some of the consequences of overdependence on licensing versus internal development and do not look with comfort toward its emergence as the major means of obtaining new technology for the firm or for the United States. (See Chapters 6 and 9.)

Technology Policy and Imbedded Technology Capability

This issue is closely related to those of make or buy, entrepreneurship, corporate research laboratories, and networking—in fact, most of the trends discussed in this section. It concerns an awakening of interest in some companies and other technology performers and suppliers (e.g., government agencies) about the need for a visible, consistent, and operational "technology policy" that can guide the decisions and behavior of its operating units and technical staff. In our studies of the R&D/production interface, software development for manufacturing, imbedded technology, and networking in many firms, we dimly see the realization that such policies are needed. However, very few companies have them in place and it is not clear whether this interest is only a passing one or whether such policies will emerge in full outline, comparable to financial, marketing, and other policies. (See Chapter 8.)

Software Development Process

Software project management has crept up on many managements gradually and they are unprepared to deal with it in an effective way. There is a clear split among organizations that are trying to handle software projects via the same mechanisms and channels they have used traditionally for hardware projects and those that are ignoring traditional project management approaches and allowing their software people to "wing it" or develop new approaches. Still others are trying to combine both approaches. Software projects clearly need their own project management methods but can also benefit from the long experience with hardware project management. This is a major challenge for both top management and technology management and needs a great deal of careful attention. (See Chapter 9.)

Sources of Chief Technical Officers (CTOs)

Traditionally, the vice president for R&D came from R&D itself, often a continuous movement up the R&D management ladder into the top technical spot in the company or operating division. This is no longer the case. What appeared as a minor possible trend a decade ago has become a major change in the source of CTOs. A large proportion of CTOs are now coming from production, marketing, and general management—often former divisional managers or product line managers. Along with some of the other trends, this one indicates top management's uncertainty about how to deal with such technical activities as R&D and their dissatisfaction with its performance in the past and present. While this provides a possible opportunity for a few select technical managers to move into such slots in the future, it has many implications for R&D career paths and for the conduct of the technology enterprise itself, some of them not very favorable. (See Chapter 3.)

2
ORGANIZATIONAL FORM OF TECHNOLOGY IN THE FIRM

2.1 INTRODUCTION AND CHAPTER OVERVIEW

Many factors influence the detailed organizational arrangement, structure, and operating modes for technology in the individual firm. However, the very nature of the decentralized firm itself, the range of technologies required to support its products, and the levels of knowledge and skills required combine to limit the alternative patterns of organizing technical activities.

This chapter presents major alternatives and discusses the issues that arise from them. Some people argue that organizational structure does not really tell us much about the way a firm actually operates. They argue that personalities and circumstances have more influence on what really happens. Certainly, personalities and circumstances have major influences on how technological activities are performed in the firm. But the arena in which these influences operate is provided by the organizational structure and the lines of communication and power that are specified or implied by the way things are formally organized. Furthermore, the organizational structure, although not a perfect indicator, gives strong clues to management's intentions about how things should be done.

A reasonable starting point therefore for examining issues of organizational form and structure and operational modes is the general pattern of deployment of organizational resources for technology.

Section 2.2 examines the structure of the decentralized firm itself and gives a very brief description of how and why it has evolved. The extent of divisionalization and decentralization is examined and then the special case of the conglomerate firm is discussed (Section 2.3.1).

Next, the forms of R&D/technology deployment are examined within the framework of the firm's overall organization and the role of the corporate research laboratory (CRL) or its equivalent is discussed. Finally, the phenomena of changes in organizational form and the "time horizon" for getting organized are examined.

2.1.1 Some Questions

- How does technology/R&D fit into the decentralized firm, other than uncomfortably?
- Why is it reorganized so frequently?
- Is there an optimum organizational form for it?
- Why does it take so long for technology to "get organized" and to begin producing results–for example, new and improved products and processes?

2.1.2 Main Body of Chapter

This contains a discussion of the organizational forms and issues and the role of the corporate research/technology laboratory. The division of the chapter by section is as follows:

2.1 Introduction and Chapter Overview

2.2 The Decentralized Firm

2.3 How Many Divisions? The Special Case of the Conglomerate

2.4 How Does the Organization of Technology in the Decentralized Firm Evolve and Develop?

2.5 The Fruitless Search for the Best or Optimal Organization Form

2.6 How Can an R&D Laboratory Be Phased Out Gracefully?

2.7 What About Basic Research in the Decentralized Company?

2.8 Changes in Organizational Form

2.9 Effects of Organizational Change on the Midlevel Technology Manager

2.10 What Is the Net Effect of Churning Divisions, SBUs and Product Lines on Their Technology Programs?

2.11 Why Does It Take R&D So Long to "Get Organized"? What Can Be Done to Reduce the Time Lag in Getting Organized?

2.1.3 Implications for Management

There is no stable "optimum" organizational form, and frequent or constant changes can destabilize operations to the point of harming the productivity of the technical staff. Unsatisfactory performance should first be addressed within the existing organization structure for technology and adjustments made, short of major reorganizations or wholesale changes in the mission and direction. When it is clear that such micro adjusting will not do the trick, then changes in organizational structure, location, and operational mode should be considered.

2.1.4 A Checklist/Watch List

- Repeated and systematic delays in getting projects moving.

- Chronic conflict and backbiting among people in the technology system.

- Appearance of bureaucratic defenses against criticism and requests for information and assistance: for example, "it's not my job" or "we're too busy."

2.2 THE DECENTRALIZED FIRM

This book is concerned with a number of factors that appear to have particular relevance for technology in the *decentralized* company, although they may also be relevant to technology in many other settings. Some of them derive directly from the very nature of the decentralized company; others appear to be accentuated by the specific circumstances of the decentralized company.

Rather than attempt, at this point, to define a decentralized company, a brief description is given below of top management's intentions in decentralizing their operations and the effects they hope to achieve through this decentralization. A partial structural description of the decentralized firm is given in Figure 2-1, where the prominent features may be recognized as a Top Corporate Management, Corporate Staff Functions (e.g., Technology Management), and a series of relatively autonomous Operating Divisions, headed by Divisional Management, which carry on the direct productive functions of the business such as production, engineering, and sales or marketing.

The exact organization structure and distribution of functions varies between firms and between industries. The salient features for our purposes are the organization of the corporation's business into relatively autonomous operating units—generally along technological or market lines—and the establishment of measurements of performance for the purposes of control, incentive, and reward.

Top management intentions in decentralizing are epitomized in this statement from an IBM announcement to its employees, explaining the company's management decentralization many years ago:*

> The new alignment of the various areas is based on products. Each of the product divisions will, within the framework of policy established by the Board of Directors

---
*See Appendix B, under IBM, section B.2.

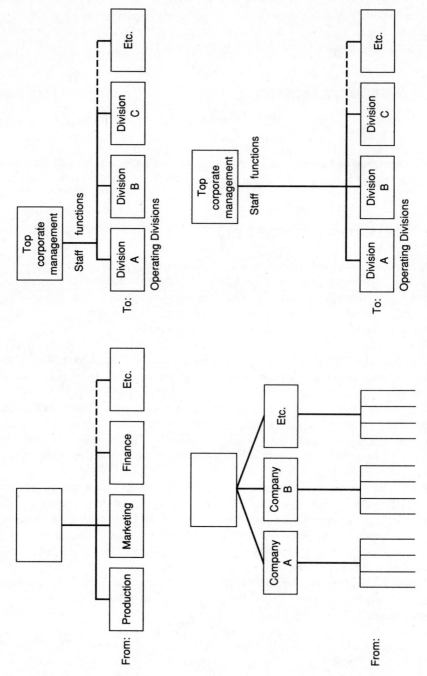

Figure 2-1. Emerging patterns of company decentralization.

24

and general management, operate almost as an individual company with its own manufacturing, sales, and service functions. Each of these divisions is equipped with special skills and product knowledge to concentrate on developing the full potential of a specific market.

Further strength is given the organization with the creation of the Corporate Staff, which, being separate from the operating organization responsible for developing, producing, selling and servicing goods, can closely examine the special areas of the business and assist the operating executives in solving problems in these areas.

An instance of extreme decentralization is described in a popular article on a company that has developed a decentralized corporate structure through acquisitions of existing companies:

> At McGraw–Edison, only the top policy decisions, company-wide in import and usually financial, are made in the head office. Otherwise, the divisions handle their own affairs. And no memoranda and visitors from the home office tell the field how to do it.[*]

A comment from a company[†] where "decentralized or divisionalized corporate form does not reflect a recent innovation" suggests the long-term benefits that may accrue to this form of operation:

> A series of operating subsidiaries focused on lines of business, with production and sales, as well as research, handled by the subsidiaries and major policy functions handled by the parent corporate staff, has been the prevailing pattern in our company for some time....
>
> Our management has found the autonomous units to be profitable, stimulating and internally competitive forms of corporate government. We believe that (our company) is a classical example of a successfully decentralized, highly diversified company with sufficient flexibility to cope adequately with economic drift across a broad range of consumer goods reaching to all areas of the free world.

Historically, we have observed two major patterns of management decentralization, with variations and combinations of both in specific companies. Both patterns tend toward the same general form, as indicated in Figure 2-1.

The speculative literature on management decentralization is voluminous. Only a few studies have been done, however, on the actual degree and effects of large-scale delegation of decisionmaking which is supposed to accompany "true" decentralization.

Very little is known systematically, for example, of the way in which major organizational changes such as decentralization are perceived and reacted to by people at the several operating levels that should be affected by the changes.

---

[*] See Appendix B, under McGraw–Edison.

[†] My source is a personal, confidential interview.

2.3 HOW MANY DIVISIONS?

Management researchers and practitioners have pondered for many years the question of how many separate and distinct operations an executive can manage and still keep control of them all and keep himself from overcommitment and breakdown. There is a substantial body of management literature, some of it going back to biblical times, on the concept of "span of control." Indeed, this issue has been one of the major ones contributing to the overwhelming trend toward decentralization of the modern corporation and its partitioning into fairly self-contained, more easily managed operating divisions.

Unfortunately for the top management attempting to arrive at an "optimum" or even reasonable number of operating divisions, the theories and principles that have been presented in this literature are rather sterile. The literature on span of control deals primarily with the "number of subordinates" reporting to a manager and fails to make many important distinctions, such as:

- The qualitative differences between the units the various subordinates represent, in terms of how much and what kind of attention they need.
- The emotional and other psychological limits the manager approaches when the number of people to be dealt with increases.
- The many nonsubordinate relationships the manager has (e.g., colleagues, clients, financiers, regulatory bodies outside the company, competitors) whose drain on his time, energy, and emotions may far exceed that demanded by his subordinates, over whom he has more control.
- The distinction between tactical and strategic or long- and short-term kinds of attention the manager has to pay the different units.

Rather than settling this question for the top manager or managers (including senior vice presidents and group executives), decentralization itself raises a whole new set of questions about how finely and into how many operating units the company can effectively be divided.

The starting logic for divisionalization can be based on such things as technology, markets, phase of the raw material to market process, and location (particularly for foreign operations). However, this overall logic does not provide answers to questions of how *far* one can safely go in dividing operations according to these criteria. According to the general philosophy espoused by many companies when they have decided to divisionalize, each "piece" should be "viable," small enough for one person to handle, and relatively self-sufficient in the principal functional areas (except legal, finance, and perhaps some others).

In general, these rough criteria have served the decentralized companies well. At least, judging from the infrequency of major redivisionalization across the board in most large decentralized companies, the initial split and the evolving split of operations have been acceptable enough to keep top management from making frequent massive redistributions.* This is not to say that the number and makeup of divisions have remained constant for most of our sample companies over the period of the study. On the contrary, there is hardly a company of the more than 200 decentralized or combination-form corporations in our sample which has not added or deleted divisions or modified the distribution of activities between them since we began collecting data from them over 30 years ago. However, the number of massive redivisionalizations in the early years of that period was very small. This lack of massive redistribution of activities among operating divisions did not mean that top management was completely satisfied with both the underlying principle of divisionalization in their company and the specific pattern of divisionalization of operations. It does appear to mean that many of them preferred other means of correcting the imperfections than redoing the whole pattern of divisionalization. The following are a few alternative means that have been employed:

- Establishing the position of group executive and forming the divisions into larger groups, based on the same general criteria that were used in the original divisionalization or other ones (see the discussion in Section 3.7 on the group executive).
- Establishing "new product" or "new enterprise" or "incubator" divisions to carry new activities through the early stages, before setting them up as full-fledged operating divisions (see Section 8.4 on technical entrepreneurship).
- Allowing or instructing some divisions to divisionalize their operations in turn. This leads to divisions of divisions, which may present their own peculiar advantages and disadvantages.
- Setting up or tolerating in some or all divisions the concept of the "product manager" or other manifestations of further divisionalization within a given division that stops short of establishing another level of fully integrated operating units. Frequently, these new units are far from self-sufficient in any of the major functional areas, personnel, or financing.

Despite all these practical alternatives to major reshuffles in divisional organization, the question remains for most companies of how many divisions (or subdivisions) they should have and how big and how self-sufficient they should be.

Number and size interact, of course, and it is clear that the fewer the operating divisions of a given size company, the larger they will be on the average. However, the upper and lower limits on size are not at all clear for many firms. Some major decentralized companies have laid down arbitrary rules for the minimum viable size of an operating division (e.g., one says it is $10,000,000 in annual sales). This is not a very good criterion, however, because it ignores many other factors, such as:

*Except for major waves of acquisitions, mergers, and disacquisitions or "redeployment of assets," which in many conglomerates often lead to several years of juggling individual units into different aggregations and configurations.

- The type of technology.
- The scale of operations (e.g., is it a watch business or a steel business).
- The number of employees (and value added).
- The complexity of the technology.
- The number and kind of customers.
- The economics of scale of operations.
- The size and nature of the division's direct competitors.
- The kinds of assets and services it needs to compete successfully.
- Phase in the life cycle of the business: start-up, maturity, end of life.

This problem of minimum size is one that has plagued the Small Business Administration for a long time, in its efforts to determine the upper limit at which a business is no longer "small" and thus ceases to be eligible for certain benefits and subject to certain regulations. So far there has been no clear single-variable answer, although number of employees is commonly used.

From an organizational viability point of view, "an operating division should either have control over or access to those services and other resources which are essential to its effective operation" (this is an example of how unsatisfying so many management principles can be). Operationally, this means that it should be able to compete on equal terms in both its own external marketplace and its internal arena. In the latter case, it faces competition for people, resources, and management attention and support.

According to this set of criteria, a large percentage of the operating divisions we have encountered in some of the industry segments we studied (notably electronics, engineered products, food, and some miscellaneous ones) have been below an apparent "minimally effective size." Although they have the resources to operate on a day-to-day basis in their specific sphere of operations, they have little ability to expand, diversify, or substantially upgrade their operations. And they have little ability to meet major crises on their own or to plan for the long run.

This situation, taken together with the (assumed) interest of the top managements in using their current divisional structure as a basis for growth as well as for current operations, often leads to the corporate central office attempting to undertake the many staff activities that relate to expansion, growth, modernization, diversification, and other nonroutine activities.

Where neither a particular division on its own behalf nor the corporate headquarters looks out for the division's future welfare, stagnation is a likely result.

At the *upper* end of the size criterion, different questions arise. Among them are:

- Overcommitment and overburdening of the divisional management in a manner similar to that which gave rise to corporate decentralization itself.
- Proliferation of staff activities and management levels within the divisions

at the cost of increased overhead, blurring of lines of command and control, and competition for skilled people in other parts of the company (divisions or headquarters).

• Overbalanced power relations with other divisions. This can evince itself in such matters as competition for resources (including money and people), bargaining over transfer prices and joint production schedules, "muscling in" on product lines and markets of other divisions.

When one or more divisions in the decentralized company gets "too big," many of the original advantages of decentralization and divisionalization are lost or at least are reduced. If one of the considerations in the original decentralization was to counter the trend toward conservatism, risk aversion, and delayed action and decisionmaking, which characterizes many large centralized organizations, then the "superdivision(s)" may restore some of those characteristics to the apparently highly decentralized company.

Again, at the other extreme, the procedure of setting up a new operating division for each new product line or new venture can lead to over proliferation of divisions and the many disadvantages of too many and too small divisions.

I do not discuss, in this book, a "magic number" of divisions or even a magic formula for deciding on a proper number of divisions, much as I would like to do so. I can say, however, that this is not a matter to be decided just once at the outset of divisionalization or during periods of crisis (e.g., merger, acquisition, major market setback or opportunity, major new product or venture possibility). This is a major organization design variable that should be monitored continually and adjusted according to the performance of various design criteria. Among these design criteria, which might give clues to unsatisfactory deployment of operating units—too many, too few, too big, too small—are these:

• Inability of small divisions to compete externally due to limited resources.
• Slowdown in decisionmaking in big ones.
• Overlap in production, marketing, or R&D in product lines.
• Missed opportunities by individual divisions or the corporation as a whole (opportunities falling between divisional chairs).
• Excessive duplication and/or overlap of staff services.
• Inadequate planning and other staff activities in particular divisions.

These and other criteria could be formulated into indicators to be monitored continually and, when they show signs of reaching unsatisfactory levels, signal the need to reexamine the deployment of operations among divisions. Some of the unsatisfactory performance might merely signal the need for slight realignments and reallocations of responsibilities among the existing divisions. Others might suggest the need for new divisions. And still others might indeed signal the need for a major reexamination of the basic principles on which divisionalization rests.

2.3.1   The Special Case of the Conglomerate

During the period of the formal study (1957–1972) and since then, many new names appeared on the Fortune 500 listing of the nation's largest industrial companies. Among those new names were a number that vaguely fitted into industry categories (particularly in the engineered products and miscellaneous groups) but that did not quite fit the traditional technological, market, product line, or organizational pattern of most of the sample companies. These were the *conglomerates*.

The number of "true" conglomerates in our formal sample was less than half a dozen, but subsequent involvement has increased our observational base to several dozen conglomerates or *very* decentralized and diversified corporations. Indeed, the *true* conglomerate, if there is such a thing, appears to be the limit to which decentralization or divisionalization can exist in a manufacturing organization without the corporate framework or superstructure becoming essentially a fiction or a financial convenience. Yet even in these extreme cases, the idea of a central or corporate R&D laboratory sometimes arises, although so far one has seldom been established.

In the "miscellaneous" type of conglomerate, where no coherent pattern is evident among the technologies or markets of the various components, the argument for any centralized or common technical activity is seldom received with favor by the managers of these operating components. Many of these people, if they survived the acquisition or merger that brought them into the orbit of the conglomerate, are fiercely independent. A large majority of the ones in our early sample of conglomerates had either been the founders or owners of the businesses that they were now managing as part of the new corporate family. In addition, many of these people had grown up professionally in firms that had the characteristics of small businesses (in operation and structure, if not in legally defined size). They were not used to having access to or being "put upon" by sophisticated or complex management systems and staff activities such as R&D, long-range planning, budgeting, market research, management science, and so on. Although many of them in the engineered products area, for example, had a strong manufacturing and product engineering base, few of them had full-fledged R&D departments comparable to those in the large chemical and electrical/electronic firms.

As a consequence, we encountered a group of new "division managers" with very little tradition of any formal R&D at all, let alone corporate R&D on the scale of the typical large chemical or electronic company. Only a few cases of corporate technology in conglomerates were included as part of the formal study. However, my colleagues and I have had and continue to have relations with additional conglomerates of varying sizes through other research projects and consulting engagements. In these cases, totaling over a dozen, we have encountered the extremes of arguments for and against the need for corporate-level technology. The degree of possible and proposed corporate involvement in these cases is seen to range from the case of a multi-billion-dollar conglomerate with one man traveling between divisional laboratories and trying to prevent duplication, to full-line corporate research laboratories (CRLs) attempting to cover the scientific spectrum underlying the divisions' highly diverse businesses.

During the past two decades, the distinction between a highly diversified decentralized company and a "true" conglomerate has become blurred. Indeed, this distinction can easily lead to a tautology such as: a diversified, divisionalized corporation that grew primarily by acquisition and merger and that *provides very little central control and staff services* might be distinguished from one that developed in the same way but that *provides a lot of central control and staff services*. We have observed a number of "true" conglomerates develop into more traditional forms of the decentralized corporation and vice versa.

Certainly, among the first activities to be centralized or provided by the corporation are financial, legal, and accounting controls. These might be followed by centralized personnel, labor relations, transportation, national advertising, long-range planning, and other activities. The idea of providing central R&D control and/or facilities appears to arise late in the consolidation or centralization process and, for our sample of conglomerates, did not seem to get very far toward the "combination" or "centralized" form for R&D that dominates many industry groups. In addition to the apparent cases of duplication of facilities, equipment, or personnel which turn up when a member of the corporate office begins to dig into each division's R&D portfolio (if it has one), the most prominent reasons for entertaining the idea of R&D activity at the corporate level include these:

- Particular divisions have a "sick" or obsolete product line or manufacturing setup that is obviously behind the general state of the art in their own industry.
- Members of top corporate management see promising lines of expansion or diversification in the product areas of particular divisions, but there are no people in those divisions who are competent or sophisticated enough to provide the technology necessary for such expansion or diversification.
- There is a pool of technical talent (often in an engineering or development laboratory) of an acquired division which could be put to work for the benefit of other divisions in the company.
- Individual division personnel and expenses for R&D (with engineering often thrown in) appear to be out of line on the high or low side for the volume of business and the technical field the division is in.
- The germ of an idea for combining the resources and talents of two (or more) divisions requires an analysis of those talents that might be relevant to such a combination.
- A routine "technical audit" of acquired businesses may turn up the need for modifications that can only be accomplished by influence or direct action from above (the corporate level).
- A talented technical individual in a division may appear too broad in scope or too prolific in ideas for his current location and may be moved up to the corporate level to flex his technical imagination or creative capabilities.

Whatever the reason, the issue of the corporate headquarters' role in the technical affairs of the company comes into conflict with the avowed or behavioral avoidance of too much operating control over and interference with division management prerogatives. Few of the conglomerates have published their doctrines of centralization–decentralization, as have a number of the functionally oriented companies who have gone through a formal transition to decentralization (see the section on the decentralized firm). One exception is this very early statement from Textron:[*]

> Textron started its pioneer diversification program fifteen years ago (1952) for two principal reasons: to avoid the cyclical effect of a single industry and to be able to participate in new growth areas in order to achieve continuing high rates of return for its shareholders. These reasons have proven to be sound.
>
> The so-called conglomerate company has come into prominence within recent years in response to some additional basic forces. One is the technological explosion, which on the one hand has been creating large new markets, while at the same time, threatening many traditional product lines with obsolescence. The multi-market philosophy permits the redeployment of resources to take advantage of new opportunities and to assure constant renewal so that return to shareholders is at a continuing high level.
>
> We bring into Textron only proven companies, with existing management organizations which have a record of accomplishment behind them, and which have the willingness and desire to continue to operate the company as part of Textron. Then we keep our operating decisions with the division management, as close to the scene of action as possible. At the same time, the corporate organization relieves the division management of many non-operating functions which burden an independent company. This gives the division president and his team more time for their primary job—to run their company as profitably as possible. Textron reserves to itself only those things which are best done centrally, or are out of the ordinary course of business, or affect planning and allocation of resources.
>
> The link from corporate headquarters to the divisions is a central organization of group executives. These men are generalists, with wide business and operating experience. They are able to provide the coordination, supervision and planning assistance necessary to assure that every unit of Textron is being operated to its fullest potential. This method is not unique, but it certainly is becoming a pattern as the most effective way to manage and control diversified operations.
>
> Thus it is that Textron is able to maintain the initiative and incentive of a decentralized operation, but at the same time give its divisions the advantage of experienced and professional counsel and services. It has worked out well in our case.

An attempt to generalize, based on their expressed motives and attitudes and extrapolating their observed behavior, might lead to a statement of the general conglomerate doctrine in these rough terms (with exceptions not infrequent): the true conglomerate is formed by a continuing process of merger, acquisition, consolidation (and, more recently, divestiture), based primarily on financial considerations. That is, unlike the decentralized companies in most of our sample industries, the

---

[*]See Appendix B, under Textron.

technology, product line, geographic location, and market structure are not the principal considerations by which conglomerate acquisition and merger programs are guided. Certainly, there are conglomerates with a major focus on consumer products, services, or heavy machinery. And there are conglomerates whose major components are on the West Coast, in the Midwest, or in the Northeast. But the dominant factor for many of them appears to be investment opportunity, independent of specific market or technological considerations. Certainly, patterns of acquisition in given market areas guide the growth of some of the conglomerates. But only in rare cases have most of the conglomerates apparently used well-defined *technological* criteria as the prime basis for selection of acquisition candidates. This situation, then, results in many of the conglomerates, at least initially, being "collections" of businesses and product lines with little in common as far as underlying technology goes. Consequently, this poses questions about the feasibility of a central R&D organization or facility such as those that are dominant in the more technologically coherent decentralized firms. One of these questions is the perennial one of how a project, started in a central corporate laboratory, can be transferred to an existing division. This transition process is beset by the many problems in the standard or more unified or older divisionalized company (discussed in Chapter 8 in the section on transition from R&D to production), plus the following:

- The corporation is a new enough entity so that there is no history of communication of technology between its components.
- The businesses and technologies of the divisions are often so diverse that no logical project portfolio for a modest-size central lab is evident.
- Power and influence relations have not yet been established between corporate staff people (e.g., technology) and divisional personnel.
- Corporate technology people (if there are any) are generally unfamiliar with the fields in which the various divisions operate and are often discouraged by divisional personnel from learning at their expense.
- The initial "hands off" policy or behavior of the corporate office sets up tremendous initial barriers to staff advice or joint division/staff activities.
- Except for some of the acquisitions by giant conglomerates or fully integrated companies with substantial R&D activities of their own, most companies acquired by the conglomerates in our sample had very small or no R&D capabilities. This meant that a CRL, where one existed, had no logical point of contact and cooperation in the new division. Typical situations we observed involved newly acquired divisions with less than a dozen people that could, by any reasonable definition, be called R&D people. This leads to conditions of limited interest, time, and capability of the divisional technical people to participate in joint ventures or to implement technical ideas initiated by the corporate R&D people.

We have been asked many times by consulting clients (usually in rhetorical fashion) why a conglomerate needs a corporate R&D effort, in the form of a coordinating mechanism and/or an actual corporate R&D facility. The arguments against its success seem persuasive, and it would probably lead to additional problems of relations between top management and divisional management. The empirical response by those conglomerates who are moving gingerly in this direction is not yet ready to be evaluated. Some of them have had modest successes, accompanied by moderate to substantial failures. Perhaps the theoretical reasons advanced in support of corporate-wide R&D in the traditional decentralized companies will have to suffice for the moment. They involve all the reasons cited throughout the book, and especially in the sections on the corporate lab, the chief technical officer, and the divisional research director. They include improved understanding of the technological base of the company's various businesses, prevention of duplication, enhanced ability to attract and hold good researchers, a coordinated bird's-eye view of the company's entire technology capabilities and operations, backup for technical crisis situations, and a platform for major departures away from the company's current operations.

In the case of the conglomerate, some of these arguments may carry more weight than others. The coordination argument can be a very strong one, because it may provide one of the few means by which corporate management can keep track of technical operations of a newly acquired company and can indirectly guide its development. The usual financial and accounting control and even strict budgetary and personnel approval procedures do not offer sufficient insight into the technical aspects of a division's operations for planning and forecasting purposes.

This motivation for having corporate involvement in the technical affairs of the divisions frequently leads to charges, by division management, of spying, too-close supervision, or interference beyond their initial understanding of corporate–divisional relations. We observed many gradations of just that—spying—and attempting to influence divisional decisions and operations. This then leads back to the central rationale for the divisionalized company, whether it arrived in that condition by internal spin-off of divisions, external acquisition, or both. If it is to be more than merely a financial holding company (which many of the conglomerates are in their initial stages), the corporate office has to provide more than a banking or financial auditing function. We have observed the gradual development of corporate staff activities in many of the conglomerates (interspersed with periods of "corporate staff bashing") to the point where some of them appear little different, to the casual observer, from a traditional functionally organized company that has divisionalized. On deeper examination, however, one finds that the lack of a common history, traditions, and personnel make rapid (or even eventual) integration unlikely.

One of the strongest arguments we encountered for a technology activity at the corporate level involves the concepts of "technical auditing" and "technoeconomic forecasting." Many of the acquired divisions have gone their independent ways for decades without systematic examination of their technological requirements, their current capability, and their future prospects. In addition, they had little, if any, tradition of technoeconomic planning and/or forecasting. This leads in many cases, to neglect of second- and third-order trends in their own industry sectors, in other sectors from which attacks on their market positions might be launched, and other sectors in which their talents might lead to logical diversification.

As an illustration of the second- and third-order trends that division managers often overlook, we cite the case of a first-generation conglomerate that, during the course of our study, developed into a modestly integrated decentralized corporation. In the second of a series of interviews with the president, stretching over a dozen years, we were told of his intention to divest the corporation of one of its currently most profitable divisions. He anticipated a great deal of resistance from the division and other executives, but he had decided it was necessary to make the move on the basis of second-order (rate of change) and third-order (rate of change of rate of change) trends in the industry which the division served. His main reason for recommending the move was the changing technological structure of that industry, in terms of production methods that were bound to become obsolete and the eventual replacement of the company's products for use in the industry by radically different types of equipment. On a subsequent (third) visit, several years later, we learned that the divestiture move had not yet taken effect (due to strong resistance from the financial people who were quite happy with its performance), but that the forecasted trends had moved up into first and second order and were beginning to become apparent to everyone.

We found many cases of staff people and corporate-level managers having such insights about segments of the company's business, but having no mechanism or insufficient leverage to bring about needed changes in direction.

Due to the still primitive state of technoeconomic planning and forecasting in even the well-integrated companies, it is no surprise that the conglomerate, with its additional barriers to integration, has made little headway in developing corporate-wide analytical capabilities. However, the need for some degree of auditing and planning may be even stronger in the conglomerate than in the more traditional decentralized company. The major reason for this is that, in the latter case, many of the executives and staff specialists have worked together for years and have a reasonably good picture of the technoeconomic environment and prospects of each other's operations. In the conglomerate, most of the top and division management group are strangers to each other and their respective areas of operation. Our discussions with managers and staff people in several of the corporate and divisional headquarters of conglomerate companies reveal a very low level of familiarity with the technical characteristics of many of the divisions. We encountered many attempts to gain this familiarity quickly and thoroughly, but few initial successes.

An elementary principle seems to be that trained and perceptive technical people are needed for this task and that the headquarters of most such companies do not have an adequate cadre of such people.

Company A, one of the half dozen largest U.S. conglomerates, had been consolidating its major divisions into several quasiformal groups based roughly on a combination of marketing and technology. One of its largest groups had a strong R&D tradition and several major laboratories. In its latest round of acquisitions, the corporation began to develop a position in a consumer product area, where there was little tradition of sophisticated R&D. As this "second-phase" acquisition program proceeded, a technique of "technical assistance" to the newly acquired divisions was developed in the laboratories of the high technology group. Immediately following the completion of the legal and financial transactions in the acquisition, a planeload of R&D personnel from the group would descend on the newly acquired company to perform a rough technical audit, a general technical scouting exercise, and a series of "can we help with your problems" meetings with personnel of the new division.

This is a new approach, not only for conglomerates but also for some of the more decentralized of the diversified corporations in our sample. Several of these were formed well before the "second round" of postwar acquisitions and mergers. In several of these cases, we found virtually no contact between the R&D personnel of major divisions or groups as late as 10–15 years after merger or acquisition. Certainly, one of the assets a major conglomerate can bring to bear is the sharing and mutual support of technology among its various components, and this is frequently cited as one of the important reasons for the merger or acquisition. However, our observations indicate that the process of establishing close relations between members of the conglomerate family is a long and uncertain one. We have found very few cases of high levels of technical integration among these loose confederations.

As some of the older conglomerates begin to add staff activities at the corporate level and begin to draw in the reins on their divisions, R&D activities, or at least a substitute for corporate R&D activities, appear. It is often one man with a variety of words such as "liaison" or "coordinating" in his often-ambiguous title. He travels a lot, if the company is geographically dispersed, and generally tries to assume a low profile when he is "out in the field" in the divisions. He has little or no staff to do background work for him or to try out technical ideas that need personnel, equipment, or facilities. His main asset is his (presumed) access to corporate management and his indirect influence thereby on capital appropriations. Most division managers tend to ignore him, unless his visit is viewed as a direct threat to their freedom or operations. His main point of contact is generally with the divisional director of R&D or engineering (if there is no divisional R&D). In some cases he finds a potential ally in the divisional R&D director, who may have been chafing for years under insufficient budget, people, facilities, flexibility, and management attention or support. The test of this alliance usually comes when the corporate liaison man is expected to produce results for divisional R&D (e.g., more budget, permission to explore certain hitherto forbidden areas, easing up of restrictions on hiring). In view of the situation relative to R&D at the corporate level in the conglomerate, however, this first moment of truth has little chance of success. The corporate liaison man who is sensitive, as well as technically competent, can build strong relationships with divisional R&D management that can lead eventually to effective integration or at least coordination of corporate-wide R&D. However, if he rushes things or tries to throw around weight (which he typically does not have), he may set back integration for a decade or more. We observed several cases where initial contacts had been so unpleasant or damaging to self-perceived divisional interest that it was years before any serious attempts could successfully be made at corporate–divisional R&D cooperation, or cooperation with other divisions in the company.

Another characteristic of many of the newer conglomerates is that they are typically dominated initially by people whose main interests lie in the business, legal, financial, or accounting areas of the firm and devote most of their energies and attention to these kinds of matters. Where they have a nominal interest in technical matters, they are often too busy to devote much time or attention to them and often postpone any action until the business and financial aspects are "under control. " For some leading conglomerates, this has proved to be a very long time and R&D has been quite low on the top management's list of priorities. Considering all these factors in the situation of the relatively new conglomerate, we see that there is little prospect that a substantial corporate-level R&D/technology activity will emerge until the acquisition stage of growth and diversification tapers off and internal growth and diversification become dominant.

2.4   HOW DOES THE ORGANIZATION OF TECHNOLOGY IN THE DECENTRALIZED FIRM EVOLVE AND DEVELOP?

Figure 2-2 presents general patterns for deployment of technology/R&D in the decentralized firm. One of three "pure" deployments is found in most decentralized firms, with variations in specific firms or industries. In general, the pure *centralized* form has only one lab or technology department (or a set of labs or departments) at the corporate level and all advanced technology work is done there.

This does not mean that all the scientists and engineers in the company are located centrally. Depending on the industry and kind of technology, there could be substantial numbers of technical people in divisional departments, groups, and even laboratories which do more or less routine technical work in support of the plants and customers of the division. However, the major process and product improvements and new products and manufacturing processes are generally worked on centrally, with the support of divisional engineering and technical service groups.

In the pure *decentralized* form, "nothing" technical is done at the corporate level: it "all" happens at the divisional level. Of course, in reality, there may be a very clever technology-oriented individual or group in a function like marketing, planning, engineering, or quality assurance, where ideas for product and process improvements are originated, collected, evaluated, or communicated to people who can actually carry them out in material form. Finally, the pure *combination* form, which is currently very common in large firms, divides responsibility for product and process innovation between corporate and divisional labs, and herein lies the tale of ambiguity and conflict, which gives rise to many of the issues discussed in this book.

The exact history of technology in each of the approximately 200 companies we have observed in a range of industry groupings is of course unique. The effects of personalities, location, the specific products and markets, size, growth rate,

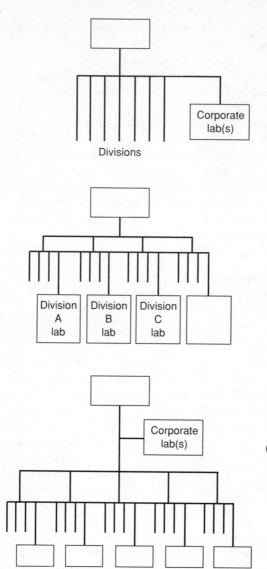

(a)   *Pure centralized* : All work on new products and processes and all work on major improvements in products and processes are done in one or more central R&D facilities. No work of this type is done within the divisions except perhaps an occasional major redesign job that approaches development proportions.

(b)   *Pure decentralized* : All work on new products and processes and all work on improvement of products and processes are done within the operating divisions. In many large companies, a given division may have a number of labs concerned with different levels of R&D.

(c)   *Pure combination* : The more general or long-range or basic or exploratory work is carried out in a corporate lab and the more specific, short-term applied work is carried out within the divisions.

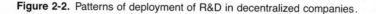

Figure 2-2. Patterns of deployment of R&D in decentralized companies.

stage of the company's life cycle, and many other factors combine to make each as unique in detail as are the life histories of different human beings. However, it became clear, soon after we initiated the formal study in 1957, that there were only a few significantly different overall development *patterns* on which the individual company histories converged. How much of this was our bias as researchers who are deliberately looking for generalized patterns of organizational behavior is hard to say. From our viewpoint, the data began to form patterns very early in the game, despite the detailed uniqueness of each case that we examined. Some of these patterns involved general organizational arrangements, funding arrangements, deployment of laboratories, relationships of operating divisions to headquarters, the role of staff groups and other organizational level phenomena. But in addition, very early in the interviewing and in-depth field study phases of the project, we began to get a strong convergence of the perceptions, attitudes, aspirations, disappointments, and other quite personal characteristics of key individuals in the firms that we studied. The most surprising set of convergences we found were among two groups of people whom one might reasonably believe were quite individualistic in their thoughts and behavior—top managers and researchers. We believe that our mixed strategy methodology (many kinds of interviews and questionnaires, repeated visits over a period of years, and "just hanging around" in some of them for extended periods) gave respondents a high degree of freedom to introduce ideas and data, free of the common constraints of a highly structured survey approach. In spite of this, many of their attitudes and behavioral patterns quickly formed into stereotypes, with only minor variations among individuals. This of course is a key methodological issue in all research on organizational behavior. We can operate at a sufficiently high level of abstraction so that all organizations look alike (e.g., they are all populated by people who have some sort of relationship to each other). We can also operate at a low enough level of abstraction, so that we are dealing with extreme detail (e.g., organization A has a president with red hair and organization B has one with black hair). The organization theorist, in contrast to the clinical psychologist or psychiatrist, operates at a level on which he recognizes what may be significant individual differences, but which allows him to concentrate on patterns of behavior that account for much of the overall variation among individual cases, *with respect to the variables about which he is concerned.* For example, if we are interested in trying to understand or predict the value orientations of the man filling the role of chief technical officer (CTO) in a decentralized firm, we may find it interesting to know exactly which university he attended, his precise specialty, his thesis topic, the people he studied under, and other details of his education. However, our theories and propositions are nowhere near good enough to use all of this detailed information in attempting to explain the problems he is having with his nontechnical colleagues. On the other hand, we can make a crude classification of people occupying this role according to science versus engineering training, amount of experience at the bench level, previous experience in industry or business, and a few other gross characteristics of their backgrounds. This analysis might allow us to distinguish roughly between those who probably will have major colleague problems and those who probably will not, given roughly the same type of situation—for example, a company-wide conflict over who should support and who should evaluate long-range or basic research.

With this methodological preamble, we shall attempt briefly to describe several patterns of the evolving organization of technology in large decentralized companies.

Let us consider first the case of the divisionalized company that had previously been centralized or functionally organized (Figure 2-1). Since the firms in our sample are among the largest in their industries (and almost all have been in *Fortune's* top 500), they typically had an R&D effort of significant size, depending on the degree of science and technology underlying their products and processes. In the chemical companies, this would usually involve several hundred professionals and in the engineered product firms it would usually involve several dozens or scores of researchers, including a large number of non-degree-holding "engineers."

When divisionalization or decentralization into profit centers, based on common technology, product, or market, is being considered, one of the major questions is: What about R&D? Arguments for maintaining the corporate lab as an organizational entity are countered by arguments that "true" decentralization requires that the division manager have available all the major resources needed to do the job and that this certainly includes technology. Figure 2-3 lists some of the issues which influence the initial and subsequent decisions about the establishment, role, configuration, and funding of the CRL.

One major pattern we encountered was an attempt (dominant in the food and metals groups in the earlier days of divisionalization) to maintain the corporate lab as an organizational entity but to modify its overall mission, specific project portfolio, and funding bases. This modification typically involved making a distinction, within the CRL's project portfolio, between work done specifically for one division and work done for multiple divisions or no current division in particular.

---

1. Relative power of corporate versus divisional executives.
2. General divisional reaction against staff overhead and interference.
3. Growth of strategic planning activity in the firm.
4. Lack of R&D inputs into corporate strategy and market planning.
5. High diversification of company product lines, technologies, and markets.
6. Increasing questions about the relevance of corporate R&D.
7. Focus on short-term cost savings for R&D should give corporate R&D a role in long-term R&D, but that is being downplayed.
8. Corporate R&D budgets are generally not growing because they are part of the discretionary budget without powerful constituency.
9. As divisional labs grow stronger, they do not perceive need for corporate lab.
10. Is the vice president of R&D really the chief technical officer of the firm, that is, part of the inner circle of corporate policymakers and decisionmakers?
11. The time lags in getting corporate R&D started and in producing results.
12. The sources and uses of R&D funds in the corporation.
13. Most corporate labs have no mechanisms for going beyond lab-scale or semiworks scale of production.
14. Who "owns" the pilot plants and/or access to full-scale production facilities for testing?

Figure 2-3. Some issues in corporate–divisional R&D relations.

As the operating divisions begin to flex their decentralized muscles and start acting as though they were indeed independent enterprises (except for legal and financial matters), they begin to become impatient with the level or quality or relevance of the work being done in the central R&D activity. For those division managers who see a real need for strong, direct technical inputs to their division's operation, the central lab seems unwieldy, distant, and not very responsive to their immediate and near-term future needs.

They occasionally snipe at the central lab, question its budget and program, and strongly hint that they could do a better job in providing technical support for their division's operations. If they have an engineering department of reasonable size and competence, they will begin to route problems, funds, questions, and projects to their own engineering group and will look with favor on the chief engineers' proposals for hiring some additional people to carry this new load.

The division managers will take every opportunity to point to the awkwardness and irrationality and infeasibility of having to depend on a central lab for their product and process support. They will give many reasons for why it does not work and, if they are being "taxed" to support the central lab, they will complain frequently and loudly that they are not getting their money's worth from it. Many will complain that it is costing them double to get their work done, since they are paying for divisional people to do it anyway, because the corporate people cannot or will not "do it right."

2.5 THE FRUITLESS SEARCH FOR THE BEST OR OPTIMAL ORGANIZATION FORM

A major conclusion of our research and experience with technology in the decentralized firm can be summed up in the title of this section. Throughout the formal study, which ambled along at an academic pace for more than 15 years, our most striking observation was the universal (sometimes desperate) search for the "best" way of organizing technology, for funding it, for controlling it, and for formulation of its project portfolio. On empirical grounds, we can say there currently is no optimum, because none of the companies studied appeared to have found one, despite many years and many dollars worth of staff work and consulting fees spent in the search. Further evidence is the almost constant change from one form or one strategy to another in most of the sample companies, and the widespread dissatisfaction of many management people with the present form or strategy of R&D in their organizations, *whatever it is*.

Managers in firms that had organized and were operating technology on a very decentralized basis were convinced that if they could pull it together into more centralized form, things would be a lot better. Managers in firms which had very centralized technology felt that things would improve drastically, if they would only loosen it up a bit and move in the direction of more decentralization. This is what they *said*. What they *did* provided even better evidence of the dissatisfaction and the search for a "better or best" solution. In almost every one of the sample companies, there had recently been, there currently were under way, or management was anticipating major changes in the way technology in the corporation was organized, controlled, and/or funded. The interesting aspect of these changes was that as many companies were moving or planning to move toward more centralization as toward more decentralization. The arguments used by many of these managers included a healthy portion of "Company X is organized that way and they are successful so how can we go wrong if we move in that direction." Unknown to them, Company X had already come to the conclusion that the way they were organizing technology needed changing and was planning to move the opposite way.

Are these attitudes and behavioral patterns irrational? Do they reflect more aimless casting about for easy ways out? In some of the situations we studied they are and do. In most cases, however, they represent overreaction or lagged reaction to situations that might have existed for some time but had not been noticed or carefully analyzed by the managers who were now advocating drastic change. Frequently, there appears to be no alternative to a massive reorganization or shake up, even after careful study (typically by outside consultants). But this "no alternative" situation may have been preceded by several years of clues and symptoms frantically signaling minor changes and adjustments such as changes in reporting procedures, funding methods, organizational structure, incumbent personnel, and coordinating and transition mechanisms. In other words, technology was not under sufficient management control in the precrisis years to allow minor adjustments to bring its operations and output in line with desired patterns or levels. This leads to another major point of this book (also a fairly obvious one): in the new atmosphere under which technology will be operating in the 1990s, it behooves both technology management and general corporate management (but especially the former) to put more effort into designing and controlling technology in the firm so as to anticipate the need for, ameliorate the effects of, and in many cases avoid major periodic changes. I shall say a lot more about this throughout the book, wherever the opportunity presents itself. For I believe that, despite the uniqueness and complexity of technology in the firm, managers can greatly improve their understanding of it and their control over it, if they are willing to devote sufficient attention and other resources to it.

But this digression left the question of why managers in our major corporations should be exhibiting this seemingly irrational and inefficient behavior of continuous, sometimes apparently haphazard, search for an optimum that does not exist. Evidence here is from theoretical, rather than empirical, sources and is the main reason why a group of organization theorists (my colleagues and I) initiated study of the decentralization phenomenon in the first place and have stuck it out (or dragged it out) for all these years.

There is a substantial body of literature in the behavioral sciences, economics, industrial management, administration, and other related fields on the behavior of people in large formal organizations. One major segment of this literature is called bureaucratic theory. Other segments deal with topics such as transfer pricing, communication, decisionmaking, control and evaluation, innovation, and liaison relations. In addition, there is a growing literature on the subject of decentralization itself. In developing our propositions and models in an attempt to make our formal studies of technology in decentralized companies moderately rigorous, we have drawn heavily on all these fields of study. A major conclusion that arises from theoretical analysis is that varying positions along the centralization ··· decentralization continuum are subject to, or themselves set in motion, a series of events, attitudes, behavior patterns, environmental reactions, and other forces which tend to lead to instability over varying lengths of time. For example, extreme centralization of R&D in Company X, with all the R&D facilities and people concentrated at the corporate level leads, within a few years, to the emergence of "smuggled" divisional R&D groups, under other names. This pattern has been observed many times in companies with centralized R&D. Likewise, a method of funding corporate technology primarily or exclusively through divisional "grants" or "taxes" has proved to be unstable in many observed instances.

The underlying reasons for the instability (and subsequent traumatic changes) lie in the variables that occupy the attention of the theorists described above. Some of these variables are distribution of power among managers, perceived organizational climate for cooperation or competition, responses to attempts to control, bargaining and gaming behavior, opportunity cost, risk taking, and investment time horizon. In other words, centralization–decentralization is a very complex phenomenon and may be examined from many different viewpoints. Our attempts to bring several of these viewpoints together have yielded many insights (which I am trying to share through this book), but no definitive answer to the question: Which is the *best* arrangement for technology in the decentralized company?

Perhaps I can soften the blow of reporting that there is no optimal way by saying that most organizational forms or formulas have their own intrinsic and general advantages and disadvantages, and that any particular form, when applied in any particular organization, will have some unique or idiosyncratic advantages or disadvantages. For example, an electronic company, run by a talented and prolific scientist (as many small ones are when they start out) may operate beautifully under tight, centralized control. Another company of the same size, in the same business, and with the same formal organizational characteristics may fail miserably under that system. So may the original company under another manager, or when it grows, or when business turns down, or when the state of the art underlying its products changes, or when the scientist–entrepreneur gets older, more tired, or less enthusiastic about direct involvement in the technical work, and so on.

To be completely platitudinous: organizations change, people change, and environmental conditions change. There is no reason to believe that the organizational arrangements that were fine for one set of circumstances will continue to be so for other, quite different, sets of circumstances. Although the influence of organizational form and formal organizational arrangements cannot be discounted, there is sufficient flexibility in many positions along the centralization ··· decentralization continuum to absorb and adapt to changes without precipitating periodic organizational crises and massive reorganizations. Periodic, *planned* changes can help offset the accrued disadvantages of operating at any given level of decentralization or centralization. However, these changes need not be frequent, massive, or destructive of morale and ongoing projects.

2.6 HOW CAN AN R&D LABORATORY
BE PHASED OUT GRACEFULLY?

From the increasing number of cases of closing up or chopping off an R&D lab that we have encountered, it appears that it is difficult, if not impossible, to perform this kind of surgery gracefully. Almost all the cases we have observed are highly traumatic, have long-lasting effects on morale, and serve as a specter hanging over other R&D units in the organization.

The reasons are not difficult to understand. Many laboratories, especially new and venturesome ones that are most vulnerable to being cut off, are associated with individual scientists or administrators who are pioneers, advocates, or champions of a new approach or a bold dash into unknown territory. If and when they fail decisively, or the whistle is blown on the venture, everyone knows whose failure it is. Such memories linger long after the champion has gone from the company or apparently recoups by some later successes. We picked up many anecdotes and case histories on these failures and they were uniformly blamed on too much risk taking, fuzzy thinking, or some other human failure, as opposed to many production, marketing, and investment failures, which were often chalked up to the "fortunes of the marketplace. " In addition to the general issues about corporate R&D labs given in Figure 2-3, some more specific factors influencing the reduction or elimination of CRLs are listed in Figure 2-4.

In our study of U.S. laboratories in Europe,[*] we encountered only a few cases of complete closing down of European laboratories, but one of them remained a cause celebre in European R&D circles for many years and is frequently mentioned when the subject of U.S. R&D in Europe is discussed. If we were to accept some of the tentative conclusions in the technology management literature about the decrease in creativity with organizational age, we would wonder why the closing down of R&D labs is not a regular process in "organizational renewal. " We discuss at great

---

[*]Forster, 1970.

---

1. Increased emphasis on cost saving and short-term R&D payoff.
2. Extreme diversification of many large corporations.
3. Emergence of the strategic business unit (SBU) and further decentralization.
4. Emergence of corporate strategic planning dominated by finance and marketing.
5. Growth by merger, acquisition, license, and other nonresearch methods.
6. Lack of evidence of the payoff of corporate R&D—whether it has really paid off or not.
7. Low level of influence of corporate R&D people with division managers or "businessmen" in the firm.
8. Poor internal marketing ability of corporate R&D people.

Figure 2-4. Particular factors influencing the reduction or elimination of corporate research labs.

length, in the section on changes in organizational form, the bad side effects of too frequent changes in organization on the ability of R&D to develop its capabilities and bring long-term projects to a successful conclusion. Let us now argue that a *planned* program of organizational *renewal*, which might include the orderly phasing out of particular R&D labs and programs, could be quite beneficial to the organization and to the members of the R&D activity themselves.

The argument (somewhat disorderly itself) goes as follows. For most researchers and R&D organizations, the processes of organizational aging and technical obsolescence start early in their life cycles and continue at an increasing rate.

Some individuals and laboratories manage to keep the rate of increase of obsolescence down to a tolerable level and perform quite well for decades, without showing the effects to nonresearch people. That is, they earn their pay or budgets in terms of apparent payoff to the organization that provides them, even though the payoffs may be much less than is theoretically possible or even reasonably feasible in their field and circumstances.

Feelings of guilt and frustration are widespread among R&D personnel about this gap between what they are accomplishing and what they know or believe is attainable with the same level of resources. Along with the guilt or frustration, they have many rationalizations to account for the gap. These include lack of a supportive organizational climate around them, lack of understanding or risk taking on the part of management, lack of time and encouragement to renew themselves and the organization technically, and difficulty of recruiting new and up-to-date talent because of either market conditions or company restrictions.

Many of the technical people who are going through the above process would like to take a step back (to retool in terms of theoretical knowledge or techniques) in the hopes of a later leap forward (in ideas and problem solutions). The costs and risks in most organizations are too high, however. It is not generally expected that a highly paid scientist or engineer will admit that he is over his head technically or in need of further training or technical rejuvenation. Certainly, many of the large companies in our sample that support graduate education for their technical people provide a better atmosphere in this respect than those who do not. But even among the former, there are very few programs that provide and very few incumbents who are provided with the several years of full-time study that are needed for a major technical overhaul or rejuvenation, especially when the retooling is in a much different field.

A major in-house alternative to massive reeducation of its technology staff is the potential opportunity an organization provides for exploring new fields via a new laboratory or a thrust into quite new and different fields by an existing laboratory. We have observed, however, that this opportunity is often missed in a number of ways. They may staff a new laboratory or new area with people from outside the company, not offering veteran R&D people the chance to participate. They may initiate the new activity with such fanfare and such high expectations for quick results that they make this a very high personal risk for which anyone can volunteer. They may insist on or allow all the traditional areas of technology activity in the company to continue at the same pace and level, thus binding the

R&D people to current projects, programs, and fields because of their experience and real or presumed expertise.

All this rather loose argument leads to the following suggestion: a program of initiating new technology areas and laboratories on a regular basis, accompanied by provisions for technical personnel to go into them on a low-risk, almost fail-safe basis, by providing job and status insurance in case they do not like it or the venture terminates for one reason or another. I could go even further and suggest that periodic forays of this type be *required* of all technical personnel, with the nonsubtle implication that they must keep studying and keep reaching out into new and advanced fields. In this way, technical obsolescence—a malady endemic to technical organizations—can be slowed (or conceivably reversed) and a nasty organizational problem—the phasing out of R&D programs and laboratories—can become almost routine.

Of course, the above suggestion is in the policy category and says little about how to do it. The specific answer for each organization, however, is lots of hard work and a willingness to experiment with alternative organizational designs and approaches to organizational change.

2.7   WHAT ABOUT BASIC RESEARCH IN THE DECENTRALIZED COMPANY?

This is one of the issues that is subject to very high statistical and definitional noise. The attempts by the National Science Foundation and others to count or measure the amount of "basic" research done by industrial companies (and others) has led to much confusion about its definition and how much is actually being performed. If we were to take some of the strict interpretations of the term and use the common operational indicators of "free choice of topic" and "freedom to publish," we would find precious few industrial firms paying more than lip service to the term. However, let us be a bit less pure in our interpretation and talk of the portions of large company technology budgets and personnel devoted to research described as nonfocused, exploratory, background, problem-oriented basic, non-mission-oriented, curiosity oriented, long-range, or other ambiguous terms. Under these headings, most of the sample companies had some sort of effort going—either a few individuals, a group, a department, or even a separate laboratory. Whatever we call this activity, it typically is not obviously or immediately related to current products, processes, or services or the operating divisions. It is often, in fact, technically disjoint from the bulk of the existing technical work in the firm. For example, many chemical companies have small groups of biologists at work in such an area; many electromechanical companies entered electronics through such groups; and many electronic and oil companies have diversified or oozed over into chemical or materials fields via such activities. Not all such groups, however, are made up of disciplines drastically different from the ones normally pursued in the company. An overwhelming majority are merely groups digging more deeply, more widely, and for longer time periods into the underlying technologies and phenomena on which the firm depends.

One thing that most of these efforts have in common is that they are typically the focus of controversy in the company. With only a few exceptions, such groups, if they are really "far out," are subject to much criticism and sniping from operating managers and even from members of other technology groups in the company. Remarks such as the following are common:

- What have they done for us in the 5 (6,7,10) years they've been going?
- Most of the stuff they turn out can't be made or sold.
- That theoretical talent should be applied to our urgent current problems.
- They really are in left field as far as company interests are concerned.
- They aren't interested in our problems, they just want to have fun.
- They couldn't design a workable product or process if their lives depended on it.
- Why didn't they stay in the university if they wanted to play those games?
- They're giving away all our company secrets in their conference papers and publications.
- We could certainly use the money they are wasting for much better purposes.

The arguments for such an activity in the large decentralized company are persuasive, despite the doubt and criticism. The combination form of organizing technology in the company is based on a sound premise, which has strongly been supported by our observations. Decentralized operating divisions generally have project portfolios that are truncated to exclude this type of activity and only a few cases were observed where "basic" research in a divisional laboratory was of sufficient size and duration and had sufficient management support to be viable and capable of performing the role expected of it. More will be said about this in other sections, especially in those discussing the pressures on the divisional technology manager. For the present, however, I am reporting that there is a vanishingly small amount of such work going on in the operating divisional laboratories of most large, decentralized companies. If it is done at all, it takes place in a central corporate lab or separate institute (e.g., some of the pharmaceutical companies or a few of the major oil companies).

The question addressed briefly here is why a large decentralized firm *should* engage in this level of risky, often costly, long-term, and frequently controversial technical activity.

The wisdom literature, especially that written by partisans of research, provides the expected platitudes on why a large firm, as well as government agencies and the society as a whole, should carry on and/or support basic research. These arguments often appeal to social conscience and to society-wide reasons such as adding to the store of human knowledge or replenishing the barrel into which you have been dipping your R&D ladle. These are certainly good reasons, but hardly sufficient to influence the resource allocation and budgeting decisions of realistic, profit-oriented managers. We have found, in our interviews and informal discussions with managers over many years, that arguments such as these make good topics for speeches but do not provide the primary basis for decisionmaking in the firm.

The main purpose of this section is to discuss what we have learned about how basic research is and can be organized in the large decentralized firm and not to make a case for or against it. However, some of the reasons and justifications that we encountered might be of interest:

- To improve our basic understanding of the materials, processes, and phenomena with which we deal.
- To improve our image with respect to the academic and scientific community.
- To help in recruiting high-grade technical people.
- To give us a window into new areas of technology before they become widely disseminated.
- Because, who knows, we might come up with a fundamental breakthrough of proprietary value.
- Because it is not too expensive on a modest scale and the efforts of one or a few scientists can provide a big payoff in terms of entry into new fields or even possibly a new product.
- Having some opportunity for doing "their own work" helps to keep basic-oriented scientists happy; its a fringe benefit.

In the overwhelming number of companies we studied, whatever basic research is going on takes place in the corporate research laboratories (CRLs). In a few cases of very large divisions in very large companies, where the divisions are in effect "independent" companies in every respect except financial, some basic work is done in those divisional laboratories. The latter case is frequently found in a large conglomerate where the current divisions used to be independent decentralized companies with a CRL. In most cases, this effort is segregated from the rest of R&D in the company and even frequently from other work in the CRL. This separation typically gives the illusion of remoteness, disinterestedness, and irrelevance to the main stream of activities in the company. Sometimes this illusion is fostered deliberately to "protect" the basic group or to endow it with prestige. More often the illusion is not only not intended but is condemned continually and presents a rationale for the periodic flogging of the basic research people and their boss, the vice president of research, for "not relating more to the business."

Our contacts with personnel in dozens of such basic groups suggest that a number of conditions are seen as necessary by their members, if they are to do their work effectively. These perceived conditions do not necessarily emerge from careful research-on-research studies of the conditions conducive to effective basic research in industry, but most of them make sense in terms of the basic research process itself and its closest model—university research. Some of these conditions are:

- Free choice of problems and, some argue, of areas or fields, regardless of their relation to present or foreseeable future interests of the company.
- Free access to the literature, to colleagues, and to other sources of information.
- Freedom to publish in the open literature as soon as something worthwhile is learned.
- A large enough group to provide mutual stimulation and a quasi academic atmosphere.
- Sufficient operating funds and supporting services to permit following up promising leads and opportunities.
- Insulation or protection from short-range and firefighting projects.

Ironically, most of these conditions are antithetical to the industrial corporation's usual mode of doing business. This leads, in most cases, to one of the following sets of circumstances:

- *Protection of the Basic Research Group by Its Patron(s) or Sponsor(s) Against Incursion of Day-to-Day Matters by Organizational, Geographic, or Other Means.* In turn this can lead to intensification of the illusions cited above and the group's isolation from the rest of the company. This can be particularly disadvantageous when members of the group do have something they would like to pass on to other groups in the company, or when they need information or the use of unique and expensive pieces of equipment (e.g., a reactor, or rolling mill, a furnace, or a paper machine) that is used in day-to-day operations.
- *Attempting to Merge the Basic Group and Integrate It Better into the Company.* This can lead to erosion of the protective shield around the group and may have a shortenng or narrowing effect on its portfolio. Another possible effect can be the loss of key basic research people who find the situation too difficult to continue in.
- *Attempting a Compromise Whereby the Basic Group Will Primarily Work on Its Own Long-Range Projects But Will also Be Available for Occasional Consultation on "Practical" or Current Problems.* This situation is inherently unstable, but some companies have managed to operate basic research groups for many years under these conditions. The cost, however, includes constant attention to potential conflicts and friction, frequent compromises, and lots of sweet talking to the researchers and their customers.

One of the worst things that can happen to a basic research group is to become too successful in solving current, practical, short-term problems. As part of their thanks, they often get criticism for not helping out more often and for longer periods. Some of them handle this problem by a rotation or other system that puts some members on the firing line for certain periods and protects the rest against intrusion. The more frequent case is that their lack of ability to solve most short-range problems becomes obvious when they make such attempt(s) and operating people are then glad to leave the "impractical long-hairs" alone, other than continually to gripe about the money and facilities they are wasting.

Are the actors in the situations described above "bad guys"? Are they naive? Are they disloyal to the company? Perhaps some of them are all of these. But more likely their attitudes and behavior with respect to basic research have been conditioned for a long time and may be based on bitter experience. In any event, regardless of the contribution to this conflict of individual values, personality, and experience, the operating people are indeed operating in a different culture than the basic researchers and view the world much differently.

This problem of the clash in values and perceptions of what is good and right and important comes up in several places in this book. It is especially prominent in our discussion of transition of results from the central laboratory(ies), if any, to operating divisions.

Suffice it to say that the formal results of our study, in support of many other less systematic observations, indicates that basic research does not fit comfortably into the industrial corporation, except under certain very special conditions. Some of these conditions are well met in certain large, diversified, decentralized companies that have a combination form of R&D; however, some of them are very rare in other companies of the same type.

Among the most pressing questions about basic research in the large decentralized firm are these: Who can afford it? Who can afford *not* doing it? How much should be spent on it? How broad should a basic research portfolio be? Almost all the large companies we studied appeared able to afford some level of basic research, as we have been defining it. One piece of evidence for this is the size of the research budgets and R&D staff. Figured at only 5–10% of their total R&D effort, all but a very few could have groups of at least 5–10 people carrying on this kind of effort. Indeed, most of them have at least that number of people nominally engaged in the kind of activity we described as "basic research" in the early part of this section. Another piece of evidence requiring a string of inferences and risky generalizations is that the productivity of R&D in many of the companies was perceived by our informants (top management, operating management, technology management, and R&D professionals) as unsatisfactory, needing much improvement. Of course, they all had their own theories and reasons for this, but among the reasons frequently given was the lack of a vigorous, large enough, high-quality, long-term basic or more basic R&D effort. In support of this argument, numerous incidents were related of the same technical problem being tackled over and over again in slightly different form in a short-term, crisis atmosphere, at a superficial level. Advocates of the more basic approach argued that even a modest long-term, basic effort could pay for itself many times over if given the chance. Many respondents argued, in addition, that their company could not afford *not* to devote a portion of its R&D budget to such basic work. The reasons they cited included the following:

• Rapid changes in the state of the arts underlying their technology.

• Incursion of new research-intensive firms into their market.

- Running up against natural laws and engineering principles that placed a limit on how much they could continue to improve their products, materials, and processes by simple extrapolation or minor modifications.
- New scientific or technical fields that threatened to wipe out the technological or competitive basis of their business.
- New materials whose properties had to be explored in terms of applicability, costs, advantages, disadvantages, hazards, and new uses.
- Environmental constraints, such as those arising from public reaction to threats to health and safety.

Added to all these practical internal reasons, there are certain outside pressures that have varying effects over time and between firms. In the 1950s and 1960s many of these large companies had to establish and maintain basic research opportunities to attract and hold high-quality Ph.D.s from the universities (and from competitors' laboratories). This influenced the establishment and expansion of many of the basic research efforts we encountered during that period. In the 1970s and 1980s, however, the situation has changed radically in a number of scientific disciplines, notably physics and chemistry, and in many fields of engineering science. Periodically, new graduates have faced a very tight buyers' market and their demands to be allowed to continue work on their (sometimes esoteric) dissertations are not being received as well as they were. There are many contributing factors that seem to be heralding a period of reduced support for basic research in large industrial firms (despite some of the statistical evidence to the contrary). The drastic cutbacks in government support of this kind of activity directly affected basic research groups in the military and aerospace companies in the 1960s. In the 1980s, increased military spending reversed this trend in certain areas.[*] However, another factor that increasingly threatens the allocation to and moral support for such activities in the primarily commercial firm, with no government R&D money, is management's disappointment in what they perceive they have been getting for their money spent on such research.

Although there is no hard evidence yet for an across-the-board retreat from a strong long-range R&D activity, there is ample evidence of holding the line, pruning out nonpromising projects, and increased scrutiny of projects and programs for near-term relevance and payoff.

All this can put the advocates and performers of basic research in industry on the defensive and it has.

For the very, very large companies in our sample (among the top 20 or 30 in the country) who are in a wide variety of markets and technologies, there seems to be no question about the need for a part of their total R&D portfolio to be devoted to the kind of basic research we have been discussing. Both top and operating managers in most of these companies acknowledge this need and do not debate

---

[*] A widely unrecognized feature of the large increases in military R&D is that the basic research share of the increases has not been proportional and that some fields of basic and exploratory research have been levelled or even cut.

it *in principle*. However, questions such as how much to spend, how long must we wait, how can we measure payoff, and how can we increase its output are frequent and widespread. Even among these top performers of basic research, there are many doubts about how to organize, control, and evaluate it; how to tie it in better with company objectives; and how to deal with the members of the basic groups and better integrate them into the company "team."

One of the most frequent issues we encounter in this area is the question of how broad a front and how deep a thrust an industrial company can occupy in basic fields—such as physics, chemistry, biology, materials science, mathematics, and psychology. Acknowledging the conditions that scientists appear to need in order to be *scientifically* productive, these managers want to know the conditions under which their scientists can be *economically* productive. Can one or two people, working in a very narrow field of a sub-subspecialty in chemistry or physics, contribute anything useful to the company, besides some papers in an esoteric journal of their specialty?

I wish we had answers to such vital questions. I can only say, based on this study and related studies done over the years by members of our Program of Research on the Management of R&D/Innovation and others, that the R&D community has developed no satisfactory means of measuring the productivity of basic research—even of the nonacademic type we have been discussing in the context of the industrial company. This is not because practitioners and academic observers of the R&D process have not tried. Quite the contrary. There have been many attempts, none of them successful so far, in providing general methods that satisfy both the practical and theoretical requirements of a valid evaluation system. Publication counting is still fashionable for measuring the output of basic research, but it fails to meet the needs of industrial managers for output measures that are more directly related to company interests.

Due to this situation in the measurement of output of basic research in industry, questions of how much to spend and how long to wait are based on other considerations in the firm. These other types of measures are based on *input* and *process* considerations, rather than on anticipated *output*. Among these considerations are the following:

- What basic[*] fields are key to our technoeconomic future (materials, processes, products, markets, general know-how)?

---

[*]I am convinced that this loose use of the term "basic research," as defined in the context of the industrial firm, is going to get me into a lot of trouble with readers, especially those who skim this section without having struggled through the earlier, tortuous "nondefinition" of basic research in industry. I hope you will bear with me throughout this relativistic discussion, since we considered basic research in our sample companies as a relative matter—relative to the main level or thrust in that particular company, in particular industries, and in particular fields of technology. For example, an appliance maker who employs two or three physicists, a biologist, and a sociologist to do "systems" studies is considered by many of his competitors as "way out in the blue sky." However, this same group of specialists might not stand muster as even an "advanced development" group in some other fields or companies.

- What is the cost of entry? Can we buy in by picking up an existing group or by buying a piece of equipment incorporating the technology, or buying a sample of the new material? Must we start from scratch and hire new graduates who are in the forefront of the field, or can we pirate specialists away from other companies or even grow our own specialists?
- What is the lowest level we can enter the field with—lowest with respect to number of people, budget, and support activities?
- What is the minimum size that will protect us against an abort, if one key person leaves or moves out of the field?
- How much attention, support, and encouragement must be given to the group to keep it going at a reasonable pace and with at least indirect connection to future interests of the company?
- What kind of leadership should be provided, and how should it be linked to other R&D groups in the company?

As indicated, these are essential *input* and *process* measures, and although they *should* be linkable directly to the probable quantity and quality of *output*, this linkage is only dimly perceived at present. Certainly the older philosophy of "getting good people and leaving them alone" is much out of fashion in industry even as a philosophy, much less as an operating principle. In most large companies, the view is that members of the basic research group need, among other things, guidance, stimulation, familiarity with the company's interests, and an occasional well-placed boot to keep them moving in the right direction. The myths surrounding a few of our best-known basic research performers in industry turned out to be just that. Except for a very few unique individuals, most of the basic researchers in these famous laboratories work under fairly close surveillance and within well-defined limits of what is or is likely to be of interest to the company.

On the question of minimum effective size of the basic group, some fields contain their own answers. In a number of cases the minimum group size is heavily influenced or even determined by a complex piece of equipment, the need for interlocking specialties that are rarely held by one person, numbers of investigators required for massive screening or experimental programs, or other considerations. Less obvious, but equally compelling in some of the groups we have encountered, is the need for enough people interested in the same set of phenomena, regardless of their specialties, to interact, to stimulate each other, and to provide internal scientific checks on each other's work. The most commonly encountered size of basic research groups varies radically with the size of the total R&D effort, the level of sophistication of that R&D and of the industry, the progressiveness of company management, and other factors.

Few of the companies maintain a complete spectrum of basic research capability, analogous to the departments in university science and engineering schools. Most of them are heavy in the primary discipline(s) underlying their technology (e.g., organic chemistry, solid-state physics, nonferrous physical metallurgy) and have only token groups in the other fields to serve as scouts and early warning mechanisms. This means that, in order for a reasonable sized CRL to provide adequate coverage of fields of *potential* relevance to the company, as well as those of *obvious* importance, they must maintain strong contacts with the university community. This then leads to the circular process whereby access to good university departments is facilitated (in some cases made possible only) by having a high-level internal basic research effort; these contacts provide the basis for recruiting the best people available and even those who are not available full time (through consulting arrangements); and these people help strengthen the quality of the internal group and thus provide access to the university. The non-R&D manager generally watches this process operating over a period of years (it can take a decade or two for an industrial laboratory to gain or regain a scientific reputation) and can only hope that something of direct and proprietary value to the company will spin off in between the seminars, colloquia, published papers, and other primarily scientific activities.

In the past 30 years, we have encountered scores of companies attempting to set up, eradicate, or drastically reorganize basic research efforts. The issues raised above were prominent in most of these cases. The frequency of appointing new managers for the groups has increased rapidly during the past 10–15 years, and the number of committees and special studies set up to look into their effectiveness also has risen rapidly from the earlier post-WWII period when they were largely left to their own devices by company management.

Despite this changing and sometimes chaotic situation, we do not see how the large modern corporation can long survive without an effective linkage to the ongoing worldwide scientific efforts in the fields relevant to their underlying technology. This need not necessarily mean a full blown in-house basic research capability. A competent, "large enough" technical liaison group, which is viewed as legitimate by the scientific community, can perform the linkage function in some areas. This device has easily reached limitations, however. If the university or government laboratory personnel perceive the company as only a taker and not a contributor to research in their field, or if they perceive that the liaison people are not working scientists and cannot participate at a professional level in scientific discussions, they may soon be conveniently away when the liaison people come to call. This situation occurs even with paid consultants who may grow tired of talking to uninformed, nonresponsive researchers in a laboratory, despite the (sometimes) high fees they are getting.

One way of overcoming this—an approach used by many large companies—is to establish basic research credentials in one or two special subfields and trade on them in contacts with specialists in other fields. Another way is by making expensive specialized equipment available or by providing summer job opportunities for faculty and graduate students. Chapter 6 discusses other mechanisms of industry–university collaboration. Whatever the technique, the price of admission to the scientific community must be paid, although the benefits are not clearly seen during the long period of waiting for discernible payoff. Most frequently this payoff is not in the form of an identifiable new product or process but comes in more subtle forms, such as increased sophistication of the company's total technology base, the avoidance of technical blunders, and early warning of opportunities or threats from new developments in science and technology.

2.8 CHANGES IN ORGANIZATIONAL FORM

There is no question that decentralization has been a successful, if not an inevitable, form of organization for the very large, diversified, geographically spread corporation. This is not the issue in this book nor is it the issue in the periodic reorganizations that such firms undergo, sometimes at an alarmingly high frequency. The practical issues seem to be *how much* they should decentralize and how they can cope with the problems that arise from and are increased by decentralization.

In the earlier days of corporate decentralization, divisionalization, and profit center establishment, there was real doubt in the minds of many managers and observers of industrial organization about whether this new form of organization was viable and efficient, as compared to the traditional functional form of organization. As firms grew larger, became more diversified, and spread out all over the country and around the world, the traditional form began to lose ground in favor of the form we have been referring to as decentralized or divisionalized. There are very few instances now where the periodic reorganizations result in a complete return to the functional form. The issue seems to be rather what degree of recentralization is feasible or desirable and which elements of the organizations should be pulled in tighter to central control and which should be allowed to operate in a looser decentralized manner.

A large percentage of the massive reorganizations that we have observed appear to accompany the "new broom" situation, when a new top manager or new top management group takes over. If they are from outside, we may expect some remaking of their new company in the image of their previous one(s). If they are from inside, they may just feel that a massive change has been needed for a long time to either "pull this thing together and get it shaped up" or "get rid of a lot of unnecessary functions and paperwork at the corporate level, so that we can get some planning and diversification done." Frequently, the reorganization and the announced reasons for it are direct reflections on the new management's predecessors: "They were too bogged down in detail" or "they didn't have control of the operation." In other instances, there is merely the recognition that "it is time for a change," with no onus on anyone for the problems of the present situation. The philosophy of change for change's sake is not usually the public reason for massive organizational changes, but some managers sincerely believe that a periodic shake-up of the existing organization can "keep people on their toes" and "keep the barnacles from growing." There may be some merit in this view, and there may also be some direct and indirect benefits to such shake-ups, but the harm they can cause is great if done without careful planning, control, and a good rationale.

There are many side effects of reorganization which do not appear immediately and which may never be greatly apparent. They may be as direct as the loss of valuable managerial time and the aborting of important projects before, during, and for some time after the change. But there may also be more subtle effects, such as decreases in confidence between top management and others in the firm, jockeying for position in the new structure at the expense of current duties and planning of long-term operations, and creation of a tense atmosphere as people wait for the next move in the game: "Here we go again, boys; this time I'll bet they recombine Divisions A,B,C; redivide Division D into its two previous parts; and wipe out the newly formed Divisions K and P." Aside from these reasons for caution in the frequency and magnitude of reorganization, there are some fundamental aspects of organizational life that must be taken into account when a change is contemplated and when its costs and benefits are being weighed. A major one of these is the natural "frequency–response characteristics" of whole organizations and of particular functions such as technology.

While no direct analogy to electrical or mechanical systems is intended here, there is a resemblance between the erratic behavior of such systems and human organizations, when the conditions under which they operate are changing at too rapid a rate.

Since this book is primarily about technology in the decentralized firm, let us use it as an example of the potential harmful effects of too frequent changes in objectives, operating conditions, or organizational structure.

A common technology change in large corporations is to set up or modify the operating charter of a corporate research laboratory (CRL). Ideally, the mission of such laboratories is described as "long range, in basic support of our current operations, and leading us into the technological future." In practical terms, it takes a lot of good people, hard work, and a long time to set up a CRL that will effectively perform these functions. How much is a long time? For some impatient managers, a year is too long to wait for discernible results. For others, who have had direct or vicarious experience in this area before, 5, 7, or even 10 years seems a reasonable gestation period. During the past 30 years we have observed many CRLs that have been in business 5 or 10 years or more and still, according to their critics, have not "produced."

This difference in time perspective and direct understanding of the R&D process and its growth and development characteristics has led to a very large number of what might be called premature shake-ups or reorganizations. They are premature in the sense that the CRL (or any other laboratory for that matter) needs time to recruit and orient new people, to organize internally, to establish projects and programs, to develop links with the rest of the corporation and with the outside technical community, and to make a few mistakes or have a few project failures before they appear to click (if they ever do).

For a moderate- or large-size CRL (50–100 or more professionals), this process almost always takes longer than their sponsors or critics expect. This leads, in a large number of cases, to intervention by the management group to "speed things up" or "shake off the cobwebs" and "get things moving." The moves they make are often wrong. They often lead to aborting activities (e.g., projects, programs, areas) that are still in the formulation stage and that are still months or years away from producing a payoff. They also lead to changes in direction of effort, which again precludes the possibility of payoff, because the effort in the original direction was just approaching a threshold or critical level. A very common class of change is illustrated as follows:

Year	Decision or Input	Effect or Response
1	Let's get into field X as a long-range diversification.	CRL begins to recruit or train specialists in field X.
2	Why is it taking so long to get organized in field X?	The best people are hard to get, we don't want to hire incompetents, and we do need a strong person to lead the effort.
3	Well, now that you've got the leader, why hasn't he produced anything yet?	He's still recruiting and training his group and getting equipment for it, but here are some of his ideas for projects and programs.
3.5	These are fantastic proposals: Where do you think we are going to get all that money?	Maybe we had better lower our sights to a more modest program in X even though we are getting some promising results.
4	Is this all the results we can expect for all this effort and money? Let's cut back on X and combine it with Y.	But we were just beginning to get some exciting results in X and if we cut it back we'll lose all we put in.
5	Do you see, we were right, X was a lousy field to get into—forget it.	We'll just keep a foot in it and won't advertise the fact that we're still working on X.
8	How come we let the competition get the jump on us in field X? Didn't we once have a project in CRL on it? Those people really do miss the boat, don't they?	We have been keeping up with X on a very low-cost basis. Now if you will just allow us to expand the effort and hire a good person to lead it....

2.9 EFFECTS OF ORGANIZATIONAL CHANGE ON THE MIDLEVEL TECHNOLOGY MANAGER

A number of problems that middle technology management faces in the decentralized company relate to the problem of "who is my boss" or "who calls the shots" or "who has the greatest influence on my future career?"

In the transition stages from a centralized corporate structure, where there is typically one R&D hierarchy, to the decentralized structure, where there may be several, the middle technology management group may be in the difficult position of having to change affiliations and loyalties: they may actually find themselves in a new organizational and geographic location or they may merely have to adapt to new lines of prescribed decisionmaking, communication, and influence.

Once the transition has been accomplished—on paper—and things begin to settle down, this middle manager may find that the problems of communicating with his new superior officer and nontechnical colleagues pose a much different problem than the ones he thought he had solved for the time being in his old location.

He may now, if he ends up in a divisional laboratory, find himself much closer to the production line and the marketplace than he was before. Or conversely, he may find, if assigned to a new corporate laboratory, serving in a so-called staff capacity, that he is more isolated from the main stream of the company's business than he had been.

These dislocations during a transition period from one deployment of technology in the company to another are more than mere once-in-a-lifetime problems for such managers. One of our most interesting observations is that most of the companies we followed over a 20-year period had recently completed, were contemplating, or were in the process of some major change in the relationship of R&D to other parts of the company. Sometimes these changes appeared superficial from an organizational viewpoint; that is, they might merely be changes in the way that the books are kept—the methods of funding. Upon closer examination, however, these changes in funding are accompanied by profound changes in organizational power and influence relationships between technology management and division or corporate management.

The striking thing about these changes is that they occur relatively frequently in the life of a company. We have observed a number of situations where fundamental changes in deployment and funding of R&D occur at intervals of 3–5 years, often in connection with a change in top management, or a changed attitude toward research in the company, or a striking success or failure in the marketplace that can be related to R&D effectiveness.

We have had the opportunity to follow several such changes over a period of years and to observe their effects on the perceptions of middle technology managements and their attempts to adjust to or anticipate the effects of these changes. Some of these effects are discussed in later chapters under various subtopics such as key players, idea flow, and technical entrepreneurship.

2.10   WHAT IS THE NET EFFECT OF CHURNING DIVISIONS, SBUs, AND PRODUCT LINES ON THEIR TECHNOLOGY PROGRAMS?

Some (many?) managers believe that shaking up the organization occasionally or even regularly is good management practice. Reorganizations, reformulation of objectives, replacement of management, and additions and subtractions of divisions, products, and product lines do get people's attention. They may even shake them out of their complacency and lead to some creative thinking or at least some new lines of thinking that are different from those into which they have fallen. Such changes can identify new opportunities, disclose below-the-surface problems that have not been visible, and accomplish other good things.

As argued earlier in this chapter, however, the frequency and magnitude of change itself can wreak havoc with technology programs and keep the whole technology effort off balance.

It is not only the direct effects of major moves like acquisitions, divestitures, and mergers that can cause trouble, but also the many auxiliary effects. Rumors of a divestiture or merger can paralyze or at least distract the technology staff (among others) for long periods before, during, and after the event, even if it does not indeed actually occur. So much time and energy can go into worrying, planning, plotting for individual survival, getting one's ducks lined up, and other nonproductive efforts that little is left over for actual technical work.

General commentary about the merits of the continuing merger and acquisition waves in the past two decades is beyond the scope of this book, although I do have some strong opinions about many of them. During the writing of this section, the quarterly list of mergers and acquisitions appeared in a Chicago business journal. The average price paid was 20 times earnings. Many of the "pieces" (divisions, product lines, and independent companies) were familiar, having been the recent objects of merger, acquisition, or leveraged buy-out by other companies. Whatever the merits of this churning, and whether or not management group A can do a better job with unit B than company C had been able to do, the impacts on the unit's technology programs, where they even survive the change in ownership and/or organizational location, can range from highly disturbing to terminal. Even where the new owners are willing to continue the technology programs (not always the case), the staff and programs often fall into much disarray during the whole change process—from early rumors to eventual settling down, if ever. This pessimistic view is not meant as a defense of the ivory tower or the fragile nature of technology professionals. It stems from too many observations of the dissolution of promising technology projects and programs because of uncertainty, personnel changes, lack of understanding by the new management, all the time lags in getting going again, and other factors.

My argument is not against well-conceived and executed business changes, such as mergers and acquisitions in the best interest of both parties. It is against completely ignoring the effects of such moves on functions such as technology, which have long time constants and which are particularly sensitive to changes in direction and levels of support, as well as the clients they must serve. Part of the planning for such major moves should be a "technology component" in which the people, projects, focus, and mission of the technology programs in the affected unit(s) are carefully analyzed and prepared for the change. Of particular concern is consideration and compensating for interconnections between the affected unit and other units in its former organization—other divisions, divisional or corporate labs, or consortia of technology activities in the company and its outside technology partners. Such planning can help avoid leaving the unit technologically "naked" after the change and unable to fulfill whatever support functions it is supposed to provide in its new organizational setting.

2.11   WHY DOES IT TAKE R&D SO LONG TO "GET ORGANIZED"?

One of the frustrations of nontechnical managers is the seemingly long time it takes for a new or reorganized R&D activity to get ready and to turn out something useful. Even where the actual elapsed time is only months or the order of a year, impatience is common while the manager waits for signs of life and productivity. Where the process takes several years—not at all uncommon for a new lab—reactions run from prodding and open criticism to direct intervention and changing management before they have had a chance to get fully organized. In this discussion we are concerned about both the kind of R&D that is tied closely to company operations, as well as that which is more remote from current operations. However, most of the comments below, attempting to account for the "organizational lag," apply to the latter kind.

Certainly the manager who has spent most of his career in marketing or production or a supporting activity such as accounting or purchasing knows that it takes time for any activity to get going. He knows that people have to be recruited or selected, procedures have to be established, and people have to learn new things. But he is typically not accustomed to the *additional* kinds of learning that are more characteristically required by technical people. Perhaps one of the major reasons for the difference (if the arguments presented below describing the differences seem credible) is that R&D is less of a "plug-in" type of activity than are many other company activities.

With respect to the degree of "plug-in-ness" of R&D as compared to other activities, I argue that the organizational learning time is more for technology than for most other functions in the firm. It takes time for a new salesperson or sales manager to learn new territory and a new product line. It also takes time for a new production person to adapt to a process that is strange to him or for an old production person to feel comfortable with a new process. However, the main difference is that the new salesperson or production people and the new sales department or production department tend to slip neatly into generalized roles that are universally recognized and accepted in the organization. There may be the usual sniffing about of the new personalities by organizational veterans and the usual tentative pecks to establish the pecking order with the new individuals or groups, but there is seldom a basic questioning of the role or the legitimacy of the function itself. In R&D this questioning of the role, legitimacy, status, and usefulness of a new laboratory is a common occurrence and has much to do with the time it takes for it to get organized.

Despite its almost nine decades of history in some major U.S. corporations, and its presence for at least 40 years in most of our sample companies, R&D is still a "strange new boy" in many companies and its role and status are continually subject to questions and controversy. Part of this is due, in contrast to the more traditional functions, to the relative lack of stable stereotyping of the researcher and his role in the organization. Many people still do not know what to expect from R&D, either in respect to behavior or output. The stereotypical caricatures of the white-coated scientist are well disseminated and accepted in the society at large. But this does not necessarily help the average employee or nonresearch manager to set his personal expectations of the particular scientific and technical people in *his* firm and in *his* area of influence. Whereas the stereotypical salesperson is expected to sell and the production person is expected to produce, the precise meaning of the terms to "research" and to "develop" are far from operational in the minds of observers. As a consequence, much of the start-up time and energy of a new R&D activity are devoted to establishing or reestablishing expectations in the minds of people with whom it must do business in the organization. Some of the specific activities (see Figure 2-5) involved in this process are described below under very loose and often overlapping labels. Taken together, however, they may help account for most of the start-up or getting-organized time and resources that are typically used by the new R&D activity.

---

1. Scouting the territory
2. Differentiating the roles
   (a) Sponsors
   (b) Clients
   (c) People who can help or hinder
   (d) Potential collaborators
   (e) Gatekeepers: access, information, resources
3. Staking out their territory
4. False starts
5. Waiting to be approached by potential clients
6. Gathering resources: people, facilities, equipment, support
7. Gathering information: about the company, markets, technology, tricks of the trade
8. Setting up the organization mechanics
9. Building internal (lab) relationships: patterns of cooperation, trust, common languages, team styles

---

Figure 2-5. Why does it take so long for R&D to get organized? Start-up activities.

1. *Scouting the territory.* Particularly in the case of new R&D activities being started up in operating divisions, the surrounding territory may be strange. For new professionals (recent graduates) and new-to-industry professionals (recruits from university or government laboratories) the operating environment may indeed be strange territory. The emphasis on schedules, fixed procedures, customer service, and other aspects of the organizational climate may be taken for granted by veterans of the organization but may be unfamiliar to the new researchers. In addition, the cast of characters may not be entirely clear to the researchers. They are not yet aware of who their potential clients, sponsors, and critics may be. They have to learn the organizational ropes, including organizational politics, which might affect their operation or even existence. As a spin-off from one series of interviews held early in our study, one of our staff members noticed a "fragmentary regularity," which was intriguing, if not statistically significant. After asking a number of researchers of widely varying seniority (new recruits to near-retirement veterans) how long it took them to learn how to operate in their particular organization, he was startled to find a rough convergence in their answers. Most of them, independent of the length of time they had been in the organization, gave a figure representing roughly *half* the time they had been there!

One reason why this initial scouting may take longer than for the ordinary organizational recruit (e.g., the new production or clerical worker) is that the formal financial, professional, and organizational status of the researcher is not consistent with his asking naive questions about the environment, a behavior that may be completely acceptable in the new clerk or production worker. The researcher has a fairly high degree of sophistication initially imputed to him, and "dumb" questions can quickly lead to a drastic loss of status. So he has to scout quietly and unobtrusively, making many small mistakes that might perhaps have been easily avoided by a formalized organizational orientation program, which is very rare at professional levels.

2. *Differentiating the roles.* In addition to scouting the general organizational environment in which he will be working, the researcher must learn to differentiate a number of informal roles that can significantly affect his work. He must identify and differentiate, either systematically or by the frequently more costly method of trial and error:

(a) *Sponsors* who pay lip service to R&D and those who will go to bat for it.

(b) *Clients* who are self-sold, those who have to be sold, and those who probably never will be sold (at least in the researcher's organizational lifetime).

(c) *People who can help or hinder* the cause of effective technology, but who are for the most part uninvolved or neutral to it initially. These may include people in accounting, personnel, information services, legal services, purchasing, building and equipment maintenance, watchmen and janitors, customer and supplier representatives, and many others.

(d) *Potential collaborators* in other functions such as production, quality control, marketing, engineering, design, planning, market research, product management. Some of these potential collaborators may have little to gain by such collaboration but may be willing to try it either for the good of the company, because they are intrigued by the possible help R&D might give them in their jobs, or for many other obvious or not so obvious reasons.

(e) *Gatekeepers* of one kind or another who control resources or access to resources that the researcher might need. In many of our sample companies the technology laboratories had miserable relations with the information gatekeepers in the organization at large—the traditional librarians, information processing people, market researchers, accountants, and others who controlled information or access to sources of information. In many organizations where good relations did exist, it had taken many years of cultivation and major behavioral and attitudinal adjustments to establish good levels of collaboration.

3. *Staking out their territory.* After scouting the territory, R&D people, especially the managers of an overall new lab activity, as well as managers of specialty areas within it, will begin tentatively to stake out areas or territories within which to operate. If, for example, personnel of the new lab feel that they can do something useful in the production process area that is currently not being done or not being done well, they may make a bid for responsibility in that area. They may find that in doing so they are elbowing aside or running roughshod over a group of process or manufacturing engineers whose territory this has been historically. Perhaps overtures to this group for collaboration may be met warmly and the R&D lab can proceed at a moderate pace to probe into problem and opportunity areas that had previously been neglected. In a similar fashion, personnel of the new R&D activity may make a bid for responsibility in the product design, quality control, customer service, standards, or materials areas. Any or all of these may have been the traditional turf of other staff groups, and the overtures of R&D may meet hostility and resistance, rather than open-armed collaboration. This may cause wise R&D people to back off. These advances, retreats, forays, and invasions of territory adjacent to the main announced mission of the new R&D activity (if one were indeed announced) take time and energy away from the main business of solving problems, turning out technical information for use by the organization, and establishing new technical fields as a base of commercial ventures.

4. *False starts* consume a lot of start-up time and energy. R&D may pick the wrong product or process for their initial thrust, wrong from the viewpoint of technical feasibility, economic justification, or political advisability. They may invest several man-months or man-years in getting ready to tackle a product characteristic (e.g., safety), process characteristic (e.g., yield), or material characteristic (e.g., performance under stress) only to find that manufacturing, cost, consumer preferences, or other factors make the potential improvement unattractive from a financial or marketing point of view. Such false starts or "learning mistakes" cannot only consume valuable calendar time and resources but can also erode the initial store of good will or tolerance toward the R&D activity.

5. *Waiting to be approached by potential clients*, rather than seeming to be aggressive, has taken uncounted calendar years by the laboratories in our sample companies. If the R&D people do not take an aggressive, hard-sell approach

to getting business or generating a stream of problems to solve, they may wait years for some of their potential clients to approach them. We encountered both extremes of initial strategy—aggressive hard-sell and passive wait-for-business. Both strategies have their dangers and we are hard put to argue in favor of one or the other extreme, on the basis of our observations. We feel confident in observing, in the special case of the corporate central lab which is aimed toward the more esoteric and basic end of the R&D spectrum, that waiting in complete passivity to be "discovered" leads to a high probability of failure in its overall mission. The reasons for this are detailed in other subsections of this discussion under headings such as "staking out the territory" and "differentiating the other company functions and roles" with which R&D must deal in order to be effective. In later sections of this book, devoted to problems of transition from the laboratory to market through the network of corporate and divisional labs and other functions, the importance of establishing and maintaining receptive relations is also stressed.

We encountered many cases where new product and process concepts lay on the shelves, in the files, on the benches, and in the heads of R&D people because they failed to market or push their ideas actively with potential clients. One assumption seems to be a reasonably good guide to selling unsolicited R&D ideas or results to other parts of the company: always assume that the potential client is too busy or too preoccupied with his own problems to tear unsolicited ideas from the arms of R&D. It is also a reasonably conservative assumption to assume that he may be too busy or preoccupied to listen to or adopt the solutions to problems that he himself asked for *at some previous time*. A familiar activity we observed in many corporate laboratories, especially new ones, was the "tour of the divisions" by technology management, with some bench researchers in tow. This is a very common orientation procedure for both parties and often represents the first tentative moves in an attempt to gain new clients or increase the flow of business from old ones.

6. *Gathering resources* in a new laboratory usually takes more time than people expect, particularly if the lab will be operating in technical areas that are new to the company. In addition to selecting (from existing company labs) or recruiting (from outside the company) the main force of professional and support personnel, there is the time-consuming job of identifying, recruiting, and orienting key technicians, first line and intermediate supervisors, and upper levels of laboratory management. In many organizations, key people and leaders are expected to recruit their own group members and, of course, this cannot begin until the leader is found and brought on board. We have observed numerous cases where this part of the process alone has taken several years for a technology that is new to the company or to industry in general.

In addition to gathering the people (which in some cases involves pirating from other company laboratories) the new lab must also identify, design, and wait for or build the facilities, laboratory instruments, and equipment they need. We found many laboratories waiting months or years for pieces of highly specialized equipment which were not shelf items.

Another aspect of gathering resources that is frequently overlooked in start-up situations is the need to gather technical, organizational, and other information as guides to the administrative and technical aspects of the laboratory's work. It is obvious that lack of information about the company or divisional products and processes has to be remedied before serious R&D can start, but it is not always obvious that there are special procedures for purchasing, personnel, accounting, and other functions that may delay getting started and that may require work to be done over for administrative reasons (often called bureaucratic red tape by impatient researchers).

7. *Gathering information* about the organization's products, processes, and materials was mentioned above in passing. It deserves more discussion at this point. It is widely recognized that new graduates in science and engineering, even (or especially) those at the Ph.D. level, come into an industrial organization with very little practical knowledge of the details of the company's business or technology. Exceptions we found were in those companies that ran extensive undergraduate co-op programs or where graduate summer employment was of major proportions. However, these programs are rare and account for a very small percentage of the new R&D employees.

Aside from the specific technical details on the properties of particular materials used by the company, the characteristics of its processes, and the technical features of its products, there are always the myriad tricks of the trade that do not appear in textbooks, manuals, trade magazines, or even company internal documents (see Chapter 8 for extended discussions of this imbedded technology capability (ITC)). They may deal with proprietary transformations of common raw materials, clever instrumentation and augmentation techniques for production equipment, shortcuts in process control, and ways of enhancing the appearance or functioning of the company's products. In addition to all these nitty gritty details, which must be learned by the members of the new R&D facility in some manner, there is typically an overall level of "smartness" or sophistication about the business and its technology that appears to be almost instinctual in the veteran employees, but a deep mystery to tyros. In those industries that cater to the whims of its customers by endless special formulations, attachments, models, machine settings, and product variations, the new researchers may be overwhelmed at first by the huge number of product variations, as well as puzzled by why their company undertakes to provide them.

It takes time and effort for the new researcher to learn enough to question company technical policies and practices (e.g., provision of almost custom products in a commodity market). Meanwhile, he has to try to understand not only the way things are done but *why*, if he is to make any significant contribution to standardization, cost reduction, or product improvement.

So far, the discussion of this issue has dealt only with the new lab whose members are new to the company, and particularly those that employ new graduates. While the time for learning and the base from which they start may be different, even experienced industrial researchers have only a limited knowledge of the technical details of other industries and technologies in which they have not previously worked. They may appear, at the outset, to be more sophisticated in general about the way things are done in practice (as opposed to the textbook), but they still may require many months of hard work to gain the intimate familiarity with the specific technical details in their new assignment, in order to be able to come up with reasonable ideas for improvement or enhancement, let alone radical change.

Again, as in the political and organizational arena, there is little formal help for the new researcher in this technical area. He has to scrounge for himself, learn by doing and making mistakes, keep his eyes open, and, if he is wise, wait until he feels confident in his knowledge before making any radical suggestions or plunging into major improvement programs. Perhaps the major advantage the researcher with experience elsewhere has over the completely new industrial researcher is that he has a bag of technical tricks, collected during his career. These may be new to his new associates and may help him establish his legitimacy as a professional with potential for pulling his technical weight.

8. *Setting up the organizational mechanics* for operation of a new lab probably does not take much longer than it does for other organizational activities. But it must be done if the lab is to tie into company policies and practices and function as an effective unit. In addition to the usual routine mechanics, mentioned above, such as purchasing, personnel, accounting, and housekeeping, there are certain special mechanics that are often time and energy consuming far beyond expectations. These are the several coordinating mechanisms for establishing and maintaining contact with clients, services, and other associated groups in the company. If they include committees, careful selections have to be made. The committee process must be established and monitored. Mechanisms for feeding information and instructions to the committees and getting information from them are needed. Where liaison relations are established through one individual, care must be taken to select the "right person" (in terms of knowledge, personality, and status). He must be instructed in the positions the lab is likely to take on certain issues and he must be carefully watched in the early stages to see if he is adequately representing the interests of the lab in interdepartmental negotiations and deliberations. The selection and monitoring of such liaison individuals constitutes one of the perennial problems in most of the labs we have encountered, both new and old. Concern with this widespread problem of liason has led us, in recent years, to undertake a major research effort into the general aspects of communication problems between groups in the R&D process, including activities known variously as liaison, interface, coupling, and technology transfer, and the emerging role of the "key communicator." The results of some of these studies are discussed in Chapter 4 (on networking) and Chapter 8 (on commercialization).

In addition to the classical liaison role (e.g., the person who keeps in touch with production or marketing or quality control), there are also laboratory roles and procedures that must be established to assure proper response to emergencies, requests for service, complaints, and requests (sometimes demands) for information, where one person is probably not able to fulfill the role alone.

In this connection, we encountered the most widespread bone of contention between R&D personnel and non-R&D personnel in the whole list of mutual problems. This is the issue of "firefighting" by R&D, as opposed to carrying out "their own" projects and programs. Many companies have kept this problem under control by a variety of means, some of them not very satisfactory, but tolerable, and some of them very expensive. These include special laboratories, distinct from R&D, for such firefighting of production and customer problems. In a number of companies these duplicate "regular" R&D labs lead to competition for people and often yield less than the best technical service that might be provided. Other organizational solutions include special sections of the R&D labs assigned to firefighting, particular individuals put on fixed or rotating service assignments, and making R&D budgets (or a significant portion of them) contingent on performing satisfactory service.

In mature, sophisticated laboratories we have encountered well-established procedures for response to client needs. Priority systems are clear and adhered to, except for dire emergencies, when "all hands fall to." Experienced and savvy laboratory directors, especially those in operating divisions, know very well which side their budget is buttered on and play the game accordingly.

All this trial and error, sensing the situation, negotiating, and assessing consequences takes time and energy. In small new labs, the director and his staff may choose to ignore the service issue initially and concentrate on the "big coup" that will yield a new product or a major process breakthrough; but in our observation they do so at their peril.

9. *Building internal relationships.* Finally, and perhaps most energy consuming, if not time consuming, is the building of the new R&D organization itself. Behavioral scientists have learned a lot in the past several decades about the social psychology of groups, including R&D groups. They have gained insights into the effects of perceptions, values, attitudes, and behaviors on group morale, productivity, and ability to survive. However, they have not yet provided enough insight into how such groups come into being and develop into effective operating entities.

We have encountered many cases of new R&D activities getting started or trying to get started and trying to reach at least an "assured survival" level. Many of them did not make it. Explanations for outright failures and expensive, time-consuming successes include all the learning factors mentioned earlier in this section. But perhaps the most critical factors, well hidden from the naked eye, are those involving interpersonal relationships of trust and confidence. Despite the many mechanical trappings of most labs, the process of problem solving still remains primarily an intellectual one. And despite the public stereotypes to the contrary, this intellectual process is heavily influenced by the emotional needs of the researcher and the people with whom he does business. The process of learning whose judgment one can trust, whose information one can have confidence in, and whose knowledge and expertise one can rely on is very time consuming and continues through the life of the organization, whenever people and circumstances change. Much early project work in a new lab can best be described as trust and confidence-building exercises or experiments. Does your colleague the chemist really know enough chemistry and its interface with physics to guide you to correct answers? Is the analytical chemist a real pro, with care and pride in his analyses? Can you get away with giving the model maker or glass blower a rough sketch and expect him to produce exactly what you need? Is there a technician you can rely on to take data accurately and completely? Are you and the process engineer on the same wavelength when you talk about yield of the process being lower than is attainable with a little effort, or will he nod and smile while you are together, but go running to his boss afterward to yell "poacher" and try to sabotage your process improvement attempts?

This is an area that students of organizational behavior have many insights about, but very little hard data and very few supportable recommendations for improvement as yet. However, it is a field that is being heavily worked and results should be forthcoming at an increasing rate over the next decade or so. Our observations of this phenomenon in our in-depth studies yielded, at least, a deep appreciation of the role of these social psychological variables in the successful establishment and maintenance of an R&D laboratory.

2.11.1 What Can Be Done to Reduce the Time Lag in Getting Organized?

It is all very well, the reader might say, for you to detail all the factors that can delay a new lab's reaching a takeoff point. We might even develop a mathematical model, indicating the combined effects of all these factors, and thus come up with a handy predictive device. But the practitioner is more interested in *remedying* the situation than in a thorough academic *understanding* of it. I shall therefore try, through a combination of our systematic observations during the study and some conventional wisdom gleaned from our direct experience with many start-up operations in R&D, to suggest some remedies. They are listed in Figure 2-6.

a. The first, and perhaps least satisfactory to the busy manager, is a psychological remedy for him, rather than an *organizational* one for the R&D lab: reduce your expectations to a reasonable level. Conversely, stretch your expectation of the time required to get organized toward a realistic period that allows for the natural processes described above to take place. In other words, consider the resources devoted to starting a new R&D activity as patient money, with an *investment* time

---

1. Reduce psychological time lag by reducing expectations.
2. Adopt an investment time horizon versus current expense or operating one.
3. Use mixed strategy in staffing: "old soldiers" and new recruits (old and young).
4. Set up orientation programs to transfer "organizational smarts" to new people.
5. Set up some low-cost learning projects to help in "learning by failing."
6. Have management (corporate and/or divisional) publicly support the lab verbally and behaviorally.
7. Do other relevant things specific to the company, industry, and technology.

Figure 2-6. How can the organizational time lag be reduced?

horizon rather than a current expense time horizon. If you try to rush the process, mistakes will be made and the net effect may be less desirable than with the patient approach.

b. Use a mixed strategy in staffing. Include several "old soldiers" who are known in and who know the organization thoroughly. They should, of course, be technically respected and liked, as well as known. If the new lab is used as a refuge for organizational outcasts, the damage may be more severe than leaving a group of all new people on their own.

c. Do not wait for the tentative and exploratory efforts of the lab people to help them learn the organizational, political, technical, and business ropes. Set up *and continue* orientation programs, meetings, interpersonal "smarts" sessions, and other ways of bringing the new people up to speed as quickly as they can absorb the information.

d. Set up some "learning projects" of low risk, where they can make their mistakes openly and not have the roof fall in on them or have the organization turn against them for the foreseeable future.

e. Have the management of the division (if a divisional lab) or the company (if a central lab) put their arms around the new lab and people in public and clearly indicate their support and intention of continuing to support it. The half-hearted "or else" attitude with which many labs in the sample companies were set up almost assured ultimate failure. If the management believes in the need for and advantages of a new R&D activity enough to devote scarce resources to setting it up, that belief ought to be clearly and widely demonstrated so that the researchers will not have to divert a substantial part of their resources to establishing legitimacy completely on their own. Of course, they must do it on their own eventually, but the initial level from which they start is strongly influenced by their sponsors' overt behavior toward them.

All of the above may sound like typical management textbook prescriptions. However, our observations were so convergent on this issue of conditions affecting start-up of a new R&D activity that I feel justified in presenting them as generalized factors.

___3
KEY PLAYERS: IMPORTANT ROLES IN TECHNOLOGY MANAGEMENT

3.1 INTRODUCTION AND CHAPTER OVERVIEW

As indicated in the beginning of Chapter 2, certain individuals can have great influence on the way technology is performed in the firm and on its outcomes. These certain individuals, in addition to the scientists and engineers themselves (the "doers"), are the various management people in both the divisional and corporate hierarchies. They include top management people—CEOs, COOs, CTOs, group executives; divisional people, including the division manager and his staff; corporate staff (e.g., finance, law, personnel, marketing, planning); and others who can and do influence the R&D/innovation process. Within the general organizational structure and specified rules of operation for technology-related activities, there is plenty of room for maneuver, manipulation, and influence on what is actually done and how it comes out. Figure 3-1 places some of these key players in their classical "solid line" or "dotted line" relationships to each other.

This chapter explores the motivations and perceptions of various key players in the technology game in the firm and tries to trace their influence on projects, programs, and outcomes.

In the context of these different perceptions and motivations, questions of trust, time perspectives, and championship of technology projects are also examined.

3.1.1 Some Issues/Questions

- Who influences the way technology is developed in the firm?
- Top management's influence

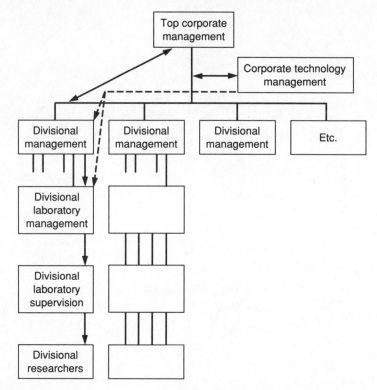

Figure 3-1. Critical junction points for influencing R&D decisionmaking in the decentralized firm.

- The group executive's ambiguous role
- The division manager and his team
- The chief technical officer
- The corporate staff
- Why are there differences in time horizons?
- Who champions technology?
- What about trust among the key players and the "doers"?

3.1.2 Implications for Management

It is not always clear whose voices and actions are exerting significant influences on the kinds of technology projects and programs undertaken in the firm, the way they are carried out, and their outcomes and impacts on the firm. Some of these influences are subtle, conflicting, misleading, unrecognized, and undesirable from the viewpoint of the firm's survival and welfare.

Although some managers—corporate and divisional—feel that the scientists and engineers they employ often ignore management's points of view and expectations from technology, they may be overlooking profound effects of their actions and words. Feedback on this is vital for a healthy and effective technology program. Terms of reference such as "second to none," "first in the market," "high-quality products," and "customer needs come first" may be too vague and may be perceived as ground rules or constraints on the firm's technology programs in a direction different from that intended by top management.

If top management really wants a strong, corporate-wide technology base and programs to support it, they must pay careful attention to the balance of power and influence of the various key players and be prepared to act decisively to assure effective organization, maintenance, and application of technology to the business.

3.1.3 A Watch List

- Continual battles over technology budgets, programs, people, and so on between corporate and divisional managers concerned with technology.
- Lack of oversight on divisional programs by the CTO: "I don't know what those guys are doing."
- Continual or frequent appeals to top management by divisional management to "get those corporate R&D people off my back."
- Refusal by divisional managers to pick up corporate research lab (CRL) projects and carry them to commercialization.
- Lack of interest in or understanding of what other divisions are doing and what the CRL is doing on behalf of "our" division.
- Semi-isolation of CRL people from the main streams of the firm's activities in the marketplace or plant.
- Obvious lack of trust between divisional and corporate-level managers in the area of technology.
- Extreme differences in time horizons between levels of managers.
- Lack of enthusiastic champions for important projects and programs.

3.2 WHY DO TOP MANAGERS COMMIT THEMSELVES TO DECENTRALIZATION?

Now, in approximately the fifth decade of the massive move by most of our largest manufacturing companies to change into a decentralized form of organization, typically along product division lines, there is no doubt that it is firmly established. There have been virtually no cases of complete return to the purely functional form of corporation in which each major area (e.g., marketing, production) is headed by one line executive. In some industries, for reasons of technology and/or market, major segments of the company (e.g., a steel or paper mill, or the consumer products marketing) remain centralized. But these are the exceptions. There will clearly be no return to the older form for most companies.

However, the perennial question of *degree* of decentralization continues to pose some real alternatives for the top manager. He can opt for a simple split up of all the operations into roughly equivalent chunks that will perform all the marketing, production, technical, and staff work for each segment of the company's business. He can withhold some of these functions and centralize them selectively (in time and place). He can allow or encourage the operating units to develop self-sufficiency in staff services and all of the supporting activities that make a modern business go. He can centralize some of these staff services. He can keep direct personal contact with and, hopefully, control over the managers of the operating divisions, or he can assign such direct contact and control to an intermediate level of executives, who in turn report to him (see Section 3.7 on the group executive).

He can monitor divisional operations closely, in fine and myriad detail; he can ask only for the "big picture" of operations; or he can select a level of monitoring and control in-between.

He can closely and personally review all results of operations, approve all major expenditures and projects, and be directly involved in all major decisions of a strategic or even midrange tactical nature. Or he can decide to remove himself from such detailed involvement and merely request data that will give him a post hoc review of such decisions and activities. In other words, he has a tremendous range of personal operating style under the overall heading of decentralization.

Since the main thrust of this book and the research underlying it relates to *technology* in the decentralized company, we do not pretend that we have a credible body of systematic data on the job of the top manager per se. During the course of the formal study and our other research and consulting activities over the past 30 years in decentralized companies, however, we have picked up some impressions and insights about the top management role and its relationship to the concept and operation of decentralization. The data underlying these impressions and insights include first-hand contact (in some cases repeated at multiyear intervals) with top managers in over 100 of the more than 200 major industrial companies in our sample.

After this buildup, I hope that the reader is not hoping for a dramatic exposé of the inner thoughts, hopes, and dreams of this handful of major management people in our society, despite the fact that some of them did indeed share some of them with us. My objective in this section, like that of other sections in the book that do not deal directly with the firm's technical activities, is quite limited. It is, in this case, to attempt a modest rebuttal to some of the massive outpouring of management literature over the past two to three decades on the "corporation of the future." Much of this literature uses the computer as a springboard and extrapolates the management information systems of the present into a vision of a computer-run company of the future.

I have no quarrel with the general picture of computers and automated information systems playing an increasing and pervasive role in the modern corporation.[*]

---

[*]Indeed, in 1986 my colleagues and I established the Center for Information Technology (CIT) at Northwestern University to do research on such issues.

I hesitate to endorse, however, the image of a single manager or top management group in the large corporation crouched over a computer console, making on-line operating decisions, either actual or simulated. I have two major reasons for stating this position:

1. The first is empirical. The state of the art in computers and management information systems has reached the point in the 1970s and 1980s where a limited degree of this "masterminding" and centralized decisionmaking on operations is technically feasible. However, I have seen very little evidence in the corporate offices of our major decentralized companies that such a mode of operations is imminent. Even allowing for the usual time lag in adopting new management techniques, I do not see a clear trend in the direction indicated.

2. The other major reason, or argument against the vision of a high degree of "computer top management" in the large decentralized company (this discussion is confined to that form of the industrial firm), is based on a series of quasilogical, commonsense arguments, which are in turn based on our contacts with many top corporate officers over the past three decades.

The keystone of this second line of argument, in my opinion, is the set of reasons why top managers agreed to or pushed for decentralization in the first place. Let us immediately dispose of the logical economic and management principle arguments, as described in some excellent research- and experience-based books on the subject (see reference to Chandler, Sloan, and Drucker et al. in Appendix B). They are good, sound arguments and are probably largely responsible for the major trend toward decentralization which we have been witnessing since World War II. I would like to concentrate on the not-so-logical subset of reasons that have to do with the psychology of the top manager, his history, and his anticipation of his personal future in the corporation. Even though the single, pivotal decision on whether to decentralize at a particular time may have been swamped by the more formal, logical, and economic arguments, subsequent oscillations along the centralization · · · decentralization continuum discussed in Chapter 2 appear to be dominated by these more personal factors.

Without attempting to play amateur psychologist, here are my impressions of some of these personal factors that influence top managers' attitudes toward decentralization and personal involvement in operating decisions. Although the average age of top managers of U.S. manufacturing corporations is probably going down somewhat, most of the heads of corporations whom we have encountered have reached their positions after long and arduous climbs through the several alternative corporate ladders that can lead to the top (staff, line, consulting specialist). The overwhelming majority still appear vigorous and full of initiative and ideas even after several years in top management (chairman, president, senior or executive vice president). However, my impression is that many of them are heaving sighs of relief to be out of direct, day-to-day involvement in operations and the marketplace. There are, of course, many exceptions. Some of those whose careers have been primarily in marketing will long for and, in some cases, involve themselves in the excitement or glamour of the marketplace, handling major accounts or keeping up relations with dealer organizations they established years before. Most of them do not, however, either *have* the time or *make* the time for such activities. One line of speculation is that they are glad to be out of the operating arena where decisions, actions, and confrontations go on at a hectic, often chaotic pace. Some of these people are just tired of that kind of activity and prefer other kinds — in the board room, in planning sessions, in outside presentations, and in small group (or two-person) "big picture" sessions. Some of them may be emotionally tired of that kind of activity and the pace that accompanies it. Others may feel less confident in their technical know-how and judgment about specific aspects of divisional and corporate operations than they did a few years ago. They are quite willing to leave it to staff specialists and the younger tigers who are now running operating divisions and staff supporting groups. Others seem to be averse to exposing themselves on specific issues in the face of limited time to study the situation and limited day-to-day involvement in the specific operation to which the issues pertain.

All of this, and what it implies, adds up to an image of the computer-managed corporation, which is counter to the one projected in much of the management literature. I do not see top managers in large decentralized companies eagerly anticipating the time when they can *recentralize* operating decisionmaking in their offices and resume day-to-day responsibility for such decisions. In fact, I see most such people glad to be rid of such responsibility and making certain that they have competent people at the operating levels who not only will take responsibility for such decisionmaking but will also be there to take the lumps if their decisions turn out badly.

One bit of evidence that I interpret in support of this argument is the minor trend toward further layering of the decentralized corporation, which is described in Section 3.7. This may be evidence of less interest on the part of top management in taking over or controlling operating decisions personally. This layering may appear to give them closer control over operations than if they had to deal with a large number of individual division managers, but it also puts them in the position of dealing with second- or third-order information and decisions (in both directions), even though some of them may be better organized and abstracted than the first-order information and decisions would be.

This discussion may appear peripheral to the central theme of technology in the decentralized company, but it is not. It is quite central. If my sketchy and nonsystematic analysis of the motives and preferences of *top* managers with regard to close control of divisional operations is correct, it has great implications for the management and effectiveness of technology in their companies.

A central working hypothesis in this book relates to *division* management's reluctance to engage in or support long-term, broad technology programs. One of the most effective counter-influences to that attitude and behavior is top management's insistence, backed by specific control and evaluation action, that longer-term, broader-scope technology programs are essential for the survival and success of the company in the future. If top management insulates itself from first-order involvement in this activity (e.g., closely following specific programs and projects), then the potentially bad longer-term effects of divisional management's very limited posture on technology are likely to be realized.

Before closing this section I want, like most forecasters of future management trends, to hedge my forecast with some conditions. As younger people come into top management jobs, with a higher energy level and a deeper indoctrination in the computer age than the current and recent generations of top management have had, we can look for more *attempts* at central control of operations. I also predict that many of the attempts will fail and that even some of these younger, more technically oriented managers will back away from "real-time" decisionmaking and control from the corporate office. The reason is that as long as the division managers are relied on to be the experts, the advocates, the risk-takers, the front-line shock troops of the corporation in its dealing with the marketplace and other aspects of the environment, it will be a reckless top manager who tries to take on this role for himself, whatever the advances in computer and information technology. I do anticipate, however, that there will be a residual trend in this direction (despite the failures) over the next decade or two, with a concomitant erosion, in some large decentralized companies, of the job and role of the division manager as we know it today. In some respects, this may spell the end of the era of the decentralized corporation as we now observe it and the emergence of still another organizational form. But we are now well beyond the scope of this book.

3.3   DO WE WANT A GENERALIST AS A DIVISION MANAGER?

The general management literature and the comments of many top managers and students of management give the clear impression that the ideal candidate for the taxing job of division manager in a decentralized company should be a management "generalist. " Cautions abound regarding the dangers of a person in such a position being too narrowly specialized and not being able to range across the entire spectrum of management specialties in his decisionmaking and other management activities. I have no intention of quarreling with that attractive general concept. We have even encountered many division managers who appear to be the broad gauge people described in the literature.

However, we have also encountered many who do not fit this general model and yet who appear to be quite as successful as those who do. In some organizations, the nongeneralist division manager even emerges as more successful than his broad-spectrum colleague. Why should these exceptions occur? (Unfortunately, this is one of the many aspects of the subject on which we cannot display impressive statistics but must operate on a combination of limited observation of cases and quasilogical argument.)

A reasonable way to approach the question is to consider both necessary and sufficient conditions for the success of people with particular backgrounds and training in the job of division manager. I will omit the many obvious personal attributes such as intelligence, personality, and decisionmaking style. Many such general prescriptions, even when brought down to certain quantitative levels, hold for most management jobs and typically involve a wide acceptable range (e.g., short people and tall people, extroverts and introverts). The focus here is on whether the generalist background most frequently advocated for the division manager is really a *necessary* condition for his success or merely one alternative *sufficient* condition of his background (in addition to all the usual attributes advocated for managers in general).

Our observation of a large number of successful division managers indicates that many are very highly specialized by their previous experience, if not by their formal education. By "highly" specialized, I do not necessarily mean "narrowly" specialized. For example, some of them are expert marketing people and have had very deep and long experience in the marketing aspects of corporate or divisional management. They may in fact be the most expert marketing people in the corporation. This does not necessarily mean that they are incompetent or ignorant in other areas, such as production, accounting, purchasing, inventory, or engineering. It may mean, however, that their major criteria for decisionmaking, evaluation, and control are derived from marketing considerations, rather than from considerations that are judged dominant in some of the other functional areas. If such a division manager has most of the widely advertised general attributes of successful managers, in addition to this high specialty, he may very effectively use the services and advice of staff and line specialists in the other areas and thus avoid blunders based on being too specialized. If he is very *narrowly* specialized, however, so that he has no general appreciation of the benefits he can gain from outside advice and assistance, then he may be subject to many classical mistakes of the type made by people with specialist blinders on. Let us argue for a moment in another direction—in favor of a reasonable proportion of highly specialized people in the job of division manager *over a period of years*. Even under the most well-prepared generalists as division managers, particular operations, decisions, and investments in short- and long-term programs tend to become imbalanced in terms of the functional specialties. A strong divisional director of production may win a disproportionate share of the capital investment funds from his division manager as a result of persuasive arguments, a close personal relationship, a demonstrated urgent need, and so on. An impressive divisional R&D director might persuade the division manager to emphasize product competition in the division's product lines over, say, price competition or development of improved distribution channels and dealer organizations.

This possibility, taken together with the high turnover of division managers (through promotion, demotion, retirement, termination, etc.), often leads to a situation of imbalance in a particular division's operations, with one or more essential areas being neglected or not receiving sufficient attention during certain periods. With an average tenure of 2–4 years in some of the companies we studied, it is possible to conceive of a given division manager position being filled with a sequence of specialists over a period as short as a decade. If we also consider the cases of the many companies that have instituted the job of assistant division manager or executive vice president of the division (in many cases as apprenticeships for the top post in their own or some other division), then we can see the possibility of the division manager *role* or position achieving the advantages of both generalism and specialty over a period of years.

Why belabor this point? One reason is concern that the trend toward training and selecting generalists for this crucial job in the large decentralized company may, strangely enough, deprive individual divisions and the corporation as a whole of the sharp, penetrating thrust of a specialist who can spot present and *future* problems in the area of his specialty that are essentially invisible to the nonspecialist. Let us use both a parochial (technology) illustration and one from another specialty.

Mr. A, the new division manager, once he has put his personal effects into his desk, met his staff, and gotten a general orientation on what is going on in the division, begins to take a closer look at some of the operations. His general business background is excellent and he is generally acquainted with all facets of the operations of businesses similar to his new responsibility. He runs through his own personal checklist of things to look out for, to probe into, and to put on his calendar for future attention. In meeting the divisional technology people, going over their current project portfolio, and discussing future plans with the divisional laboratory director, he gets a nice warm feeling that this is one area that probably will not cause him any trouble in the foreseeable future. Within the year, his optimistic assessment turns out to be an expensive one. In reality, as it turns out in the postmortem, the divisional lab director is a nice guy and a good salesperson of his program but is in fact technically obsolete and also largely out of touch with the second-order changes going on in the marketplace. His program looks beautiful on paper but is full of projects based on very weak technical assumptions and inadequate market analysis.

Would an R&D specialist have spotted this right away if he had been the new division manager? Quite possibly. Would a marketing or production specialist have known better than to rely on his own judgment about another specialty and called in a corporate technology specialist or an outside consultant to make a technical audit of his division's R&D organization and program? Again, this is a matter of speculation. It is clear, however, that the new division manager was neither expert enough to second guess his technology manager nor sufficiently modest about his own technical competence to call for outside help.

Mr. B. came straight out of a marketing career into the job of division manager. His first major effort, even before arranging his desk and meeting all the folks, was to defer to his nagging suspicion that the division's marketing setup was not all that it appeared on the surface. He hopped onto a plane and visited the headquarters of the half dozen sales territories serving his division's product lines and the half dozen other offices shared with one or more other corporate divisions. Then he returned home and had a good cry, followed by a massive cleaning out and shaping up of the marketing and distribution system serving his division directly. He also initiated action to confer with the division managers of several other divisions that shared marketing facilities with him. Would a generalist have spotted the weaknesses in his marketing organization? Probably not so readily. Would a generalist have called for help in auditing his marketing organization? Maybe and maybe not. Again, let me protest that I am not deliberately playing devil's advocate on the concept of the generalist as a division manager (or the division manager becoming a generalist, whatever his prior experience and tendencies have been). I am seriously questioning the assumption that generalists can do the best job of running an operating division in a decentralized company and that this (being a generalist) is the *only* way to succeed. We have seen many cases of success where the division was preeminent in *either* marketing, production, or technology and adequate in the others. But I wonder how many cases there are statistically in which a division manager and his division are not outstanding in *any* of these major functional areas and are still highly successful in the marketplace.

It is clear that there are multiple ways of being successful in the marketplace. Balance is one of them, but being outstanding in a given area is also one of them, and we have been impressed with the performance of a number of division managers who are strong advocates and champions of strength in the area of their own specialty but who do not neglect the other areas.

3.4   THE DIVISION MANAGER AND HIS TECHNOLOGY TEAM

One recurring observation we made about the many division managers we have encountered may help to account for a major share of the conflict between them and R&D/technology laboratory members, where such conflict is above a normal level. This is the inability or unwillingness of many of these division managers to differentiate the roles of specialist groups serving their divisions to the extent that members of these specialist groups think is proper. That is, a significant proportion of the division managers we encountered did not assign as specific and as specialized a set of roles to their supporting staff groups as the specialists (particularly the professional members of such groups) preferred or thought necessary to their professional operation. This difference in perspective might, for identification purposes, be called the *team* view versus the *specialty* view of supporting staff services. It might help us to account for much of the pulling and hauling, grunting and groaning, and open conflict that we found associated with the "firefighting" duties assigned to or imposed on the technology lab in the division or to that portion of the corporate lab supposed to support divisional operations.

Under the *team* concept, the division manager certainly respects and appreciates the special talents of professionals such as accountants, lawyers, engineers, and scientists. However, this appreciation is often subordinated to his current priority, importance, and urgency rankings of problems that need attention. When he is in a crisis situation, he is not very interested in the fine distinctions between chemists and chemical engineers, or the finer distinctions within his group of chemists. If the problem is important and urgent enough and chemist X can solve it, or at least "get it off his back" for the moment, then X is the person to jump in and work on it. The fact that chemist X is in the middle of a series of long-term experiments, aimed at a possible new product 5 years down the road, is almost irrelevant in the division manager's hour of need. Indeed, if X does a good job of bailing him out on a problem, the division manager may reward him by telling him to forget the long-term project and to stand by for more firefighting. The same process was observed with respect to *specialty groups*, as well as individuals. If the division manager (and his divisional functional management people) finds that one of the R&D groups, normally working on *indirect* support of operations, has talents and is available, with a little coaxing or threatening, to work on *direct*, immediate problems, he sees little wrong with institutionalizing the arrangement or at least with calling on the group frequently for such service.

The rationale is clear in the minds of the divisional management group, and especially of the person charged directly with profits in *this period*. "Everyone in the division is a member of our divisional team. Our first duty is to survive into the next accounting period. Anyone who can and will help us do that is a good team member. Anyone who won't is really not a true and useful member of our team."

Coupled with the pressures on virtually all the division managers we encountered to perform in the short run, this attitude leads to several strains, if not outright conflict, between divisional management and those members of the divisional lab with a longer time horizon and a preferred broader scope of activity. Some divisional technology people who cannot reconcile their project portfolio, time horizons, and work schedules to divisional requirements may (and do) request transfers to a corporate laboratory if one exists. Alternatively, they may adjust to the situation and give up or drastically modify their ideas of making major, long-term contributions to divisional or corporate operations and settle for more or less routine development and service activities. The divisional researchers we encountered who neither abandoned ship nor made the adjustment constituted a core of disappointed, often bitter people with potential for (1) doing "great things" if the environmental constraints were lifted or (2) being a source of discontent and carping in the division if they were not.

More will be said about the reaction of divisional technology personnel to the constraints imposed by their divisional environment and the attitudes of the divisional management. For the present, however, I would like to return to my assertions about the division manager himself.

We generally found divisional managers in the companies of our sample to be a group of attractive, dynamic members of the industrial community. They differ in size, shape, age, education, and in many other respects. By and large, however, they are intelligent, alert to their environment, willing to take certain kinds of risks, and very, very conscious of the time and financial constraints under which they operate. With many individual exceptions, of course, they come across as quite a different group than the functional executives we encountered in the same companies. In other sections on the role of the division manager, I discuss sources of such executives, their backgrounds, and some of their general perceptions of technology. Here, however, I want to continue with this theme of his ability to differentiate specialist roles in his organization.

The division managers we have interviewed and interacted with were almost all members of this country's giant corporations, most of which have strong technology bases in their materials, products, and production processes. As a consequence, few of these division managers were, in general, negative to the need for and benefits of good technology in helping them and the overall corporation carry out its objectives. However, when we began to probe into where this technology should be done, who should pay for it, and how closely it should be tied to current operations, we found strikingly similar (on some issues, converging) attitudes, which reflected reservations, doubts, and outward hostility to the way that technology was being performed in their divisions and corporations. The overt expression of dissatisfaction with their own divisional R&D was carefully expressed, however, by those who appeared to be the strongest proponents of the team concept of divisional operations. If the particular division manager had been in office for more than a year of two (presumably time enough to mold it to his image), he was less likely to indicate high levels of dissatisfaction with his own divisional lab. Exceptions to this were instances where he could blame the interference of corporate technology management or the general attitude of the R&D people in the company who wanted to "play around in their laboratories instead of helping the company to make a buck."

Most typically, when questioned closely about the role of their divisional laboratories in meeting divisional objectives, most division managers expressed some dissatisfaction with the pace, focus, and responsiveness of the lab to his expressed needs "which must be obvious to everyone on the divisional team."

One recurring circumstance, which I believe nicely illustrates the division manager's team concept as it applies to R&D, is the occasional success of the division in introducing a new product that is distinctly different from the ones the division has worked with before.

In the several cases we encountered of introduction of a radically new product by an operating division, the divisional R&D or product development lab (where one existed) played a major role. Whether the new product originated in the division or the idea or fully developed prototype came from the corporate research lab, from another division, or from outside the company, the divisional lab typically took on further development as its major task. Except for the maintenance of their regular service to existing product lines (e.g., factory and customer service and continued surveillance over materials), the major efforts of the lab personnel were devoted to getting the product ready for production and the marketplace and keeping it going after market introduction. In those cases where the product or idea had not been fully developed when it was selected as a major focus of the division's activity, this frequently meant overtime and extreme pressure to get it out the door and making money for the division.

Even though many such projects had rather short time horizons (several months or a year or two), the aftermath of such projects typically had repercussions on the divisional lab for several years after the initial thrust to get it on the market. We encountered a number of such cases in industries characterized by fierce product competition, based on competitive product features, differences in formulations, and products with a regular "new model" cycle. There are many special activities that characterize the preparation of a new product for production and marketing, including purchasing materials, preparing plant facilities, adopting or acquiring production equipment, market testing, establishment and monitoring of warranties, packaging and labeling, setting up marketing channels, preparing advertising copy, estimating costs, planning production schedules, helping customers to get ready to use the new product, and recruiting and training (or retraining) employees.

Few of these activities, other than the purely technical ones, require the attention and time of people trained in science or engineering whose major company role has been R&D. However, several major factors contribute to the situation of the major energies and time of divisional R&D people being sucked up into the vortex of a new product crash program.

Since this topic has arisen in the context of the division manager's view of his divisional personnel as relatively undifferentiated parts of this team, this aspect is considered first. When he makes a personal commitment to such a new product crash program, the division manager typically pulls out all the stops and throws the full weight of his resources into the project. Ideally, he would like to have the "best" people available, in terms of specialty and quality, for each special phase of the project. Usually, however, he does not have access or at least direct access to even the best people inside the company. They are nominally available to him as staff consultants, but he typically finds this availability more nominal than actual. Moreover, these people are not under his control, even if he pays dearly for them in interdepartmental charges. They have other duties, other interests, and other loyalties. They also may be very clever in their specialties (e.g., patents, advertising, market research) but they do not know his style and they do not know and probably do not care much about the specific ways in which his division does business. So, faced with a myriad of special tasks and problems in connection with the new product, he looks around at his divisional resources to find people who can perform them. Many of the specialized tasks he automatically assigns to existing specialty groups, such as production control, plant layout, quality control, cost accounting, and purchasing. Frequently, these tasks are an overload on an already overloaded department and he has to keep flogging them to be sure their energies, time, and enthusiasm are shared with the "new baby."

He surveys and resurveys his resources and somewhere in the process concludes that his divisional R&D people, although not necessarily the most qualified in each specialty involved, are the people he can spare the most to push the project through. He sometimes concludes this reluctantly, because he knows that in order to do this, they must give up some of the longer-term efforts they are working on and thus perhaps he will miss future cost savings or profit potential.

But it is really no contest. Based on the time perspective that has been forced on him by circumstances and/or that has made him a successful (or at least surviving) division manager thus far, he opts for the bird in the hand and throws the R&D members of his team into the new project.

This can be a marvelous opportunity for many of his technical people who have been "hiding out" or confined to lab activities for most of their careers. Here is an opportunity to get into the business end of things, share some of the excitement of bringing a project to a climax, and, incidentally, demonstrating their potential for general management responsibility.

As a consequence of all these factors, with the R&D people often willing members of the crash project, we have encountered many such new project ventures staffed temporarily with highly intelligent and educated people whose nontechnical contributions range from excellent to inept. Much of the ineptitude results from plunging into highly specialized aspects of business where the plungers completely lack knowledge of the tricks of the trade and skill in meeting new, specialized circumstances. Examples of this abound in blunders arising from pricing and advertising, equipment design or purchasing, dealing with contractors, test marketing, use of market statistics and economic forecasts, product styling, setting up marketing and distribution channels, and many others.

The "undifferentiated team" approach to recruiting them into the new project (often organized into a temporary subdivision of the operating division) casts technical people in a variety of roles for which their education, experience, and temperaments may make them unsuited. Our observations of several new product programs revealed scientists and engineers agonizing over decisions on pricing, styling, and other aspects of marketing for which they had neither the experience nor the "feel."

In most of these new product crash programs that we have encountered, the question of who would end up on top of the venture, if successful, was never far from the minds of the key project people. In several companies that are well known for generating new operating divisions around their new product lines, it was taken for granted that a member of the project team had a good chance of becoming manager of the new division, if one were formed. In a few of the more research-based companies in our sample, a significant proportion of division managers had reached their positions from an R&D background via the new venture route (see Section 8.4 on technical entrepreneurship).

From the viewpoint of the manager, if he retains the new product or benefits from having seen it through the venture stage, this may appear all to the good. In addition to the end result, he may even have some spin-off benefits through identifying hidden talents for general management and other activities among his hitherto cloistered technology people. The material and emotional benefits to these people may also be high. And certainly the corporation may gain from a new product line or business area. Despite all these apparent benefits, I would like to add a bit of a cautionary (sour?) note to this discussion and point out some less obvious side effects and longer-term consequences which all three groups (lab, division, and corporate) may come to regret later on.

These other consequences have to do with the functional integrity of the divisional laboratory and its capability of service to the division and corporation. With a glamorous new product effort absorbing the energies of divisional lab management and some of the best technical people in the lab, which it typically does, the lab effort becomes bimodal. That is, normally (or ideally) divisional labs distribute their efforts along the continuum of technology activities in full support of current and future divisional operations (from service to plant and customer, through product improvement, into general explorations and keeping up with the state of the art in the division's materials, processes, and markets). With the push on to get the new product out and the need to keep up the service end of their portfolio, the likelihood of all other activity being curtailed or stopped is very high. If the new product push lasts only a short time—for example, a few months—and the project people return to normal division lab work, no harm may be done.

However, if the new product succeeds, it is often the case that many of the divisional technology people follow it through its birth pangs, into infancy, and puberty. Some of them cycle back into R&D when the glamour and excitement wear off, when the choice jobs in the new organizational entity are allocated, or when they or other people discover that their talents and interests are really more in technology than in general business. However, this process is long and uncertain. What *is* certain is that there are many technical matters associated with the new venture that can absorb many man-months of their time in ironing out details, writing manuals, getting up the production "learning curve," establishing process and quality control, instructing customers on use of the new products, looking for the initial cost-reduction opportunities, and so on. Where the product is of the formulation type (an intermediate chemical, a paint, a food product, a polymer, an additive, a piece of equipment, or a subassembly that comes in many possible configurations), there is an endless stream of modifications and adaptations that must or can be made to suit specific customer or market segment needs.

All this takes time, and before the division manager realizes it, if he ever does (because he is often promoted following a major success), the divisional laboratory has lost a good deal of its expertise, organizational memory, and momentum on other new technical ventures.

The rebuilding process is often a slow and painful one. We encountered several cases of divisional laboratories having never recovered from a great new product success. That is, its former technical leadership in the division's broad areas of business and technology may be lost for a decade or more.

I do not go into as much detail on the cases (much more numerous) where the new product venture fails after many months or several years of frantic effort. However, the smell of failure or having been in over their heads does not help the morale of the project people, and sometimes it is hard to readjust to the more leisurely pace in R&D after the hectic pace in the marketplace.

What does all of this add up to in terms of a reasonable strategy for the division manager with respect to technology and pursuing new ventures? Should he abandon his team concept and confine his specialists strictly to their specialties? Of course not! He has neither the resources nor the temperament that will allow him to do so. However, it is clear that unless he plans a bit beyond current crises and crash programs and makes provision for the *maintenance* of his key specialty groups, he is likely to find the division depleted of critical capabilities for survival and future effective operations.

The situation described above with respect to some division managers' (not *all* or even *most* division managers') "depletion" of human resources for current operations is quite analogous to the similar use of material resources—particularly production equipment—which can be a perennial bone of contention between divisional and corporate management. In this case, the sacrifice of production equipment through curtailing or omitting preventative maintenance and "rest periods" and through running equipment above capacity and fatigue limits can lead to a sorry mess for the division manager or, most probably, his successor several years later.

3.5   LINE VERSUS STAFF: SPECIAL TREATMENT FOR ACQUIRED DIVISIONS/OPERATIONS?

This book is not about heroes and villains in the arena of technology in the firm. It is true that some (many?) line managers and technology people hold mutually unflattering stereotypes and even caricatures of each other. Some of these are best not reduced to words or pictures.

Despite differences in values, styles, objectives, time perspectives, career goals, backgrounds, and experience, most such people seem to work together reasonably well in the firm. That is, projects are initiated, approved, funded, staffed, worked on, completed, and implemented. Some of these tasks are done well or splendidly and others not so well or miserably. What I am focusing on in this book is how to improve the overall R&D/innovation process for the benefit of the firm and with less stress on its individual members. Sometimes improvement is seriously inhibited by negative stereotypes and the expressed attitudes and behavior that flow from them. This book is exploring ways of modifying some of the stereotypes at least to the point where they do not interfere with effective use of technology in the firm at any level—corporate, divisional, or strategic business unit. Even short of that goal—of modifying the stereotypes—some ways are presented of operating effectively in spite of them. "We know these guys are really not interested in our problems and are trying to sabotage our work, but let's take that for granted and try to work with them on a basis that meets their needs as well as ours." "If we can't improve cooperation by increased understanding of where our adversaries are coming from, then we can try to make the best of the situation and work around these potential obstacles to accomplishing our objectives."

Let us look at the corporate versus divisional staff situation from the special viewpoint of the heads of newly acquired divisions, which may have been independent companies before the acquisition or may have existed in essentially the same form, in approximately the same kind of company, and were involved in a simple swap or transfer. Let us deemphasize those acquired divisions that were "sick boys" and bought by a "company doctor" who plans to shake them up and shape them up. The focus of this discussion is on relatively successful operations that were previously independent companies or relatively "stand-alone" divisions in a highly decentralized environment.

Presumably, the acquiring company viewed the acquired unit's operations favorably (or it would not have bought it) and wishes its success to continue. *However*, the new company, in order to pay for the acquisition or just on general grounds, is in a cost-cutting and "lean and mean" phase. It has been and continues to trim staff at both divisional and corporate levels and expects the "new boy" to do the same.

But, wait a minute. One of the secrets to the acquiree's success has been a good set of supporting staff groups and excellent relations between marketing, technology, data processing, operations, and other specialist groups. The people have grown up together and have just been congratulating each other about the fact that they contributed significantly to the high price paid for their unit. Now the order (suggestion?) comes down from the new corporate bosses and their staffs that there must be a decrease in "head count" at the (new) divisional headquarters.

For some staff services, such as legal and financial, they can see the justice or wisdom in such a policy, since there is a lot of duplication between some of their staff groups and those in the new corporate headquarters. But for technology? "No! That is a ridiculous idea; it means breaking up a winning team and robbing the division of its eyes and ears in our fields of special interest. It also means starting the whole process of team building all over again and losing our ability to maneuver in the marketplace when technical threats and opportunities arise."

The division manager may protest, but, after all, he *is* the new boy and he does not want to start rocking the boat right away, at least until he has scouted and calibrated these new people. Besides, he may have gotten a good personal deal as a result of the acquisition—stock, cash, bonuses, perqs, a substantial salary increase, perhaps a percentage override, and a golden parachute in case things do not work out.

He knows the corporate staff people cannot really mean that he has to eliminate much, if not all, of his staff support, especially in the technology area. So he may go along and make some reductions in his staff, allow some of them to be transferred to the corporate offices or other line operations, and allow some to retire early (see Section 8.3 on imbedded technology capability).

In addition to personnel, it now seems the capital investment funds and even working capital are getting tight, perhaps because of the acquisition (which may be part of a package or sequence of acquisitions). His cash flow is eagerly gobbled up by corporate management and only a dribble is given back to him—on a detailed justification basis—for "renewal" investments in people, facilities, equipment, and technology base.

Is he worse off than before? Perhaps, perhaps not. He may now have a large CRL to use for support of his technology, products, and longer-term plans. When he was an independent firm or a stand-alone division, he did do some longer-range planning (his bankers and board of directors made sure he did), but for technology he was pretty much on his own.

But the CRL is in a remote location, staffed by strangers, and seems to be preoccupied with a long list of their own projects. They sent a task force to visit his division soon after the acquisition was consummated and beautiful music was made between his technology people and their technology people. The specialists seemed to get along fine. But now he has lost some or most of those specialists in his own division and may even have lost the lab he previously had. So who is now going to interface with these CRL people and launch projects for the direct benefit of his division?

And besides, who is going to pay for "dedicated" projects in CRL if they do agree to undertake some for him? The "tax" he now must pay out of his operating revenue only pays for *general* support of the CRL. Specific projects for his division cost extra. He now joins the "old timer" division managers in bad-mouthing CRL and vowing to go out and get whatever technology he needs from more responsive, outside sources (assuming he can get capital funds for this or divert some of his cash flow in that direction).

The above is admittedly a stereotype of a particular type of situation. But it is a very representative stereotype of many situations we have encountered in decentralized companies. Whether the division manager is new or has been in office a while, he is expected *both* to keep his products and plants up to date technically and to conserve cash and personnel by not allowing a "bloated" technology activity to exist in his division.

If the above scenario sounds bleak, consider the new division manager who was the original founder of the acquired company and who was the inventor or innovator of the company's products and the secret to its success. His tolerance level for such interference and stonewalling by corporate staff is much lower than that of a professional management person who arrived at the top management level in the acquired company through many small steps involving a lot of compromises. The entrepreneurial division manager typically will not take such interference and frequently leaves the new organization early on. He takes his money and goes to play golf or set up a new enterprise where, again, he can be his own boss.

I have worked with many such people in my 25 years in the venture capital business and find a very poor fit between them and the large, supposedly decentralized corporation. More will be said, in some detail, on the topic of entrepreneurship in the large firm (Chapter 8). And still more will be said about the role of corporate and divisional staffs in the technology area.

Where do the influence and power really lie with respect to technology in the firm?

We have looked, in a more formal manner, at the motivations, perceptions, and behavioral patterns of several sets of key players in the technology game inside the firm. Let us now look at some concrete manifestations of these factors on technology policy and decisions. The following are some vignettes drawn from my consulting experience and research observations at the interface between some of the key roles in the firm.

- "Get off my premises and don't come back until I personally invite you." Was this a plant manager threatening a union organizer or an OSHA inspector? No. It was the culmination of a long series of insults and conflictual encounters between the president of an operating division (on whose premises they were) and the corporate chief technical officer (CTO). For over a year, the newly appointed CTO believed he had the backing of the chairman and CEO to "whip the divisions into shape technically." He had tried the mister nice guy approach and, in vain, requested a meeting on several occasions with the division president and his divisional head of technology. After the above incident, the CTO appealed to the CEO and asked for his intervention in allowing the CTO to audit or at least review the division's R&D program and its manufacturing technology program. The CEO came down on the side of the division president, saying "it's his business and if he is not interested in your help, that's the way it is."

- *Division Manager:* But what's your job description, Mr. CTO?

  CTO: It includes coordination and review of all technology programs in the company.

Division Manager: Can you change a division's technology budget?

CTO: Well, not really, not without the agreement of the division manager.

Division Manager: Can you reject, recruit, or appoint people to managerial or key technical positions in the divisional lab?

CTO: No, that's really a divisional responsibility.

Division Manager: Can you terminate or change the direction of inappropriate or ineffective projects in the divisional labs?

CTO: Of course not, that's their prerogative.

Division Manager: Well, what do you mean by "coordinate and review?"

- *CTO:* We have this great training program in management of technology for the operating divisions. It should really help you a lot in improving divisional technology programs.

  Divisional Technology Manager: That's O.K., thanks, but we take care of our own training needs here in our division.

- *CTO:* We're planning a series of seminars and workshops to help acquaint divisional technical people with some new technology such as microprocessors, lasers, computer-aided design, and so on and would like to invite you to send some of your divisional R&D, engineering, and manufacturing people.

  Divisional Technology Manager: Sorry, we're too busy. And besides, who's going to pay for their travel and hotel? We have no funds in the budget for such things.

- *Division Manager:* Please, Mr. Chairman, get those corporate R&D people off my back and out of my hair. How do you expect me to meet my quarterly profit goals if my people have to waste time listening to their blue sky schemes and impractical ideas?

  Mr. Chairman: O.K., I'll tell the CTO to keep his guys away from your guys.

- *Division Manager:* Look, I'm paying 20% of the budget of your CRL out of my division's revenues, and I want a substantial percentage of your efforts devoted to servicing my plants and customers. We have a lot of start-up problems with several new product lines and if I don't get some of your best people assigned to help us out over the next year or so, I'm going to take those funds and buy myself some outside talent to do the job.

  CRL Manager: I'm sorry my people haven't been as responsive as they should have been. I'll see that a good group is assigned right away to work with your manufacturing engineering and technical service people.

- *Division Manager:* I think this high-tech acquisition is going to be very good for my division's product lines and manufacturing technology.

  CEO: Don't you think you ought to ask the CTO and his specialists to look it over and give you an evaluation of the technology of that company in terms of it being up to date technically and cost effective?

  Division Manager: I really don't have the time for that kind of analysis. The deal is hot and we have to move now, if we want to close on it. Besides, the CRL people would analyze it to death and they really don't understand our business in the first place.

CEO: O.K., give me your proposal for the capital investment committee meeting next week and good luck with it.

- *Chief Planning Officer:* I know you're the CTO and supposed to be responsible for making sure the divisions are up to snuff on their technology plans, but you can't say those kinds of things about Division A's business plan just because you think they don't have the technology base to carry them out. The CEO and division president would have a fit. Please go over it again and be a little more subtle this time.

This is a small sample of the interactions and conflicts that we have observed and participated in when push came to shove between corporate technology staff and divisional management and other managers. The dominance of divisional executives over corporate staffs is, of course, not universal. It is certainly not generally true in the cases of the legal and financial corporate staffs. And it is also not universally true for the corporate technology staff. However it is "mostly" the case for technology and many other corporate staff functions, such as marketing, personnel, training, and operations research. The jury is still out with respect to data processing and information systems. That situation is still being debated and fought out between corporate and divisional people throughout industry. For the corporate technology function, however, including the corporate research lab (CRL), the balance of power seems generally to have been resolved in favor of the operating divisions. Wait a minute, one may say, there are lots of good examples of close cooperation between CTOs, CRLs, and the operating divisions. Yes, there are some excellent examples, based on a strong and highly credible CTO, a division manager who desperately needs help, a very visible champion role by the CEO, or a modus vivendi arrived at over a long period of give and take.

Most cases of "about even" influence have come about because (1) the CTO and his people clearly understand who brings the revenue into the firm and keeps it going and (2) the divisional manager and his people clearly understand that they cannot keep doing it without the help of some smart and technically up-to-date people at the corporate level. These two groups have teamed up, at least temporarily, to meet the challenge of the "monsters in the marketplace."

3.6 WHY DO DIVISION MANAGERS AND CORPORATE TECHNOLOGY PEOPLE DIFFER ON TIME HORIZONS FOR PROJECTS?

An initial observation stimulated this long-term (since 1957 and continuing) study of managing technology in decentralized organizations. It was that managers of subunits (divisions and departments of large decentralized organizations) seemed to make their time and other resource investment decisions on a rather short-term basis. That is, they appeared to turn down projects (particularly in technology) whose anticipated payoff stretched beyond the near future. For some of them, this near future ended within the current fiscal or operating year; for others, it stretched beyond the current year, but not much further than a year or two beyond. Of course, like many other unsystematic observations, this was a very oversimplified view on our part. Certainly, in many parts of the operating (divisional) manager's investment plan, there were specific projects that would definitely take several years to complete. Among these were new plant and new production line construction, development or exploitation of raw material sources, development of new marketing organizations, some institutional advertising campaigns, new data processing systems, and some personnel development programs. Upon closer examination (i.e., as we formalized our observations and began to talk to division managers in decentralized companies and to talk about them with their colleagues in the corporate offices and the technology activities of their companies), other factors than pure time to payoff emerged as possible explanations of their investment strategies.

Since our major interest was in their attitudes and decisions about and their behavior toward technology, we attempted to analyze their "project portfolios" in terms of time horizon and other characteristics. With respect to R&D, at any rate, there appears to be a clear interaction between time to payoff, uncertainty of outcome, and degree of departure from current activities (e.g., product lines, production technology, material base, market applications). So far, we have not been able to develop a credible and general analytical model that will provide the coefficients or weighting of these factors (and others) in project selection and portfolio construction decisions (see Chapter 7). However, leads uncovered in our research and consulting on technology in decentralized companies have formed the basis for some of our current research on these very questions.

In any event, these preliminary casual observations plus the early systematic observations, once the decentralization study was begun in 1956, led us to formulate the working hypothesis that has been the basic stimulus for the overall study. This can be stated generally as follows: unless significant pressures to the contrary are perceived by the division manager, he will tend to prefer a technology portfolio that is relatively short term, narrow in scope, and relatively inflexible.

With, of course, striking exceptions, we have found no reason during the course of the formal study and our related observations and experience in decentralized companies to reject this as a fair characterization. A major thrust of this book is to explore:

• Why division managers tend in this direction.
• What the consequences of this tendency are for technology in the division and in the corporation.
• What countermeasures are and can be used to mitigate this tendency, without significant undesirable side effects.

The division manager's first duty is to survive. One may question whether, in some cases, his personal survival may be at the expense of the survival or at least the effectiveness of his division and of the corporation. This question led us, in the early 1960s, to probe into the basis for division managers' compensation and how closely their personal reward structure was tied into divisional and corporate success. Little quantitative evidence was found at that time of the degree to which division manager compensation was directly tied to personal and divisional performance. However, the general impression was gained throughout the study that, regardless of the specific quantitative details of the compensation system, there was no question that poor performance of his division, *in the current time period* (month, quarter, or year), spelled potential personal trouble for the division manager.

An extreme illustration of this, in numbers, was provided by one company president who, when asked if he *really* tied the division manager's bonus directly to divisional performance, said: "Three years ago division manager A earned a 33% bonus; two years ago he earned a 17% bonus; last year he received none; and this year he is gone."

Many top managers whom we interviewed said they did not believe in such direct bonus payments. They claimed that the division manager was rewarded primarily by salary for his performance, considering factors beyond his control as well as his own efforts. In these cases, the major sanction was to demote or fire the person who did not produce. In other cases we have encountered intricate profit-sharing arrangements, based on formulas that included both divisional and overall corporate performance (profits, return on investment, or other measures of return).

Overall, we found that division managers are controlled in large decentralized corporations by a very small set of criteria (typically, but not exclusively, return on investment or assets or changes in such ratios) and that they are subject to a very small set of sanctions (typically, removal from their positions).

This situation appears to be typically American, however, for our limited examination of European and Japanese counterpart organizations and managers does not appear to follow this "few criteria, few sanctions" practice. Our limited observations with Japanese decentralized corporations (derived from my occasional visits to give seminars in Japan and from Japanese executives visiting the United States) indicate a quite different approach to incentives and sanctions for division manager performance, with removal from office or discharge a much less likely event.

Certainly, the position of the division manager in a highly competitive market sector (e.g., one product line or a small set of related product lines and, for many, a narrow range of customers) makes day-to-day or month-to-month survival a major preoccupation. For many of them, in such a narrowly prescribed situation, there is a limited scope of action, beyond fighting for a larger market share (at the expense of one of their competitors) or squeezing their costs down. For the division manager in a fairly static market, with relatively inelastic demand and a fairly rigid cost and price structure, there is virtually nothing he can do, in the short run, to increase his share of the market other than vary his advertising (if his product line requires much advertising) and some minor aspects of his product (such as the "rechrome" strategy in some segments of the appliance business). He may, as some do, make a major impact with a significant product change in the intermediate or long term. But this gets him back to the situation where he must spend highly certain "current" dollars for a highly uncertain advantage in the long run (when he may not even be division manager or at least manager of *this* division).

This is a fairly oversimplified view of the external competitive situation for many division managers, but it does serve to illustrate the external pressures they may be under. Now let us consider what may be an even greater set of pressures—those coming from internal competition within the corporation.

At a minimum, the division manager must compete for personnel, facilities, capital, corporate services, and other scarce resources in the corporation. However, the most competitive situation for many division managers is for the attention of top management. In some of the very large decentralized corporations we studied, the number of divisions ranged from 20 to 100. Prior to the emergence of the role of the group executive (one of the trends that emerged during the course of the formal study—see the next section), the lineup of division managers to get some of the CEO's time was incredible. Some division managers had difficulty getting a few hours a year to discuss their long-term strategies, problems, aspirations, or plans, other than at formal stand-up review presentations.

Except when a division was in serious trouble or had made a dramatic coup and needed immediate, massive attention by top management, the allocation of top management time to individual divisions for quiet, long-term planning was found to be very low in most large companies with many divisions. In many of these cases, the division manager did not want for certain types of attention from the various *staffs* at the corporate office who wanted to know why he was doing things this way or that way and why didn't he conform to these policies and those procedures. Competition for the CEO's *personal* attention to his perceived problems appeared to be one of the most severe forms of competition that the typical division manager experiences in many large decentralized companies.

Under these circumstances (and others discussed elsewhere in the book) of external and internal competition, it is small wonder that we observed a strong tendency for division managers to suboptimize, to concentrate on the short term, and to emphasize the relatively certain aspects of technology as well as other functions.

This tendency to concentrate on the short run, more certain, less innovative kinds of projects is also evident in the attitudes and behavior of many division managers toward supporting activities other than technology—planning, information systems, facilities, maintenance, management sciences, and human resource development.

3.7 THE GROUP EXECUTIVE

Among the major changes in the decentralization picture in the 1960s and early 1970s was the emergence of the group executive. Some organizations have had them for decades, but this role has only recently emerged in the majority of large decentralized firms. Typically, he is not directly responsible for the operations or the P&L of one division or line of business. He is typically responsible for two or more divisions, which are often grouped according to a rationale such as technology, market, class of products, age, size, location, or history in the company.

In some organizations, his role is clear: he is a member of top management, increasing the ability of top management to pay closer attention to operations in terms of monitoring or control. In other organizations, his role is equally clear: he is merely a "superdivision manager," with direct responsibility for the operations and profitability of more than one division. In many companies with group executives, however, the role is still a muddy combination of these two roles and this can lead to great confusion and conflict.

The emergence of this new role has posed several sticky questions in the realm of decentralization:

- If the group executive is merely a superdivision manager, to be controlled and evaluated as a division manager, then how can he be an integral part of top corporate management—a member of the CEO's inner circle?
- How is decisionmaking shared by the division manager and his group executive? Some forms of this relationship appear to divide their functions along lines of strategy versus tactics, with the latter being reserved for the division manager and the former for the group executive. In other cases, the relationship is viewed as a facilitating one, with the division manager retaining most of his original freedom of decision but now having a "helper" or "communication channel" or "staff strategist" in the form of the group executive to facilitate relations with corporate management and other divisions.

Where is this person to come from and what qualities should he have? The most common source of the group executive is from the ranks of division managers. This poses a problem in his mode of operation, his loyalties, and his relations with former peers. One problem encountered in a number of companies has been the tendency of group executives who were formerly division managers to continue to act in that capacity relative to their former divisions and to place relatively more emphasis on their former divisions, due to more intimate knowledge of its people, operations, and problems and, perhaps, due to a need for greater comfort and personal security in the early stages of their new role.

In the 1960s, many large decentralized companies experimented with (tried out) the group executive concept. Some of them appointed one or two and waited a while to see how they would work out before appointing more. Others started the new approach with a bang—combining all their operating divisions into groups and appointing group executives to head them. In several cases, which we have followed for a number of years, there has been complete or almost complete turnover among this set of executives in under a decade, independent of retirements or promotions. A typical reason for the turnover has been that "he (or they) wasn't the right choice and had to be replaced." Few of the companies that have made commitments to this new organizational layer have reconsidered the basic premise on which it is based and have either abandoned the concept or done a thorough job of organizational design to assure its success.

In addition to the basic questions of apparent role and prior experience raised above, there are a number of issues that may not seem important at the time the group executive position is established and the first incumbent appointed, but that can become crucial later. Among them are the following:

1. How many such groupings are viable, before a whole new layer of decentralized organizations appears and is itself in need of coordination? That is, let us consider a decentralized corporation with, say, 15–20 operating (product) divisions operating in half a dozen markets and based on 3–4 distinct technologies. Overall organizational, marketing, and/or technical logic or differences in size and age of divisions may indicate that six, or four, or some other number of "groups" makes sense. It is very difficult (if not impossible) to optimize the number of groupings according to all the relevant criteria. So each "magic number" represents a compromise. Our observations are that, although these compromise groupings appear to remain relatively stable in some companies for many years, there are continuous pressures to regroup according to some other criteria than the ones in force. When the number of groups approaches the same order of magnitude as the original number of divisions (e.g., 5–6 groups, 10–15 divisions), many of the conditions that gave rise to the perceived need for the groups appear with respect to the groups *themselves*. Hence, we have observed a third-level grouping, where a series of "senior" or "executive" vice presidents are appointed to oversee the activities of the group executives and to relieve the top corporate management of the details at the operating group level. And so on.

My feeling is that this multilayering of the divisionalized corporation, which has appeared to work well over a significant period of time in some giant companies, has gotten out of hand in many other decentralized companies which have not spent sufficient time and effort thinking the concept through and really *experimenting* with it, rather than doing it on a cut and try basis.

2. Can the group executive really pull the switch on one of "his" division managers and order something to be done or undone? Many of the group executives, particularly those we encountered in newly established positions, did not really know and were hesitant to try to find out. This situation is treated in the organization theory literature under such headings as "power" and "legitimacy of authority. " If he does try to pull the switch, and finds that the wires are not connected, he can have lost any future possibility of exerting influence on the division managers. If he temporizes, when faced with a need to decide or act, he may also let the apparent power of his office erode and slip away from him (and possibly his successors). This set of circumstances is very intimately related to the prior and obvious power positions of the division managers in his group. Some of them may be very independent (and very successful) managers. Others may also be independent but may not be so successful. If he plays the game of "pecking" those who are not powerful (either by virtue of their past success or by

virtue of their personalities and/or their historical relations with top management), he may win some battles but may lose the overall objective of establishing a legitimate, routine level of control over all the divisional operations in his group. Frequently, he is left no alternative but this jousting, since top management may merely appoint him, publish or announce a vague job description, and throw him to the wolves. For some appointees, these may be sufficient conditions under which to establish their legitimate and effective role. For others it may be (and has been) sheer disaster. These latter find influential, successful, or well-connected division managers circumventing them, getting concessions directly from the CEO, or just ignoring their orders or "suggestions." This is particularly bruising for a group executive who is given authority over a particular division because the division manager is a maverick and the top management wants him off *their* necks.

3. Does the group executive have a staff? Some of the military metaphors used in the above section suggest that, like a military commander, the group executive, in order to be effective, may believe that he needs a "general staff. " He may perceive the need for accountants, personnel people, technical advisors, planners, marketing specialists, financial specialists, and so on. He may indeed attempt to set up a miniature corporate office in his group headquarters (many of them do). But this raises many fundamental questions involving specialist staffs throughout the corporation. What do we mean by decentralization and the necessary control that accompanies it? How many staff groups can the company afford at the division, group, and corporate levels? Even if these groups can be supported, what is the logic and how are problems of coordination and division of labor to be settled? How are these groups kept from falling all over each other and interfering with, rather than complementing, each others' functioning? We have observed such consequences of lack of careful organizational design and poor organizational design involving many staff specialties. The particular focus of this book, of course, is on technology, but we have also closely followed related activities such as information and data processing, systems analysis, operations research, personnel and training, long-range planning, market research, and management science (which some professionals in the field believe encompasses many of the others).

Of course, some group executives (and some top managers) in decentralized companies prefer to operate essentially without staffs. They may have a secretary, an accountant, and (in the case of top managers) a lawyer or two. We have encountered this style in many of the giant conglomerates and are hard put to assert that the system does not work well that way. We can see and have been told about the many longer-term deficiencies that mode of management can create, but there are no clear effectiveness criteria on which to compare the performance of such companies with that of other more staff-oriented companies. The size of corporate (in some cases including group) central office staffs that we have encountered in the top 200–300 industrial companies in the United States ranged from half a dozen people to 10% of the firm's total workforce. This in itself is worthy of a deep study. The several superficial studies in the literature which use only statistics of staff sizes do not penetrate into the essence of the question as to whether large staffs are essential at the corporate, group, or even the divisional levels.

In the 1980s, as well as in earlier periods, staffs at all three levels have become a "happy hunting ground" for corporate raiders and buy-out specialists who are trying to service the massive corporate debts involved in the acquisition process. Many staffs in the "10%" category soon find themselves down to a skeleton "turn out the lights" crew.

3.8 THE GROUP LABORATORY

We have encountered only a few cases of "group laboratories" since the emergence of this new group executive role, but it is possible that for reasons of economy and technical efficiency, this organizational form will become more common. It is tempting, for example, to attempt to eliminate or reduce the redundancy in separate divisional labs working in the same disciplines or applied areas by regrouping these activities into common group-wide technology services. This issue arises in the specific case of mergers and acquisitions (see Section 3.10 on champions for a discussion of that). It also arises as a natural outgrowth of changes in markets, products, technology, and competence of existing divisional and corporate labs.

This concept of "group" laboratories and its underlying logic are strikingly similar to the concept of "regional" laboratories, which we have encountered in our research over the past 30 years on the organization of science and technology (S&T) in developing countries. Due to the limited resources of most developing countries and the striking similarities or overlap in their apparent needs for science and technology, a logic has developed among international funding and advisory organizations that a good solution is the regional, multicountry laboratory. Consequently, many regional laboratories of one type or another (single purpose, multipurpose, basic, applied) have been set up in Asia, Africa, Latin America, and the Middle East. Some appear to have succeeded, in that they have continued to exist and are performing useful services. Many others have come apart at the seams, and others continue to operate, but not as truly "regional" labs.

A good theory on the success and failure of the regional lab concept and of specific applications of the concept is yet to be developed. However, we have many clues as to why they run into difficulty. A major source of difficulty, of course, is the problem of submerging parochial national interests in such an international enterprise. For countries that already are rich in S&T resources, their participation in a particular regional or cooperative lab may not be very important to them in terms of pride, control, direct benefit, and level of expenditure. Thus, for example, several highly successful regional or international laboratories or major projects have been operated in western Europe for the past few decades in areas such as physics, atomic energy, and aerospace. Recently, however, some of these ventures, and several other major international R&D projects, appear to have become either "too important" to individual countries or "too unimportant" to permit their continued cooperation on the former "give it the money and laissez-faire" basis.

When we examine regional technology institutes and major projects among the developing countries, we find that virtually all of them are highly important to the member countries from the viewpoint of their monetary and other resource contribution, national pride, interests in specific fields and projects, and mode of operation. Some of the evidence for continuing difficulties comes from hassles about who will lead the lab or institute, what proportion (and status) each nation's professional contingent will constitute, what languages will be used (if there is a language difference between countries, which is fairly common outside Latin America), and perhaps most important of all, where the buildings will be located and the funds expended.

In the case of industrial corporations and the group laboratory, some of these problems appear only by analogy (e.g., nationalism and differences in language), but some of the basic issues also exist: location (near which division's plants or centers of operation), project portfolio (whose specific interests and product lines will be emphasized), management (where will the director and his staff come from), control (who will call the shots), and support (on what basis will the participating divisions contribute or be taxed).

In several large decentralized companies, where operations are clearly bifurcated—for example, into two sets of divisions or groups with entirely different technological bases—the group labs are, in reality, a set of corporate research laboratories (CRLs) with responsibility for only a part of the company's total technology. We have observed this to operate fairly well, except in the instances where advancing technology and need for diversification have driven (rather than drawn) the two labs closer in technology base, without cooperation and joint projects being set up between them. This can lead to, for example, poor electronics being done in the engineered products group and poor mechanical engineering being done in the electronics group. We have encountered several cases of this, which are illustrated by caselets elsewhere in the book.

Some situations are not simple bifurcations but involve several group labs. Each of these is intended to service a logical set of operating divisions of different sizes, different stages of technical maturity and sophistication, and different positions in their respective markets. For example, some of the divisions may be technical leaders while others may be followers; some may be monopolists or oligopolists, locked in nose-to-nose combat with technologically superior competitors. The divisions may have quite different time horizons for product development and technical service. Under some of these circumstances, the group laboratory can become a nightmare for its director, the group executive, and the division managers and their technology managers who are supposed to support and benefit from it.

3.9 POWER AND INFLUENCE OF THE CHIEF TECHNICAL OFFICER (CTO)

This section is about the individual in the corporate office who is considered the top "technical" executive in the firm. Almost all very large, technology-based firms—

decentralized or not—have such a role. Generally, it is formalized with a title and occasionally someone informally occupies this role. Most of the exceptions are among the extreme conglomerates that are run with a very small corporate office. That consists mostly of legal and finance people and one or more senior, executive, or group vice presidents.

Such an ubiquitous role has been a rather recent development for many companies that did not have a very strong science and technology tradition historically. They managed to watch over technology matters through their set of divisional labs or a small isolated CRL. In some of these companies it is not clear whether it has become fashionable to have a CTO, along with a chief planning officer, a chief information officer, a human relations or human resources executive, or whether the role has resulted from careful organizational design and/or necessity.

The general move away from the earlier title of vice president R&D (VP/R&D) or similar "scientific" designations to chief technical officer (CTO) seems to reflect more than fashion. Many firms are quite concerned about their *entire* technology picture—R&D, manufacturing, product design, technical intelligence, patents, materials, quality, standards, regulation, energy cost, environmental impact, and related matters. In some firms the job of CTO—often only a few years old or only into its second generation of incumbents—is intended to be much broader than the traditional role of VP/R&D. However, not many of these incumbents have been successful in gaining sufficient control, or even adequate information and access, to make these added functions part of their domains. Many of these technology-related functions already have deeply entrenched individuals or groups who strongly resist being superseded or pushed aside by a corporate officer who is viewed as less experienced than they in their particular areas.

Top management's increasingly frequent search outside the firm's own R&D or technology ranks for a chief technical officer at the corporate level often stems from frustration or dissatisfaction in their previous dealings with technical specialists in the company. Where the CEO inherited or (in his view) mistakenly appointed a classical "scientific type" to the position of CTO, he may have had a series of disappointments with such people in attempting to use advanced technology in the firm's businesses. Or he may have had a series of personal disappointments at "not being able to communicate" with "that person" or "such people."

Whatever vague or specific feelings the CEO may have about the incumbent in this important corporate position, it might be described as a very unfruitful relationship for the CEO and, perhaps, his fellow corporate and divisional officers. He may be receiving complaints—subtle or overt—about the impracticality, ambiguity, unresponsiveness, lack of savvy, or foot dragging by corporate technology people when asked for information, assistance, or material results (new products and processes). Added to this frustration may be an even more personal reason for dissatisfaction. The CEO may feel very uncomfortable in talking with his CTO or even being in the same room with him. This latter feeling may result from personal antipathy, stereotyping, or specific personal experiences in school or elsewhere with "scientific types."

Whatever the reasons, this sense of discomfort can be a very strong, if not completely conscious, reason for finding a replacement from among the "real businessmen" or managers in the company who "really understand what I am trying to do, what directions the company is going in, and what our time frame for action is." Although current statistics are not readily available, there has been a notable increase in the past 15 years in the number of CTOs coming from other areas of the firm than the traditional R&D activities. One indication of this growing dissatisfaction with the traditional type of CTO as compared with the newer, *non-R&D* type is the increase in examples of the latter, as compared to the former, who are now included in the inner circle of top corporate management. Where the new style CTO was previously a division or group executive, he may even be welcome, initially, in the inner circle of the firm's division managers, where there is such a "cabal." (In some firms, competition for career paths and resources is so fierce between division managers that they are seldom found consorting with each other.) After the initial welcome, based on his previous line assignments, the new CTO may still be suspect because of his current assignment. It is amazing (but not surprising) to see his former peers "turn against him" after he seems to begin to assume the values as well as the formal duties of the CTO and become a critic of their divisions' (stated or intended) technology policies and capabilities. While I do not want to go so far as to say that "the role makes the man," it is clear from a wide range of contacts with such executives that the effective ones will gradually or even abruptly "change their stripes." They will take positions vis-à-vis division management that the firm's and division's technology base is being neglected and must be strengthened. From being "one of us," they become "one of them" (corporate staff nuisances) as viewed by divisional management. They are then subjected to many of the same difficulties as their technology-career colleagues who arrived at the CTO job from a strictly R&D or technology background.

Depending on the personalities of these non-R&D CTOs and on the special circumstances they find themselves in, they *can*, and many do, fulfill the hopes of the top management that appointed them, by exerting more influence on divisional management than their nonbusinessmen colleagues can and do. "Value switching," at least on the surface, is not unusual. Classical examples are (1) noncommissioned officers who receive battlefield commissions and who are suddenly not the same people their buddies liked and esteemed and (2) graduate students in a production management class who, when assigned arbitrarily to the roles of "research," "production," and "marketing," begin almost immediately to attack their counterparts as "narrow minded" and "too specialized."

Of course, as in many aspects of the subject of this book, there are exceptions. In some industries, such as food processing, there is a long tradition of oversight or control of quality and R&D by a single corporate officer. In some of the sample food companies, in fact, the title of the CTO was actually vice president of research and quality control. In some process industries, the vice president of engineering (or, in an earlier day, the chief engineer) had responsibility for production technology, equipment, and facilities, as well as product development. Many of them still do. In still others, the regulation issues are so important to the firm that the CTO has this as a major part of his job description. In other firms, regulation issues are deemed more of a nuisance and given to the CTO to handle so as to relieve other executives of the need to worry about them. The CTO of the future, if I may wax poetic for a moment, may indeed have much broader responsibility for technology in all its manifestations than most incumbents currently have. From an ideological viewpoint, someone in the firm has to be looking out for *all* aspects of technology and its implications for the survival and success of the firm.

In this potential transition (it is by no means clear that the decentralized form and philosophy of many firms will permit such broad powers in the hands of a single corporate executive), many battles will have to be fought and much ambiguity clarified by the CEO and the top management group in order to allow the broad-range CTO to assume his responsibilities and do his job effectively. One example, from one of the sample companies, may suggest some of these difficulties, which are discussed in some detail in this section in connection with today's CTO. The vice president of research and manufacturing (sic!) in a large machinery firm had the usual oversight responsibilities for engineering, product development, and research as well as line responsibility for a corporate research and engineering lab. As a result of some significant product liability suits against the company, he began to look into the "total cost of warranty, rework, returns and allowances, and liability," involving both internal and external (legal costs and settlements). He was amazed to find that such costs had never been consolidated before or even considered as a "controllable" cost in the firm. The total figure was huge, relative to operating profit and other cost items. Much of it was also addressable through improved technology (some of it already on the shelf in industry or on hand in the firm) and through just plain focused attention. The screams of rage from the sacred cows and the oxen being gored drowned out his mild voice in the corporate headquarters and, after a brief attempt at getting top management's attention and support, he withdrew quietly to nurse his wounds and "attend to his own business." Many other examples abound, many of them directly related to the focus of this section—the power and influence of the CTO.

In order to examine this important issue systematically, we must first look at who these people are and where they come from. Their organizational backgrounds, educations, chronological ages, organizational ages, titles, and lengths of time in the CTO position vary widely. However, a few general patterns can be used to describe the origin of most of the men in such positions.[*]

- Career R&D men, who had come up through R&D in their present organization. Many of these had been directors of the corporate research lab (CRL) or a large division lab. They typically have advanced degrees in science or engineering (or both).

- General management men, with little or no recent experience in a lab, who were appointed to the post (often newly created or newly vacated by an R&D-experienced man) from an executive position in another functional area. These men rarely come from outside the company; they have typically held a num-

---

[*] Again, as in the preface, I apologize for use of the male pronoun. There are some outstanding women in the CTO role in industry, but males dominate by an overwhelming majority (more than 95%).

ber of general management positions in the company over a long period, one or more of which (often "way back") may have been in R&D or engineering.

- Outside R&D men who are appointed directly to the top technology job from a position in a university, a government laboratory, or a research institute. Some of them have had direct industrial experience before; some have not.
- Outside R&D men who held similar positions in smaller companies, or positions at one level down in an equivalent-size organization.

The practice of hiring technology executives from the outside has changed radically over the past three decades. Earlier in this period, we found the original person who had started the laboratory, or his protege, still running it in the late 1960s. This was remarkable for many very large decentralized companies, and it was in contrast to other functional areas in the firm. These were among the largest firms in U.S. industry and some of them had had a substantial R&D activity for over 20 years. Through a combination of rapid growth, great success, and (in some cases) lack of a clear path from R&D into top management, many of the chief technical officers or VPs of R&D we encountered during this period had originally set up or been the first appointee to the top technology job in the company. During that period, however, many of them had not yet been appointed officers of the corporation (e.g., corporate VPs).

The situation has changed radically since the 1960s, however, again due to a combination of factors. One is that the ranks of this first or second generation of R&D executives, who either founded or presided over the rapid expansion of industrial R&D in the postwar period, have been thinned by retirement. In some instances, the third generation of top R&D executive has reached the top through natural succession. However, other factors have played a major role in the dramatic change among the top technology management personnel in the largest U.S. companies. Among them are the following:

1. Changes in the earlier gentlemen's agreement not to raid each others' laboratories, which dominated the situation in several major industry segments for many years. Of course, some other industry segments had made a practice of doing this from the beginning. Despite confidentiality agreements, employment contracts, and other formal barriers, the backgrounds of some of the top technology people in those industry segments looked like a roster of their current company's competitors. In those industries that had not made a practice of such wide-ranging search for outside talent for their top technology jobs, the choice was generally limited to either the next R&D person in line in the corporate office, the head of one of the divisional labs, or someone pulled from one of the lower ranks in R&D who fit the specifications for a new (although internal) broom. As external recruiting of top R&D executives has become more common, the alternatives have broadened considerably.

2. The periodic increase and subsequent phasing down or out of major government R&D programs, laboratories, and companies that depend primarily on government contracts. In particular, the aerospace industry and the related gov-

ernment laboratories and university laboratories which took a licking in the early post-Vietnam War period became major sources of top technology executives for primarily commercial companies.

3. The gradual disillusionment with the output of industrial R&D which began appearing among top industrial managers in the 1960s and which had reached crisis proportions in many companies in the 1970s and 1980s. Top managements have been looking in different places and for different types of people to help turn their technology operations around. Many more are either going outside the company, outside R&D, or both, although in some companies, the outside, non-R&D appointment is still fairly infrequent. These simple categories—inside–outside and R&D–non-R&D—are not intended as a cavalier categorization of people occupying this important position in a decentralized company. It is merely intended as a starting point for general discussion of the subject of power and influence of the person in the position and specific illustration of some of the problems that we have observed the top technology officer encountering. A major set of questions that guided segments of our interviews and unobtrusive observations in decentralized companies were:

- How much actual power and influence does the top corporate technical person have?
- Is this power and influence limited to strictly technical matters, or does it carry over into other functional areas of the firm (finance, general business, marketing)?
- Where does this power and influence come from; how much comes "with the job" and how much does each new incumbent in the job have to earn on his own?
- How does he exercise his power and influence?
- Specifically, how much power and influence does he have with respect to top management, divisional management, divisional technology, other corporate functional areas, and customers and other outside organizations?

The temptation here is to go off into a long discussion of the theoretical meanings of the terms "power," "influence," and others, such as "authority" and "leadership," which are treated in a huge literature in the general behavioral sciences and specifically in organization theory and management theory.

If there were unanimity on these concepts or even relatively good agreement, it might be worth the trip. However, there is not much convergence in this literature and, furthermore, much of it is at an abstract level, far removed from the day-to-day behavior of people in organizational situations such as the one being discussed in this section (the power and influence of the top technical person in the large decentralized company). Therefore, I will forego a rigorous, theoretical treatment of these terms (power, influence, etc.) and refer the reader to some of this literature.

For present purposes, let us consider power and influence to be closely related. It is the attempt to distinguish between them that leads to the ballooning of the theory literature. I shall focus, very roughly, on the ability of the CTO to "have people agree to things the way he sees them, and do things he wants done." If all other managers in the company already see things and are already deciding issues the same way as the CTO, then a discussion of his power and influence over them would be extraneous. In this best of all possible, but unrealistic, worlds there would be no differences in perception and no conflict over objectives for technology, its organization and method of operations, the projects and programs it works on, its role in the company, its sources of funding, and the value of its results. Unfortunately, this is a very unrealistic picture in all the sample companies and, probably, in most organizations. There are distinct differences in viewpoint on some or all of these matters. Conflict—ranging from mild, latent levels to fierce, open levels—is quite common. The sources of such disagreements and conflict are discussed in many places in this book. Among them are differences in values, time horizons, operating styles, behavior patterns, and professional orientation. The concern in this section is with the ability of the CTO to overcome or circumvent those differences that interfere with the ability of technology-related activities, such as R&D, to perform effectively in and contribute significantly to the organization. Among the means at his command, some of which are more under his control than others, are:

- Aspects of his personality which have to do with persuasiveness, instilling of confidence, ease of dealing with him, apparent integrity and competence (as perceived by others), and other communication and human relations skills.
- A personal reputation for competence, fairness, and reliability which may have followed him from his previous position or organization.
- The reputation of the corporate laboratory or department or his position (prior to his succession to the post) for effectiveness, responsiveness, and "right-headedness."
- The formal charter that describes his role and formal powers.
- The overt or implied support of him, his position, and his activities by top management and other influential people.
- Formal control or authority over budgets, personnel matters, project portfolio (of divisional labs), work load in the CRL, new areas of investigation, and other key aspects of company-wide technology.
- His corporate status—officer or not—and whether he is a member of important corporate committees or the unofficial inside group of top management.

Some observers (academicians and practitioners alike) place a great deal of emphasis on personality and other personal characteristics in attempting to explain the degree of power and influence of the CTO. Others place primary emphasis on the formal aspects of his job and his assigned status. My view is that there are many combinations of these several factors that can lead to success (or failure) of the CTO to wield power and influence effectively in the company.

One of the most common comments we have encountered, both in the formal phases of the study and in informal contacts with hundreds of technology managers (through seminars, consulting, friendships) over the past 30 years, is that the CTO must indeed depend on informal influence and persuasiveness to further the interests of technology in the company and to assure its effectiveness. We certainly have no systematic data to present to the contrary. However, our observations clearly indicate that the CTO who has to rely *primarily* on such informal personal qualities is at a decided handicap in dealing with, for example, division managers in the decentralized firm. We have found that very few corporate technology executives (vice presidents or otherwise) have direct, clearly mandated authority to decide on allocation of R&D funds to divisions and within a division; hiring, promotion, or termination of divisional R&D personnel; or final "sign-off" approval on specific projects and programs to be undertaken by the division in its own lab or in the labs of outside contractors. Typically, the CTO does have control over these matters (except the amount budgeted) for his own corporate research lab (CRL), if the company has one.

Some CTOs have ultimate control over which division-sponsored projects and service activities they will perform in the CRL, the degree of control varying with the method of funding divisional projects in the CRL.

Consistent with that part of the theory literature that deals with "apparent versus real" and "informal versus formal" authority, power, and influence, we have encountered a significant number of instances that clearly demonstrated the interaction between formal and informal power and influence. For example, we have observed top corporate technology executives who had certain of the above powers formally delegated to them, but who were unable to exercise this power effectively. Conversely, we found many who had virtually no formally allocated power over these matters, but who exerted a dominating influence over them.

In general, however, we found that the CTO has little direct power or control over major decisions concerning divisional R&D. In the few cases where we followed attempts by a CTO to wield such power, over a period of months or years, the issue was generally decided by whether he really had the backing of top management to force decisions on the division manager and the divisional R&D staff. In other words, informal and personal influence and persuasiveness aside, major decisions on the divisional R&D, when there is a sharp difference, generally go in the direction favored by the person who actually controls the divisional R&D budget, personnel decisions, and project portfolio decisions. This person is typically the division manager in most decentralized companies. I can anticipate a storm of protest to this highly oversimplified statement from our many friends who are R&D managers and many more R&D managers who have neither been nor probably will be friends, after reading this. I plead guilty to this oversimplification but argue that this is the only way that we can explain the many instances we encountered of exercise of raw power in disagreements or open conflicts between the CTO in many of our sample companies and divisional management. These instances are of many types; here are three brief illustrations:

1. Division manager A proposes to appoint Mr. B as new manager of divisional R&D. The CTO violently objects on grounds of B's technical incompetence. Mr. B is appointed.

2. Division manager C requests that the members of the CRL who are being funded to work on long-range problems of his division be switched over to short-term service work and minor product and process improvement work. The CTO objects. Top management says: "It's his money; he has the right to say how it is to be spent."

3. The CRL comes up with a potential new product for division D. The new product has great promise for intermediate or long-term sales and profits for the company. The manager of division D says his people are too busy to fool with it now and, besides, he's not going to pay for the next 2–3 years of development costs until he is sure that it will be a moneymaker *for his division*. If there is no other mechanism than transition to an existing operating division, the project is put on the back burner in the CRL or dropped.

These only illustrate one side of the issue—the side that says that the locus of "real" power has a decisive influence on the outcome of "yes–no" type decisions involving conflict between corporate and divisional R&D. We could cite many contrary cases, in which a strong, persuasive CTO swung decisions his way by wielding informal influence over division managers over a wide range of issues. The point to be made here, however, is that much of the mythology of the influence of the CTO in the decentralized company involves his *informal* power and influence, whereas it seems clear that *formal* power over the allocation of resources is frequently decisive in the crunch.

This issue brings us back (as many others raised in this book will) to the basic assumptions on which the decentralized firm is organized and operated. In its extreme, this set of assumptions delegates to the division manager "total" control over the operations of his division, except for capital funds, some legal matters, and some necessity to avoid conflict with corporate policy in such matters as industrial relations, national institutional advertising, overall pricing and behavioral standards in the marketplace, general ethical practices, and a few others that involve general guidelines rather than specific decisions or courses of action.

We encountered very few instances among the companies we studied where control over matters pertaining to technology tactics and day-to-day decisions involving divisional technology programs was retained or assumed by the corporate office, either in the hands of the CTO or the top management themselves. The most common situation is one in which such matters are left to the division manager, subject to general overall (and frequently quite vague) guidelines. The CTO is then expected to use informal influence, consistent with those guidelines, in persuading the division management to "think and do right."

We have encountered many manifestations of the power of division managers over the technology related to their divisions. Even in those companies that have apparently centralized their entire R&D operations, division managers who pay for R&D either directly or indirectly (through a sort of "tax") can typically get their way in matters pertaining to their division or can circumvent the system by going outside the company for technical help. When called on this issue by the CTO or by top managers themselves, strong division managers will counter with "this is my business and I must do what needs to be done to be sure it is successful."

In those divisionalized companies that have designed the degree of decentralization far short of the "in business for myself" stage, all bets are off. Where the corporate office dominates short-term decisions in the divisions as well as longer-term strategic ones, the CTO may very well have an almost "line" relation over division managers in matters related to R&D and other technical activities.

Indeed, we have encountered a number of cases where such is the case overall, or merely with respect to a particular division or divisions which are in trouble. In the latter instances, a "staff" CTO has "gone into" the division and started throwing weight around (with top management's backing) in order to get the mess cleaned up. In several instances, where the particular CTO has had prior substantial experience in production and other non-R&D functions, his temporary raw power may apply across the board of divisional operations. We have seen this happen in several instances where the CTO is the only staff corporate officer who is available for troubleshooting or who had the necessary background and credibility to do the cleanup job.

Where does this rambling discussion of the "real" versus the "apparent" power and influence of the CTO take us with respect to improved design and operation of technology? At the least, it clearly raises the issue of how much power top management is willing to or can grant to a person in this position vis-a-avis the managers of the company's operating divisions. If top management really wants their corporate CTO to look after the entire technology program of the company, to assure its technical vitality, to have the various labs complement each others' work, to keep a balance between short-term and longer-term work, and to provide the company with ability to respond to technical threats and to capitalize on technical breakthroughs and opportunities, then it either will have to give him the formal power he needs to resolve major conflicts or must be prepared to play the continuing role of arbitrator between the CTO and divisional management.

In addition, it may develop a set of criteria through which top management can monitor the relations between the corporate and division activities related to technology. This could provide a sensitive control device for pulling in or letting out the reins on their division managers with respect to technology matters.

Short of the exercise of raw power in resolving conflicts, how do the successful and unsuccessful CTOs operate with respect to division management and other functional areas in the firm? This is where the double dichotomy of background comes in. Let us first consider the "inside–outside" dichotomy. Obviously, each category has its own set of advantages and disadvantages, or practice would be essentially uniform. There had been a strong tradition of promotion from within the companies that have been active in large-scale R&D for a long time. The expected candidate for the new or vacant top technology spot in the company had traditionally been the director of research, if such a number two position existed at the corporate level, or the director of one of the larger divisional labs. This tendency is still dominant, but there are indications that the hiring or appointing of outsiders is increasing.

When the new CTO is an outsider, he has two sets of handicaps relative to an insider. The first set includes all the procedures, people, historical detail, informal channels of communication and decisionmaking, and other aspects of his new environment, which he must learn and make judgments about in order to operate effectively. The other set includes all the things that other members of the organization have to learn and make judgments about relative to *him*, so that they can operate effectively from their viewpoints. The process of "enculturation" takes time, patience, courage, and other precious resources of a person eager to dig into a new job and begin to show some results. This is a process that is not required in the case of the insider appointed to the CTO position, unless his previous organizational location and functional responsibility were so different that people have to recalibrate him all over again and that he has to explore and establish a whole new set of relations in the organization. Of course, the outsider may have a distinct advantage in not having to live down past mistakes, blunders, and bad relations which the insider may drag with him into the new job. In addition, the outsider may not have to live down the "junior" image that often sticks to people promoted to a high level from within the same organization or the "protégé" image that sometimes obscures the individual qualities of a person who has been in the shadow of his predecessor for too long.

In the approximately 40 decentralized companies we examined closely in our detailed field studies[*] in the 1960s and 1970s, 7 of the 20 CTOs who had recently been appointed were from outside the company. Some companies had had several incumbents in the job up to the time of our visits, and others had only recently established the job, with the first incumbent still in office. The proportion of outsiders has risen dramatically over the past decade as some companies appear to be playing "swap" for their CTOs as well as for advertising, planning, and information officers. For the outsider who comes from another firm in the same industry, his new position and organization may be similar in many ways to his old one, except for a possible rise in status. In addition, if he were active in professional and trade associations in the industry, he may know a lot of the technical people in the new company and even some of the nontechnical corporate executives. It is unlikely, however, that he will know many of the division managers and even less likely that they will know or have heard of him.

Thus, the process of getting acquainted, courting, trial and error must be initiated by the outsider upon his arrival in the company. If the position he comes into is well established, the process can be shortened quite a bit, with only personal relations between himself and his colleagues, subordinates, clients, and sponsors to be developed. If, however, he and the job are both new (six of the seven cases mentioned above) then he must engage in many of the activities described in Section 2.11 entitled Why Does It Take R&D So Long to "Get Organized"?

Where the outsider is a person of some reputation in the R&D community (such as a prominent government laboratory director or a well-known university scientist), the path may be somewhat easier for him in terms of getting integrated into the R&D aspects of the job. It is not very likely that a scientific reputation will do him much good, however, in establishing working relations with divisional managers. Quite the contrary, their stereotypes of "esteemed scientists" may get in the way of cordial and easy relations.

[*]Some lasting several years.

Of the seven outside people whom we followed subsequent to their entry into CTO jobs, one was promoted to president, two were kicked upstairs into "nonportfolio" staff jobs, two left the company within a few years, and two were still on the job approximately 10 years later.

All but one of them (one of those who remained in their positions) ran into classical difficulties in trying to serve the corporation and the divisions. They all encountered strong opposition from operating divisions and most of them had trouble justifying their operating results (from the CRL) to top management. We followed four of the seven companies from a distance since the original field visits to them (which took place in the period 1956–1964) and have observed continuous thrashing about in three of them, attempting to find an "optimal" way of integrating corporate R&D into the rest of the firm.

These sparse data do not, of course, give us a direct comparison with firms who appointed or promoted inside people to the job of CTO (13 of the 20). In the six firms whose inside CTOs were no longer in the position 10 years later, three retired and three were replaced for other reasons. We cannot really say that these firms had less churning in their R&D organization, because our subsequent follow-up of them happens to have been less systematic than with the firms that had appointed outsiders. Observations in other firms (outside the formal sample) suggest, however, that insiders have not fared significantly better in terms of management satisfaction and stability than their outsider counterparts. This seems to imply that the insider–outsider factor is not the dominant one in the perceived success of the CTO.

Now let us look at the R&D versus non-R&D dichotomy for appointments to the CTO job. Available data are not plentiful or clear enough to indicate a decided trend toward the appointment of people with *no* significant amount of previous R&D background to the position of CTO. However, the number of such cases has increased significantly in the past decade.

The major motivation of top management in appointing a person without an R&D background was stated in various ways to us by various top managers and others. In summary, these seem to converge on considerations such as the following. The CTO should be a man who:

- can bridge the communication gap between R&D and operations or general management.
- can serve as advisor to top management and take responsibility for all technical matters in the firm.
- talks and thinks the language of business.
- has faced operating problems and appreciates time and money constraints.
- has a proven record of company loyalty and effectiveness, who commands respect in the company and can work with people.
- can straighten out the mess in R&D and turn it into an asset rather than a liability.

In other words, some of the basic lack of trust in or discomfort with R&D people, which is reflected in the attitude of many top managers toward R&D,

leads to appointing someone to the top technology post who is "one of us"—the practical business and management group and not one of those "blue-sky types."

But these non-R&D inside appointments (they are typically from inside) are not home free in terms of getting control of the situation and meeting management expectations. Despite his personal qualities, such a person generally faces initial (and sometimes lasting) hostility from within the R&D community. This is often based on stereotypes of the hard-hitting, nonintellectual business executive who cares little for science or technology in itself and is often fed by resentment of the inside people who were passed over in his favor.

If such a CTO quickly demonstrates an appreciation of the "important" aspects of the R&D process (i.e., does not blatantly violate any sacred cows) and shows his ability to improve the resources and conditions of work for the labs, he may gain grudging respect from his staff. In a few of these cases we have followed, however, the idea that he is "not one of us" and "doesn't really understand R&D" seems to lie close to the surface for years after his appointment.

I am not sure where this recent trend toward appointment of non-R&D people to head up technology is going. It is very attractive from top management's viewpoint, since they have a person they know they can trust, understand, and work with. It remains to be seen whether the communication and understanding gap which such an appointment almost inevitably creates further down the line (e.g., between the CTO and his subordinates) is a big enough disadvantage to cause a slowing or reversal of the trend. One significant additional factor, which has not yet been recognized by many top managers, is that removal of the top corporate technology job from the set of rewards for which many R&D people in the company strive can have a serious effect on incentive, initiative, and risk taking. Historically, a number of R&D people who have made it all the way to top management have traversed through the job of CTO. If that channel is no longer open to them, the decentralized corporation may be closing off an important source of the needed scientific and technical inputs it needs at the highest level of management policy, decisionmaking, and planning.

Perhaps the principal method available to the CTO for building long-term good relations, acceptability, and influence with operating divisions is the use of the CRL (if one exists) as a combination training ground and "reverse farm club" for the divisional labs. Several objectives can be served by a constant flow of people from CRL to divisional labs and an occasional return flow (which is much less common).

One of these objectives, of course, is to provide well-trained and scientifically oriented people to divisional labs so as to establish and maintain a reasonable quality level for their technical work. Another, perhaps more political, objective is to assure that CRL alumni, who are sympathetic to the viewpoints of CRL and have certain "proper" scientific attitudes, will sooner or later be running the divisional labs and interacting with CRL on policy and program matters.

In some organizations this has worked admirably. The divisional R&D directors are by and large loyal alumni of CRL. They can communicate with, engage in cooperative projects with, and also sympathize with CRL viewpoints in such matters as quality standards, priorities, areas of program emphasis, and evaluation of specific people and projects. They often form common cause with the CTO and members of CRL in bargaining with their own division managers. On the other hand, sometimes this strategy backfires. Some of the divisional R&D directors we have encountered are indeed alumni of the CRL. But they are bitter alumni and go out of their way to oppose or passively block CRL programs and ideas. A proportion of these cases may involve specific personality and personal antipathy problems. We have little to say about that from an organization theory viewpoint, except to call attention to the CTO who naively assumes that all his protégés will be faithful, happy fans when they are no longer under his direct control.

I am concerned here with a possibly more systematic cause of some of the divisional–corporate struggles that we have observed within the company's technology/R&D community. This may be the very process that generates or selects out members of CRL to become directors of divisional labs.

Certainly, with few exceptions, the division manager has the ultimate (sometimes the complete) say in who becomes his divisional lab director. However, many division managers (almost all in some companies) rely heavily on the CTO or VP/R&D to recommend or approve nominees for the job. Whether the division manager chooses the nominee himself or relies on advice from the CTO, it is clear to all concerned that he is generally not looking for an outstanding scientist and R&D specialist as much as he is looking for a practical person who is, of course, technically competent but will serve him and his division first and the cause of science or research a distant second.

Almost by definition then, potential candidates from CRL are not necessarily the most outstanding scientists in the lab. It is true that a certain percentage of rare individuals combine outstanding administrative and business skills needed for the top divisional technology job with outstanding scientific skills needed for success in basic-type research. However, we have encountered very few of these paragons among the hundreds of divisional technology managers we have had contact with in the formal aspects of our study; in our teaching, training, and lecturing activities; our consulting; or our professional and personal contacts. In fact, the selection procedure seems to work against outstanding researchers becoming heads of divisional labs in many companies. This general statement does not necessarily apply to those leading firms that are highly science-based (e.g., pharmaceutical, aerospace, scientific instruments, state-of-the-art electronics). But in most firms, the choice of divisional technology managers heavily weighs administrative and business ability as dominant selection criteria, subject to acceptable levels of technical competence.

One possible consequence of this selection bias can help explain much of the conflict that appears after the former CRL member becomes head of a divisional lab. The argument goes this way: depending on whether the CTO is himself a management, business-oriented individual, there may have been real differences in values, style, standards, and interests between him and his former subordinate who is now a divisional lab manager. If, prior to the appointment, these differences were latent—that is, held in by the subordinate—they may suddenly break loose when he is no longer under the thumb of the CTO. His resistance, even opposition to some of this former boss's ideas, may shock the CTO and perhaps lead to higher levels of conflict and hostility than if the same behavior were manifested by a stranger. I am certainly not professionally competent to and have no wish to get into a Freudian analysis of this relation, but many aspects of parental or sibling rivalry may be relevant to this situation.

In any event, the successful transfer (from the CTO's point of view) of a CRL person to head up a divisional lab depends on the former working relations between him and the CTO, the degree to which they share the same values and see company and divisional problems the same way, and the interpersonal relations between them. The CTO would do well to analyze carefully these major factors and to attempt to extrapolate them into the new situation, before automatically assuming that such an appointment will automatically serve his interests and the interests of corporate-level technology.

One final point should be made on the choice of a member of CRL as head of a divisional lab. Such appointments may have very salutary effects on relieving the upward mobility pressure on the CTO. Traditionally, the path of advancement for most CRL people is toward the top corporate R&D post, which many incumbents hold until retirement. This effectively reduces the possible lines of advancement of almost all CRL members, since many of them are too specialized or too high priced for "equivalent" non-R&D jobs in the company. The outlet of divisional technology director or manager provides a major channel not only into higher-level technology jobs but also into non-R&D jobs in the division and, for some, into the corporate office itself. Many people have traversed this route and it is one of the major benefits to technology that decentralization has provided.

3.10 WHO CHAMPIONS AND PROTECTS TECHNOLOGY IN THE FIRM?

The logic of having an advanced technology program and laboratory(ies) in the modern corporation seems clear. It is also clear that the initiation of large-scale R&D, corporate central research laboratories, major long-term research programs, and other nonroutine aspects of technology are subjects of controversy in large organizations, with many advocates and many opponents. We have encountered very few companies with substantial (or even quite modest) corporate central research laboratories (CRLs) in which there were not vocal groups of critics of the CRL and serious questions arising from time to time about whether its existence was justified. Where some of these critics are people with substantial influence in the company—for example, division managers of large and successful divisions— the fate of the CRL is not at all certain in the intermediate and long run.

There had been only a handful of cases (not more than half a dozen in our formal sample companies) of a complete close-down of a CRL in the period of 1950–1970. However, the continued existence of many had been seriously questioned and moves made to curtail their resources and scope of operations. In the 1970s and 1980s a significant number of CRLs have been in jeopardy, for a number of reasons (see Figure 2-4). It is not easy to guess how many will eventually be shut down, but many have been cut back severely and an increasing number have been closed down.* Why should this be the case, if economic and business logic indicate the necessity of corporate-level technology in large decentralized companies? One obvious explanation is that the a priori logic that led to the establishment and continued support of a CRL in many companies has not been justified by subsequent results. Another is that some divisions have become relatively self-sufficient in technology or that if they have to "buy" research services outside the division, they prefer to go all the way and buy from an outside consulting laboratory over whom they believe they can maintain better control (than over a CRL). Without, in this section, going into the reasons why the future of the CRL in many decentralized companies is not very secure, I can say that there are many factors and many situations and events which can influence the survival and fortunes of a CRL. Some of these are discussed in the section on the CRL (Chapter 2) and in other related sections.

In this section, let us examine a major countervailing influence, which not only was responsible for the continued support and expansion of CRLs during the 1950s and 1960s and into the 1970s, but was also responsible in many instances for their original establishment.

This is the faith of one person or a small group of people in the corporation who strongly believed in the need for and the potential benefits from having a CRL. We encountered several of these people. Sometimes they had been trained in engineering or science, sometimes they had worked in R&D in earlier stages of their careers, and sometimes they merely had faith, supported by no material evidence, that their company should have a CRL. Where this person or these people constituted top management, there was little doubt of the outcome: a CRL would be established and protected. This does not mean that it would be effective or valued by other members of management, but it does mean that they are not in a position to destroy it as long as the boss is its champion and as long as his faith does not waver. Where the champion or advocate is not in the top power position, the struggle is more difficult and the outcome continues to be in doubt.

Certainly, I do not mean to imply that there are assassins lurking in the shadow of every CRL, waiting for an opportunity to blow it up or chop away at its foundations. Far from it. When times are good, many of the critics, especially those who are division managers, take a philosophical view and "write off" their contribution to supporting the CRL as "another damn tax on doing business" and they do not worry about it very much, except when it frustrates their plans or refuses to go along with projects that are specifically relevant to their divisions. It is when the crunch comes, in a variety of guises, that the fate of the CRL reemerges as an issue. In addition, there are events or periods in the life of the corporation which bring the question of CRL survival and prospects to prominence. A number of the caselets presented in Chapter 1 and elsewhere in the book deal with such events or transition periods. At this point I merely list and comment briefly on classes of

---

*More than a score in major companies during the 1980s.

events or periods which may serve to raise (and sometimes settle negatively) the fate of a CRL or any corporate-level technology activity.

1. *The Pre-birth Period.* This is an extremely vulnerable period for a potential CRL, since a great deal of inertia must be overcome in order to get it going. People and conditions may be displaced, funds may be diverted, space and services must be shared, and so on. In addition, there is really no evidence that it will work or pay its way (how can there be?). Not surprisingly, some of the strongest opponents may be managers of existing divisional laboratories who may not see or may all too clearly see the consequences to them and their laboratories of the formation of a CRL. There is, of course, competition for the job of CTO, VP/R&D, or director of research among the divisional R&D people, and many such jobs have been filled from within the company. However, many of the sample companies went outside for the top corporate technology person, thus, perhaps, increasing the resentment or opposition of the people already on board. Any new activity in a company will require some displacement and some inconvenience and loss of position or resources for some members of the organization. However, a new CRL may carry additional threats, due to the uncertainty of its role and the probable path it will take.

2. *The Early Years.* We have followed a number of these nascent CRLs for a decade or more in a number of decentralized companies and in some cases for over two decades. We have predicted a decreasing probability of success as opposition, indifference, fear, or disdain kept the CRL from achieving a critical mass or developing into an effective, ongoing resource for the company. Many mistakes can be made in the early years of a CRL, particularly when it is staffed by outsiders—for example, Ph.D.s in science who are brought into a company that had previously employed only development engineers in their R&D. Some of these mistakes are technical—for example, choice of the wrong project or approach or material with which to experiment. But many are also political—for example, ignoring advice from "old soldiers," failure to probe deeply enough or even to be interested in the technoeconomic nature of the businesses the company is in, wasting time on nonprofitable or "dead horse" product lines, failing to respond to requests for advice and firefighting in the plant or market, giving the impression that they are smarter than the people who have been in the company, and just plain not integrating into the organization.

3. *Specific Goofs.* Many of the above events and actions on the part of CRL personnel can be lumped, by neutral observers and critics, into the "getting started" or "learning the ropes" process (see Chapter 2) and can be forgiven and forgotten eventually. However, some of them really hit the fan in a big way. We encountered endless stories about major mistakes made by the CRL which could not be passed off as part of the learning process. Some of them were perpetrated by mature CRLs and some by young and immature ones. They included major marketing blunders, missed opportunities, products that failed for technical reasons, major design goofs in new plant or new process construction, picking the wrong fields, and not coming to the rescue of a divisional product or process in technical trouble.

4. *Business Recessions*. When profits go down and the cost-cutting process begins, technology, especially long-term R&D, is a likely target. Since the budgets of most CRLs are not part of any larger operating budget (other than G&A expenses), it is highly visible and attracts a lot of attention in terms of cutting back. Until the 1969–1970 country-wide downturn, few CRLs had been eliminated or drastically cut back for pure cost-saving reasons. In the latter part of the 1960s, and into the 1970s and 1980s, however, the high rate of growth many of these labs had experienced was suddenly cut or reduced to zero in many companies. Caught between the inexorable rise in the costs of conducting R&D and the reduction of budget increases, many CRLs, in effect, have been cut back or eliminated. The effects of cost cutting hit the CRL in a double way. First, its own budget is a prime target for direct reduction, or at least no increase in the face of steady increases in the cost of doing research. Second, in companies where the CRL is supported, in whole or part by fee-for-service work sponsored by the divisions, an even more significant cutback occurs. The division manager who is hard pressed to conserve funds will look around for the projects and activities that are "postponable." It is not surprising that R&D will be one of the most postponable things to hit his eye. If he has his own laboratory, he will probably cancel or cut back any work he has "contracted out" to the CRL and put it in his own lab, if he still wants to continue it. If the cuts he deems necessary include some of the work already being done in his own laboratory, he may try to pass some of it off on the CRL, even though it is clearly not part of their intended mission in the company (e.g., product or process improvement work or customer service). Either way, the CRL may become the victim—through direct cutback of support, having to switch its efforts to pick up former divisional development and service work, or in other ways.

5. *Changes in Top Management*. Many changes in the status of a CRL, and in some cases corporate-wide technology, have followed changes in top management, through natural succession, merger or acquisition, or other means. See the discussion of top management in Section 3.2 for impacts on technology of change in the incumbent(s).

6. *Merger, Acquisition, or Major Reorganization*. Such major organizational events are often a time when fundamental questions are raised about the role, size, and future of a CRL. When two companies merge that each have a CRL (or several, as did a number of the companies studied), then chances are good that one, both, or all will undergo significant changes in organization, size, mission, constraints, or personnel. In a few cases, the merger of two giants, formerly in quite different fields of business and/or technology, will leave both CRLs intact. We have followed several of these cases for many years, watching for signs of a merger of the two, elimination of one, the creation of a third super CRL, or other change. Some of these have occurred. Most frequently, however, one of two things occurs: the CRLs are merged in some fashion or other, or they are in effect relegated to the role of superdivisional or group laboratories, without overall corporate responsibilities.

One of the major side effects of the wave of "megamergers" and acquisitions of the mid-1980s has been relatively unnoticed by the business community. Their attention, naturally enough, has been focusing on the astounding (multi-billion-dollar) financial aspects of such merger/acquisitions. This side effect has been the potential impact on long-term technology capability in the newly formed firm, and in particular the CRLs of the previously independent companies. More than a dozen such megamergers have occurred in the late 1970s through the 1980s and perhaps more are possible in the next decade.

One of the rationalizations for such mergers, other than the strictly financial objectives of raiders and buy-out specialists, is the view that many large firms have been inefficient, top heavy, unproductive, and increasingly uncompetitive. The "dead wood" advocates believe, and in many cases quite correctly, that a lot of fat has grown in the corporate offices of the previously independent firms and that a shaking out, cutting back, and substantial reduction in corporate staffs is long overdue.

"Lean and mean" is the fashion for corporate headquarters of the 1980s and massive reductions are under way through early retirements, layoffs and redeployment of personnel to operations. For many corporate staff groups, including technology, out-of-pocket savings can be realized immediately and the promise of more efficiency attained through "getting the attention of those bureaucrats who have been feeding off the company overhead for so long." This discussion is no defense of overstaffed, inefficient, obsolete corporate staffs. It is, however, a cautionary word about cutting too fast, too indiscriminately, and in too bloody a fashion for the future effectiveness of the technology functions in the firm. Here are some cases of what happens to corporate technology groups (not to mention other corporate groups in planning, marketing, manufacturing, human resources, management science, etc.) when the axe falls:

- The soon-to-be-owner of a conglomerate that he was in the process of buying out gathered the corporate staff survivors in the company auditorium a few weeks before the deal was consummated. He announced that "ninety percent of you are not going to be here when I officially take over." And they were not. Included in the massacre was the complete corporate technology group of three people (remnants of a previous period when their number was much larger). In recent months that technology group had been actively engaged in trying to revitalize the level of technology in the operating divisions, some of which were under severe competitive, regulatory, and advancing state-of-the-art pressure.

- The CRL of a "top 100" corporation, which was acquired by another "top 100" firm, was disestablished because the newly formed firm did not want "two corporate technology centers" (CRLs) even though the existing ones were in quite different fields of technology and served quite different markets (part of the rationale for the acquisition in the first place). The new company had inherited from the acquired firm a large number (scores) of divisional engineering groups, many of which were falling behind technically in their own markets and many of which were well below "minimum effective size" to do anything but routine technical service work on current products and manufacturing methods.

- Acquisition of a major firm in the same industry led to the complete elimination of the acquired firm's CRL, which had been considered one of the "good" labs in the industry.
- Firm A acquired firm B in the same industry. Both had CRLs, several thousand miles apart. It was decided to close down the CRL of the *acquiring* company because the acquired company had recently built a showplace R&D center that was a matter of pride to the CEO of the acquired company, who remained a major shareholder of the new company. This was done despite the fact that the two CRLs (and their parent firms) had been in quite different segments of the market and that their underlying technologies, despite some overlaps, were somewhat distinct.

There have been other such cases. On the surface, such events may not be disasters. In fact, they can lead to immediate gains in cost reduction and attractiveness of the combined firm to investors. The firm can now be viewed as more efficient because it got rid of a lot of the duplication of corporate functions and "pretty, but useless" laboratories. In the longer run, however, such cavalier (and some of these moves appear to be just that) "deorganization" of major technology assets can cut deeply into the combined firm's technology capabilities and, ultimately, its competitiveness.

The factors that sometimes precipitate cutting back or eliminating CRLs, which are presented in Chapter 2 and elsewhere in the book, may indeed lead to a deliberate "design" decision. It may be concluded that there is no need for two major corporate technology centers, that the firm's operating divisions are capable of looking after their own needs for technology (by either "make or buy"), or that the surviving CRL—usually the one belonging to the acquiring company or the larger partner to a merger—will be able to serve all the divisional and corporate technology needs. Some of the people, groups, and projects/programs from the disestablished CRL can be brought into the surviving CRL and, after a brief adjustment period, perform as well or even better than they did in their previous environment.

A counterargument in favor of doing this very, very carefully, over a longer-than-usual time period, or for not doing it at all (eliminating one of the CRLs) is not as obvious. This line of counterargument, or "special pleading" as some may view it, has to do with all the factors that contribute to or detract from the effectiveness of a CRL in the first place. Figure 2-5 in Chapter 2 and the supporting text describe the many actions, decisions, changes in perceptions, and relations that are involved in starting up a new long-range research program or CRL. The time and difficulty of some of these stages may be lessened for the surviving CRL which is taking over some of the responsibilities for the terminated CRL. That is, its own internal start-up activities may be speeded up or eased. However, the whole set of activities that require getting to know and establishing working relations and trust with the client divisions of the former CRL may still be time consuming and costly. In addition, as we have seen in many such cases, the transition may never be effectively accomplished at all. "What do those chemists know about our machinery products" or "their electronikers are so far behind the ones in our former CRL that they'll never catch up" (even though the divisional people may have been harsh critics of their own former CRL).

Aside from these parochial arguments about the internal workings and relations within the technology group itself, there is a major factor often overlooked in the enthusiasm for cost cutting and staff bashing. It revolves around the general view that any technology-based firm, and certainly very large ones with highly diverse interests, must have an "over-the-horizon" and "side-looking" view of the science and engineering underlying their current and potential technology base. It involves the necessity that this upward and sideward view be specifically related to the current and likely future technology needs and interests of the firm. Certainly, very competent consulting and technical forecasting analysis can be bought from outside by the pound or by the hour. We have seen file cabinets and shelves full of it in planning and executive offices of large firms. But much of this material is on a somewhat abstract level—based on industry-wide considerations or for the "hypothetical" firm. Effective use of such material has to be related to the specific objectives, plans, policies, and circumstances of the firm and its component businesses. Generally, such material is "too rich" or too time consuming for direct use by divisional technology or management people. Translation, adaptation, and tailoring are needed for the firm's purposes. Traditionally, corporate technology staffs and CRL people have performed this intermediate role. Some have done it poorly or in a mediocre way. Others have done it superbly, charting the way for the firm into new fields of technology and new business opportunities.

As in the case of ideas for technology projects, there has to be a *need* for the idea and a *means* of bringing it to fruition (see Chapter 6). In a similar manner, there must be a receptivity for new technology and a good understanding of where it might fit into the firm's current and future operations. This takes time and dedication to learn. In some cases, the surviving CRL has people who are already somewhat familiar with the acquired firm's technology and markets and are "quick studies." In many cases there are no such people or not enough of them with enough slack time and interest to allow them to engage with "those other guys' problems." This situation can, and frequently does, leave the acquired firm's divisional technology people "on the beach" with no long-term and sophisticated technology group looking out after their interests, again no matter how much they complained about their former CRL's responsiveness and competence. Add to all the difficulties of any CRL in serving its constituent divisions—discussed in Chapters 2 and 3 and elsewhere through the book—the additional one of the "we–them" barrier, and the potential for technical disaster has been established.

Recall that the reason for mentioning all these factors relating to change, crisis, or instability which might affect the future of a CRL is that the role of a *champion* of technology and the CRL may be crucial in helping it to survive and come out of these crisis situations in fairly good order. During some of the classes of events described, quick decisions are often made which result in the reduction of a CRL's potential to contribute to the future welfare of the company or even to its survival. The presence and influence of a highly placed champion at such times may be crucial for the CRL. In a number of the cases we have observed, such a champion appears in the role of the chairman or influential board member whose advocacy of technology and the CRL goes back for decades. A timely word or full-scale intervention by such a champion may turn the tide and save the CRL from damage or oblivion. Without such a champion in the top management or on the board, it can be very vulnerable indeed during some of the transition and crisis periods discussed in this section.

3.11   THE ROLE OF TRUST IN INTERPERSONAL RELATIONS IN THE DECENTRALIZED COMPANY

Interpersonal trust is clearly a major influence in the relations between members of any group or organization. In a small, face-to-face working group, the members can soon assess the degree to which they can trust fellow group members. In a larger organizational context, however, getting a good reading on this important aspect of interpersonal relations is more difficult and the amount of data used is generally much less than in the small group. As a consequence, we observe members of different parts of a large, decentralized organization using stereotypes or single incidents as the basis for deciding whether they can trust members of another functional area or another part of the organization. A marketer (or researcher) has certain rules of thumb about how far one can trust a researcher (or marketer) to give you the straight dope, to help out in a pinch, to look out for your interests, or to behave in other ways deemed important by marketers (researchers). Frequently, these judgments are global, applying to all or most members or a typical member of the other group with, of course, exceptions for specific people in research (marketing) whom the marketer (researcher) knows and whom he knows he can trust. Some of these global attitudes and evaluations of the trustworthiness of others appear to be almost instinctual or "inherited. " A marketer (or researcher) who has had virtually no contact with a member of the other specialty may still have a fairly clear picture of how much and in what circumstances he can trust such a person.

   Throughout our interviews and other contacts with members of technology functions and other areas in the decentralized firm, we continually encountered statements and attitudes that relate to the degree of trust between the particular individuals or groups and other individuals or groups who are supposed to be cooperating in the support, conduct, and application of technology. It is clear, however, that in addition to global types of judgments about their trust of others, our respondents had subcategories of trust which applied to specific situations and classes of behavior. Among these were:

Trust of the other	To do what I would do
	Not to do anything dishonest or unethical
	To help when needed
	To be discreet
	To do a competent and reliable piece of work
	To look after my interests and the interests of the company (as I see them)

In the relations between corporate technology and the operating divisional management, for example, many of these aspects of trust come into play. If divisional people develop a perception that they cannot trust corporate technology on some or most of these dimensions, then relations between the two groups are not likely to be very conducive to cooperation. This may come into play, for example, in selecting projects and areas to work on and in getting the results of research applied.

The principal source of data on which members of an organization can base "trust" decisions between themselves and others—for example, members of operating divisions and corporate technology people—is the set of statements and behaviors exhibited by the other party that appear to have relevance to their basic values and their attitudes.

Occasions for the gathering of this type of data may be infrequent, but they can leave a lasting impression. For example, consider this illustrative dialogue between one of our interviewers and a member of one of a decentralized company's operating divisions:

Question: How are your relations with corporate technology?

Answer: What relations?

Question: Well, do you exchange technical information and cooperate on projects?

Answer: You must be kidding. Those guys aren't in the least interested in our problems or in cooperating with us.

Question: Can you give some examples of that?

Answer: It's obvious. They have never come around to find out how they can help in our real problems. All they do is send those unreadable scientific reports through the internal mail, which haven't got the slightest thing to do with our business or our technical problems.

Question: Have you tried taking the initiative with them—that is, requesting technical assistance or visiting them to see what they are doing?

Answer: What good would it do? They're too busy working on their own hobbies. And besides, they are full of textbook nonsense; they don't know a thing about how to make our products and design our production lines. The couple of times I did call them about a plant problem, they struck out or they came up with a nice theoretical answer that was too expensive to apply and that would have taken too long anyway.

Question: Will you call them again, or go visit them, if another opportunity arises in the future—that is, would you give them another chance?

Answer: I might, if I'm really desperate, but what I'd really like to see is some evidence that they really care about company problems and our problems in particular and not just the blue sky stuff they are working on.

Now consider his opposite number in the CRL, responding to similar questions:

Question: How are your relations with the operating divisions?

Answer: Not very good. They don't seem to trust us and they don't seem to want us nosing around in their facilities or their affairs.

Question: Can you give some explanations or examples of that?

Answer: First of all, most of the division technical people are not very well trained, and the division's management people have no technical training at all. When we try to help them, we can't get even simple technical ideas across. On the few occasions when they were desperate for help (they only call us when they are desperate) and did come to us, they wouldn't sit still for an analysis of their problem. They just wanted a quick and dirty solution. On one occasion the trivial production problem they brought us (which any well-trained engineer or scientist should have been able to analyze) was really symptomatic of a whole set of related problems, reflecting a basic error in design of their production facilities. It just violated two or three well-known physical laws. But they weren't interested and, in fact, they got mad when we tried to point this out to them and tell them that a much deeper study was needed.

Question: Do you think you could really work with them and help them in the future?

Answer: Sure we could, if they were willing to listen and to do a little homework on some basic scientific issues. They've also got to quit thinking they can get away with technical shortcuts and tricks in making a complex production process operate efficiently. Oh yes, and they must also get over that snotty attitude that just because they are apparently contributing directly to company profits in the current accounting period that they are better and more loyal employees than we are.

3.12 WHO WATCHES OUT FOR DIVERSIFICATION AND WHAT IS THE ROLE OF THE TECHNOLOGY GROUP IN DIVERSIFICATION?

According to the textbooks and general exhortations in the technology management literature, R&D plays a vital role in leading the company into new areas. As a partisan of the effective use of R&D in the firm, I would like to be able to report that this is generally the case in the companies that we have studied and consulted with. I cannot. In some of the industry sectors, such as chemicals, pharmaceuticals, and electronics, this was very much the case in the 1960s, when industrial R&D in some fields was still performing beyond the expectations of their sponsors and clients (typically the division managers). Since the late 1960s and beginning of the 1970s, the situation changed for many of these companies, including some of the high-technology companies in the three industries mentioned. Increased competition, rising expectations in the face of falling performance, and the severe depletion of the scientific bag of tricks from which many industrial R&D programs got their new product and process ideas have taken the initiative away from R&D in many cases. Diversification by merger, acquisition, joint ventures, purchase of technology, expansion abroad, and other methods have pushed R&D into the diversification background in many companies.

In their place in the diversification game have appeared new staff functions that report to various other executives than the CTO. These include new venture groups, development departments, long-range planning groups, merger and acquisition (M&A) groups, and product and market planning groups. Characteristically, these new groups are paper and pencil activities, without laboratory facilities, but with many contacts outside the company and a healthy travel budget. They are typically found at the corporate headquarters level but also exist at divisional level (for very large divisions) and at the group level, where such a formal level exists (see Section 3.7 on the role of the group executive).

Certainly the R&D groups in the divisions and especially the CRL, where one exists, are involved in studies *related* to diversification. They all claim or acknowledge that it is a vital part of their role in the company and that they have the competence to study the technological aspects of diversification. In spite of this, most of the R&D groups in the companies we have been involved with had only a peripheral role in this important activity. These are some of the reasons we encountered:

1. The divisional laboratories are too specialized in the current technology of their division and too closely tied to short-range work directly in support of current operations and technology. In addition, where they are located in narrowly specialized product divisions (e.g., tied to one small set of chemical compounds or one narrow market niche), they do not have the incentive or the encouragement from their divisional management to wander far afield. This is especially true when their division is successfully producing and marketing its current line of products. Paradoxically, it is also the case when their division's products are in difficulty and their direct help is needed to bail a product out.

2. The central labs are usually too far removed from the marketplace to have a direct input to diversification. Most of them, of course, are continually inputting ideas for new products and services, technological forecasts of changes in products and services, and analyses of emerging fields of science and technology that could lead to new businesses for the company. Most of them, however, are lacking in

major ingredients of a full-fledged, effective diversification activity. Typically, they do not have experienced marketing people on the R&D staff. The same is true of production people, financial people, and, strange to say, engineering people who can estimate plant and equipment costs, who can provide realistic start-up dates and costs, and who can successfully anticipate learning curve, quality, and other operating problems. They are seldom equipped to handle the business and financial negotiations that are required when buying into a new field and protecting the company against legal, financial, business, labor, and interpersonal difficulties. All this makes many of the corporate R&D people (in the CRL) sound pretty inexperienced in the field of business, especially in the initiation of new enterprises. My general impression, although we did not make measurements directly on that, is that many of them are. More important, they are perceived that way by a large percentage of top managers whom we interviewed and discussed this with. (See Chapter 8 on technical entrepreneurship for more discussion of this.)

During the course of the formal decentralization study, we did observe a number of sharp exceptions to the situation described above—R&D departments that were fully equipped for and experienced in doing the work of getting their company into new markets and new fields. In all these cases, the R&D labs involved worked in and were imbued with an atmosphere of entrepreneurship, diversification, and market development. Some of these are well-known examples of companies that have grown from within and spun off division after division, based on products invented or developed in corporate labs. The general attitudes of the professionals and managers in these labs are noticeably different from those observed in the traditional CRL—the locus of whatever "basic" research is going on in the company.

Many of the companies we studied have been experimenting with one-person or few-people groups to get them into new fields. We observed very few cases of success under these conditions. The reasons appeared to include lack of a size big enough to do more than one thing at a time; lack of command over resources (people, money, facilities, equipment); lack of diverse enough skills to handle the various aspects of diversification or entrepreneurship on a professional level; lack of a stable organizational platform from which to work (many of them were "do or die" efforts, with no clear career path for those who failed); lack of historical relations with groups they had to look to for cooperation and services; and, perhaps most significant, lack of clear wholehearted support from top management, as indicated by their failure to provide the resources and conditions needed for the job.

Sometimes these individuals succeed, thereby amazing everyone, and, if their talents warrant it, end up heading the new enterprise or acquisition. More often they fail, thus making it more difficult for their successors to work up a head of steam and overcome the problems to which the earlier groups succumbed.

During the study we encountered many such people and, we hope, learned to distinguish to some extent between those who "had it" and could do the job, barring events beyond their control, and those who were "losers" in this difficult field.

Only a few of the ones who appeared to "have it" were members of CRLs. One type of R&D person trying to push a diversification effort might be typified by a young Ph.D. chemist we encountered very early in the study. He came to visit us at M.I.T. in 1957, from one of the very large chemical companies. He gathered together a group of faculty members on the unspoken assumption that there might be some consulting work in it for some of us (consulting was very scarce in those days, when people were not as concerned about the organization and performance of R&D). He said he had a charter from the corporation to diversify it into "any field that promised to be profitable, regardless of technology or market." This was in the morning. By late afternoon, after careful (and somewhat brutal) probing of the actual terms of his charter (aided by the early "happy hour" in the faculty club that day), it turned out that his *actual* charter was quite different. He was constrained to find some new markets or product configurations that would exploit idle capacity in a very narrow area of the company's business. Furthermore, he had not done and was probably not capable of doing any rough economic analysis of what it would take to add even a modest percentage to the division's return on investment and what capital was required or available.

I do not mean to put down R&D people as important contributors to diversification. But in many of the companies we studied, R&D generally seemed to have dropped the ball they could be carrying. Part of the problem is, as indicated, lack of experience and training in the business skills necessary for successful entrepreneurship. But a more serious problem is attitudinal, especially among members of corporate laboratories who express little interest in the business and development aspects of the process. This is of particular importance in many large decentralized firms, since growth from within or buying into new technologies is vital for their continued health. The problem is that their contribution *is* needed, since many of the nontechnical groups that try diversification continue to make technical blunders by purchasing technical lemons or committing the company in areas where they are unable to produce technically.

As implied in Chapters 6 and 8, a major factor in successful diversification is its reporting focus. In effect, this reflects its ability to command top management attention, the cooperation of people all over the company, and significant amounts of capital (initial and continuing). In very few cases have we encountered diversification efforts within R&D which could command all these critical resources in sufficient quantity. In many of the sample companies, R&D was already in a defensive position with respect to these resources in relation to its *ongoing* programs. It had little residual power over such resources to apply to diversification efforts. In addition, there was, in many cases, a history of ideas for diversification from R&D that had not worked out. Some of these were specific new products; others were complete new market concepts or product (or service) lines.

The position of R&D with respect to diversification is closely related to the general power of R&D in the company and this, in turn, is directly related to the power and influence of the top technology executive—typically, the CTO or VP/R&D. More is said about this subject in Section 3.9. It is important to note here that R&D has difficulty playing an important, let alone a leading, role in corporate diversification in the absence of strong, indeed dominant leadership. Sometimes this leadership is gained by virtue of the personal esteem for a distinguished scientist who heads R&D. More often, it is only available to a person who understands and operates effectively in the managerial–business culture in addition to the R&D culture. This may explain the increasing trend, mentioned in Chapter 1, of appointing general management or business types of people to the job of CTO, in place of the traditional road up from director of research to VP/R&D or CTO.

3.13   SOME ISSUES RELATED TO THE INTRODUCTION, ORGANIZATION, INTEGRATION, AND CONTROL OF NEW STAFF SPECIALIST ACTIVITIES

In several different contexts throughout this book, the subject of staff specialty groups is discussed. It is mentioned in the context of the division manager's job, the job of the group executive, the relation of R&D to other activities in the firm, transition from lab to market, and other places. This frequent reference to specialist staff groups reflects my belief that they are essential to the very existence of the modern corporation, whether it be centralized or decentralized. This may seem a bit tautological, since one of the hallmarks of the "modern" corporation is the existence and effective use of such groups. If we can take this as given (beg the question?) without having to belabor the general philosophical questions of whether such activities *are* essential in the modern corporation and what percentage of the firm's employees or resources should be devoted to them (a subject of a small body of nose-counting literature), we can proceed to discuss their role in the large decentralized corporation. As part of the POMRAD research portfolio, we have been engaged in systematic studies of the problems of introduction, organization, and use of specialist staff activities for over 30 years, on a very broad front (including my own Ph.D. dissertation at Columbia University). The theoretical depth at which these studies have been and are being conducted (see Appendix B for a list of papers resulting from these studies) do not permit spinning off, at this time, a few handy conclusions or principles on their care and feeding.

We have followed the birth and death of some of these groups (e.g., operations research, management science, technical intelligence, quality assurance, long-range planning, information and data processing, organizational development, program planning budgeting) in more than 200 organizations in U.S. industry and the federal government, and approximately 20 foreign countries, looking for generalizable organizational design features, measures of effectiveness, problems, and solutions. Based on this work, some of which is still in progress, I comment briefly on their relevance to and problems of integration into the large decentralized company. A set of issues related to specialist staff activities, particularly as they relate to the large decentralized company, is given below. Some of them summarize or repeat arguments that have been presented elsewhere in the book. Others are based on the findings of our formal studies of "The Introduction and Utilization of Operations Research and Management Science (OR/MS) Activities in Organizations" and other relevant POMRAD studies. Still others are based on observations and other data collected in the course of the decentralization study itself.

- Where should specialist staff activities be located in the large decentralized corporation?
- How can a balanced portfolio of direct service activities for company (internal) clients and longer-term backup research and technique development be achieved and maintained in these specialties?
- How can such activities be funded?
- How can they be controlled and kept from becoming technically obsolete or too far removed from immediate company interests?
- What career paths can be devised so as to attract and hold good people in these specialties?
- How can the efforts of multiple staff specialty groups be coordinated and integrated so that they do not duplicate, overlap, or interfere with each other's activities and contributions to the organization?
- How can their results be quickly and effectively transmitted to clients for implementation?

As a preliminary to discussion of these specific issues, I would like to mention briefly the relation of this subject to the organization of technology and relate how we became involved in this long range OR/MS study.

In the mid-1950s, industry's interest in the new specialties of operations research and management science (OR/MS) was beginning to result in the establishment of internal groups to perform these activities (up until that time most companies had relied heavily on outside consultants and many still do). We observed that many of these groups had great difficulty in getting started, surviving, and performing to the satisfaction of company management. This led, over the next 4–5 years, to attempts by our research group to formulate some propositions and models that might help us to understand better this phenomenon and to seek means of overcoming the many difficulties it involved. The place of the project in our own research portfolio seemed logical for a number of reasons. One practical one was that the Department of Industrial Engineering and Management Sciences, which has been the locale of our POMRAD research program since 1959, has a major commitment to graduate training and research in the fields of operations research, systems analysis and design, information and data processing, and other specialties that make up the field known as *management sciences*. We were among the first two or three departments to break from the traditional industrial engineering curriculum and to develop a full program in support of the management sciences. As teachers, we felt we had to know something about the roles our graduates would be expected to occupy and the problems they would encounter in attempting to perform satisfactorily in those roles.

The logic of initiating this project (really a subprogram, consisting of more than a dozen distinct projects) in our program of research on R&D/technology management seemed clear and grew even more clear as the OR/MS project expanded, diversified, and shifted its main center of operations into the newly formed Kellogg Graduate School of Management at Northwestern, under the direction of Mike Radnor.

R&D, OR/MS, and many of the other technical staff specialties that have been introduced into the firm in the past few decades share many similarities of history, reception in the firm, search for support, opposition and indifference, need and difficulty of demonstrating effectiveness, organizational location, influence and power, balance of portfolio, selection and development of personnel, and so on. In addition, many of these similarities appear to have been evident in older, well-established staff activities in the modern firm, which appear to have weathered the initial stages of their existence—for example, quality control, industrial engineering, personnel, and controllership.

Suffice it to say, from both an organization theory viewpoint (in which we seek general levels of understanding and ability to predict) and an organizational design and control viewpoint (in which we seek better ways to organize and operate these activities in specific situations), that most of these staff specialty groups, old and new, have a lot in common.

In addition to this general argument as to why specialists in technology management should also be involved in these studies of other management staff activities, we find a high functional interdependence between the R&D activities in the decentralized firm and the many other specialist staff groups who may be clients, collaborators, providers of service, and competitors to R&D.

3.13.1 Some Comments on the Classical Distinction Between Line and Staff

I have no intention here of attacking head-on the emerging concept of the "blurring of line and staff" which has become increasingly prominent in the management literature. The rise of staff specialists and staff groups of all kinds makes it clear that the traditional distinctions need some modification. In the many studies of decisionmaking that our group and many other academic researchers have undertaken in recent years, it is clear that the pattern of decisionmaking has changed drastically. There is no doubt that staff advisory, analytical, and service groups wield an immense and increasing influence on the decisions made and the programs undertaken in the large corporation. What has not been made clear in the literature and indeed has been obfuscated by the endless invocation of terms such as "responsibility" and "authority" is the "real" decision structure of the firm. Perhaps this reality is an illusion and decisions sort of ooze out in many different ways, without actually being "made" by anyone. Certainly the abortive attempts by many academic researchers, including ourselves, to get a nice clean picture of the "locus of decisions" or "the characteristics of and influences on *the* decisionmaker" suggest a high degree of diffusion of the influence, power, authority, or responsibility that goes with decisionmaking. Maybe we have just not been clever enough to smoke out the real decisionmakers and the real decisionmaking process. But we are still working on it in several of our major research areas. In these areas we recognize that individual influence is a determining factor in decisions to select projects, enter new areas, attempt to adopt an innovation, and so on. In our studies of the adoption of new pieces of technology from outside the firm, for example, we certainly encounter the influential staff groups whose analysis and advice weaken or strengthen the chances that a piece of new technology will be adopted. But we also encounter the "line" executive who says "forget it" or "I want that piece of technology ready by the end of the year."

In the decentralized company, data of the latter kind are much in evidence. We find the decisions and influence of the division manager, for example, dominating the opinions, suggestions, and advice of staff groups which he himself may have set up and/or which he certainly supports. Most of the division managers we have encountered listen to staff advice and many even follow it, but there is no doubt whose decision it is. As a typical division manager might put it:

> I appreciate the advice and factor it into my decision, but it is *my* division and *my* job that are involved here, and *I* am the guy that top management comes to when things go wrong. I call the shots and if the advice or recommendations of my staff people don't feel or smell right, I follow my intuition. After all, I was put in this job and am being very well paid to make the decisions and take the consequences if they don't work out. This means that I make the decisions and I live with the results. Many of these staff people are pretty brave when it comes to making recommendations for bold, risky moves. But some of them are very reluctant to put their jobs or incomes on the line if the recommendations don't pan out.

Perhaps, at some levels of the corporation, it is difficult to say whose decision it really was, particularly when interlocking mechanisms of staff groups and committees and task forces are used to process decisions. At the division manager level in the truly decentralized company, however, we have had no difficulty in distinguishing staff from line management.

There is a segment of the organization theory and related behavioral science literature which deals with the problems of professionals in organizations. Among the issues discussed in this literature are the reference groups that a professional (a member of a socially recognized group of specialists such as lawyers, accountants, or scientists) uses as standards for his behavior, the values that guide the decisions and behavior of the professional in his day-to-day work, and his relative loyalty to or identification with the organization he is working for and the profession of which he is a member.

Some of these issues are believed to have a significant influence on major career decisions made by such people—their choice of job and organization or the decision to remain in or leave a job or organization. In addition, however, there is evidence, at least among scientists and engineers in R&D, that these considerations may also influence their on-the-job behavior and decisions. We have observed this phenomenon in two related aspects of the R&D process—the selection of projects and the transmission of ideas that may lead to new projects. We have found clear evidence that R&D personnel evaluate ideas and projects according to criteria that appear to relate to such factors as scientific versus engineering orientation, managerial versus nonmanagerial company roles, chronological and organizational age, amount of higher education, phase in their career, and their own past history. We gathered such data in our studies of operations research and management science (OR/MS) groups, and we suspect that we might find similar regularities among other highly specialized professional staff groups (e.g., information and data processing, long-range planning, economic analysis).

What are the implications of this for the sponsor and client of staff specialist services in the decentralized firm? One is that he may find a conflict of values and loyalties within and among some of his staff groups, where particular incidents find the professional torn or at least uncertain of whether to do something or decide something the "company" way or the way that his colleagues in his professional specialty would approve. In R&D, such conflicts frequently arise in connection with questions of product quality and safety, depth of attack on a research problem (quick and dirty or the proper, scientific approach), or choice of a commercial project to work on versus one with scientific potential that might enhance one's reputation in the profession.

Until recently, such issues were not paramount in connection with staff groups other than R&D. Exceptions were professional accountants in an internal auditing function, in-house doctors and lawyers, and the occasional in-house psychologist who had to make moral and ethical distinctions about what was good for his individual client (an employee being counseled or tested) and, if there were conflict, what was good for the company.

With the modern corporation's dramatic increase in the staff specialized services it supports, we can expect such potential conflicts to increase among some of the newer professionals. No "solution" is suggested here, except that this is a matter to be considered in the organization, location, reporting relation, project selection procedures, and other features designed into such groups. They can make a significant difference in the quality of service provided by these groups, the rate of turnover of their members, and their morale. It can also have a significant effect on their loyalties, in terms of rendering assistance in emergencies and standing by in troubled times. We have seen a number of instances of such staff groups abandoning ship or withdrawing from a bad situation when particular clients for whom they felt no loyalty were in difficulty.

4

NETWORKING OF TECHNOLOGY AMONG THE OPERATING DIVISIONS

4.1 INTRODUCTION AND CHAPTER OVERVIEW

Under the best of circumstances, cooperation between autonomous and often competing divisional units within a decentralized organization does not occur easily. Self-interest and focusing on specific short-term unit goals dominate the behavior of division members and do not encourage general sharing of technology with other divisions, unless they are highly interdependent with respect to markets, products, or technology. Divisional development or engineering labs, charged with supporting their own divisions' products and services, are not likely to spend a lot of time and effort trying to help other divisions in the company solve *their* problems or get *their* products out the door, in the absence of some strong incentives.

When resources are tight, and that is the usual case in a "lean and mean" operating division, there is not a lot of slack for long-term cooperative activities with other divisions that have only marginally relevant interests or even for short-term activities that do not directly benefit "my" division. Unless there are strong pressures toward cooperation in technology between two divisions, the likelihood of its occurring is slight. Generally, time and resource pressures soak up any slack that might be used for such excursions and it takes a pretty strong-minded divisional researcher to resist such internal pressures and play the "good guy" for another division.

If that is the case, why bother even to worry about it. In the "well-run" decentralized company, each operating division is evaluated on its own performance and presumably has the incentives and resources to generate, acquire, and use the best available technology for its divisional mission. But this presumption may not reflect the actual situation in many, if not most, divisions. In this chapter, we examine the assigned and perceived mission of the divisional technology lab and see that time and resource pressures do not always (in many cases, seldom) allow the divisional technology people and their management to develop, acquire, and apply the best available technology to the division's needs.

For a number of reasons, technical cooperation or "networking" between operating divisions may be highly desirable. In many cases it may even be necessary for effective operations on a company-wide basis as well as for the benefit of individual divisions.

Major reasons include economies of scale, sharing of equipment and facilities, common technical problems and opportunities, and lack of a corporate-level technology group, such as CRL, to look "over the horizon" and "to the side" of the company's current technology. In this chapter some of these factors are examined and a case is made for improved interdivisional technology networking. The actual patterns of networking can vary widely and there is no ideal or optimum way of doing it that has emerged from either theory or practice. There are "better" ways and "not-so-good" ways of networking and some of these are presented. The exact pattern for a given company, as with so many other organizational issues, needs to be tailored to its specific needs and circumstances.

One major approach discussed in some detail in this chapter involves special organizational roles—often informal and not to be found in organizational charts or manuals. These are the roles played by key communicators (KCs) in transferring ideas, information, and technical assistance between operating divisions and bringing technology in from outside the firm. This is not a heroic role and very often, instead of rewards, KCs receive criticism for being nosy, pushy, or "out of their territory." Despite this, a strong pitch is made for establishing such roles between operating divisions (and other parts of the firm) or strengthening them where they already exist in a less-than-effective form.

4.1.1 Some Issues/Questions

- What forms of technical cooperation can exist between operating divisions?
- To what extent do divisional development and engineering labs cooperate across divisional lines?
- What keeps them from cooperating more—or at all?
- Is the "mission statement" for a divisional lab a barrier to cross-divisional cooperation?
- What kinds of incentives might increase technical cooperation and technical networking?
- How do key communicators facilitate networking?
- How are "group technology" and large-scale system development affected by the level of cooperation and networking?
- What patterns of networking are feasible among diverse divisions?
- What specific steps can be taken to improve cross-divisional networking?

4.1.2 Main Body of Chapter

This contains a discussion of the networking phenomenon in general and the specific case of interdivisional technology networking.

4.1.3 Implications for Management

It is not "natural" for divisional technology people to cooperate freely and exchange information, advice, and technical assistance. In order to accomplish this, both corporate top management and divisional management have to make special efforts to encourage this, using both positive and, where necessary, negative incentives.

4.1.4 A Watch List

- Interdivisional projects "dumped" rather than transferred.
- Significant duplication of effort between divisions.
- Excessive contracting out for technical assistance when the needed skills and capacities may exist in other divisions.
- Use of less-than-best skills available in a given division when a higher level of that skill exists in another division.
- Absence or very low level of technical personnel transfers between divisions.
- Excessive recruiting of technical personnel from outside the firm without a search inside for required talent.
- Absence or low level of sharing of specialized equipment or facilities.

4.2 MISSION OF AND PRESSURES ON THE DIVISIONAL TECHNOLOGY (R&D) LAB

Whether or not the firm has one or more corporate labs—CRLs—the responsibility of the divisional technology director (sometimes a divisional VP or CTO) is clear. He is responsible for assuring the timeliness, cost, and quality of the division's technology. He often shares this responsibility with other divisional specialists— the director (or VP) of manufacturing, the chief engineer, the director of quality control, if there is one, and others. However overlapping this responsibility is, the divisional technology director will get his lumps if the division gets into trouble because of its products and manufacturing technology. Sometimes these troubles have to do with customer complaints, such as warranty matters, difficulty of use, or product liability. Sometimes the trouble will come from the division's plants, which are having difficulties implementing manufacturing methods or materials or other technology the divisional lab has developed or brought to them in a "middleman" role from other parts of the company or from outside.

Whatever the source of the problem and whoever's "fault" it may be, division-al technology is expected to play a major role in fixing it and bailing the division out.

Aside from such almost continuous or routine problems with customers and the manufacturing plant(s), the divisional lab is also supposed to provide a constant stream of improvements in divisional products and manufacturing methods. The product improvements are generally aimed at capturing or maintaining market share, satisfying particular customers with custom design, providing marketing and sales with new features that will help differentiate the division's products from those of competition, and/or reducing costs and increasing margins. Many of the projects are aimed at cost cutting and materials substitution in existing divisional products, improving quality, or correcting environmental health and safety hazards and other "fix" kinds of improvements.

Every now and then, the division manager, perhaps egged on by the marketing people, will ask divisional technology to come up with a substantially or radically new product to broaden out the product line, meet a major competitive threat, or replace an existing product that is on its way out. Even though it was "known" many months or even years before that such a new product was needed, the demand to get one developed is often imposed in a crisis mode with a very tight time schedule. After all, "you guys know our products, our markets, and our manufacturing capabilities, why should it take so long to pull something out of the hat?"

As if the above tasks were not enough, divisional labs are also expected to be ready to "mobilize all hands" to meet specific customer or plant emergencies— even those with only marginal relevance to the lab's main mission. "Dammit, I know you're busy on lots of other things, but our best customer is making noises about straying off the reservation if we don't help them clear up the problem they are having in their plants with our material (or product, or equipment)."

Divisional people do not argue with that kind of "request" when it comes from the division manager or one of the other functional top managers in the division or their surrogates. As implied in Section 3.4 on the division manager and his team, when the DM itches, everybody scratches.

Do divisional R&D people go to pieces or dig in their heels under such pressures as we have implied some CRL people appear to do? Not generally. They may be overworked and mentally, as well as physically, exhausted by a combination of time pressures and sometimes "unreasonable specifications and features for the new product," but that's the name of the game out here on the firing line—"we are not a bunch of ivory tower prima donnas like some people in the corporate labs."

Pressure builds even more as the time for plant start-up or market introduction nears. Tempers may get short and technology may also be "shorted" in order to get the product out on schedule or on cost. The rewards for a successful new product introduction by divisonal personnel can be great. Unlike some of their corporate technology (CRL) counterparts, divisional technology people have a number of clear career paths outside R&D if they wish to try for them. A successful new product launch, given a clear and effective contribution by the lab, can be a ticket to promotion within the division up the manufacturing, marketing, or general management line.

Of course, if and when the product fails, such ambition may be thwarted or set back a few years. However, there appears to be more room for forgiveness within the divisional "family" because of the team approach promulgated by some division managers.

What has all this discussion of time and other pressures been in aid of? Of course, the closer you get to the factory and the marketplace, the tighter time and resource constraints are going to be and the greater the expectations of success. These considerations are introduced here as a preamble to addressing the central question of this chapter: Why should divisions engage in technical cooperation with each other?

A quick answer to this question might be: "There would have to be a very good reason for them to do so and even then, they may not feel they have the time and resources for such cooperation." Before assuming that I am condoning this attitude and giving up on technology networking among divisions, let us look at some reasons why individual divisions need and, in return, should give help to their fellow divisions. Then we look at some mechanisms for such networking, either in direct self-interest or in the spirit of "bread upon the waters."

4.3 ISSUES/PROBLEMS OF DECENTRALIZED TECHNOLOGY/R&D

Figure 4-1 lists a number of issues or problems of pure decentralized technology. Many of them exist even in the "combination" form of technology organization,

---

1. Time horizon
2. Narrow scope of work/technology spectrum
3. R&D career path
4. Critical mass
5. Fluctuation of support
6. Lack of networking for scientific and technical information (STI), technology transfer, and technical assistance (TA)
7. Opportunities for exploiting nonmainline developments
8. Size of the idea pool
9. Gaps and overlaps between divisional territories
10. Limited role models for "far out" researchers, projects, ideas, and approaches
11. Domination by plants and/or marketing
12. Lack of technical audit, peer review, or second opinions on technical issues
13. Not invented here (NIH), rivalry, and mistrust
14. Arena may be too limited for "hot shot" R&D group or individuals
15. Tendency toward obsolescence
16. Limited slack time for renewal, exploration, multiple approaches, and "wild cards"
17. Operating culture may not be supportive of R&D

Figure 4-1. Some issues/problems of pure decentralized technology/R&D, subject to specific circumstances such as size, technology base, geography, company history, personality, organizational climate, and competitive environment. This list is not in order of importance.

where there is a CRL. In either case, an individual divisional lab or technology department must deal with many of these issues while attempting to carry out its assigned mission or extensions of that mission into longer-range and broader technology programs than those with which they are accustomed to dealing.

Of course, the existence or severity of these issues or problems depends on the specific circumstances of a division, such as size, technology base, geography (degree of isolation from other company technology groups), company and divisional history (how they became decentralized), personalities involved, internal organizational climate, competitive environment, age of their product line(s), and modernity of their manufacturing facilities.

Subject to such particular circumstances, the following factors can influence the degree to which a division can provide its own technology support for its operations without help from elsewhere in the firm.

1. *Time Horizon.* This describes the expected time (on average or at most) for projects to be completed and for the results of projects to pay off. Generally, as indicated earlier, it is fairly short in the typical operating division. There is little time for side explorations of potentially applicable science or technology and the tendency is to go with off-the-shelf items that suppliers can provide or that divisional people can "polish up" for a specific application.

2. *Narrow Scope of Work/Technology Spectrum.* Divisional development work and the related technology are generally focused on a fairly narrow band around the specific current needs of the division's products and manufacturing processes. Possible improvements or new ideas that are peripheral to that focus seldom get a good hearing and the resources to pursue them.

3. *R&D Career Path.* Most divisional labs, even in some very large companies, are of modest size (surprisingly so for some giant corporations, which may have a lot of 5- and 10-person divisional labs). In such a lab there is little room for significant advancement in R&D itself. Promotion and other career opportunities must typically be sought in other technology or nontechnology functions in the division. Cross-divisional transfers are not common and few divisional R&D people move into or back into the CRL. As a result, most divisional labs are not likely to attract many people with a high commitment to science or advanced technology and whose technical breadth can provide depth and strength in advanced technical fields.

Even where the divisional lab has several dozens or scores of professional people—engineers and scientists—a few years of narrowly focused, short time-horizon development and lots of firefighting are likely to dull the technical edge of many of them to the point where they concentrate primarily on the things they know and with which they are comfortable.

Larger divisional labs do in fact have specialists in a relevant range of technology underlying their businesses, but such specialists are less likely to be operating at the scientific or engineering cutting edges of their fields as compared to their colleagues in the CRL. Thus, advancement very far as a narrow technical specialist is not very likely.

4. *Critical Mass.* Given a range of technical specialties and a generally modest size, divisional labs are seldom able to assemble a "critical mass" of people to blitz a complex or big problem that requires different facets of a given specialty or a broad range of specialties.

This means that divisional technology people are frequently out of their strong areas of competence and out of their depth even in their own fields when they attack other than familiar problems or try to sally into totally new areas.

5. *Fluctuation of Support.* The actual size of divisional labs, in terms of professional and support personnel, does not vary widely in the short run. However, when crises occur or times are bad, this is an area of discretionary funding which the division manager feels he can cut without immediate harm to his operations. Divisional technology people are not generally laid off in such circumstances but are more likely to be reassigned to other activities and asked (ordered) to put longer-term activities (such as radically new projects) on hold while they pay more attention to cost cutting, customer technical service, and other activities with a quicker payoff. If economic conditions become or remain severe, layoffs or permanent transfers may indeed occur.

6. *Lack of Networking for Scientific and Technical Information (STI), Technology Transfer (TT), and Technical Assistance (TA).* Depending on shared markets and common technology interests, some pairs or groups of divisions (especially when they are under the same group executive) may help each other technically in one or more of these three modes. In some rare cases, groups of divisions may even share the support of a technical information center or library. They may, often under duress, cooperate in accessing new technology and share their skills and experiences. In the general case of autonomous, specialized, and highly differentiated divisions, this is not common and each division is typically on its own technically. Exceptions may occur for assistance they can buy from outside or from the corporation, through a CRL. In other words, networking among divisions is not the typical operating mode for most firms, with very specific exceptions such as a given division being a supplier to another division(s) in the firm.

7. *Opportunities for Exploiting Nonmainline Developments.* In view of the size, time horizon, and constraints on scope common for most product divisions, it is unlikely that the divisional technology group will be allowed or encouraged to "follow the rabbit" into "potentially" new areas as spin-offs from mainline divisional products and markets. Exceptions, of course, are common when they fit into the DM's plans or hopes for a broadening or deepening of his product line and market segment(s). He may send the message to "look for new applications of our manufacturing technology or capacity" or "new variations of our products to keep our current customers happy" or "loss leaders that will help us get a toe into a new market." Such thrusts, in the highly fragmented firm where one

division is responsible for "big blue ones" and another for "small blue ones" or "big green ones," have the potential for getting those toes stepped on. We have been involved with several such companies where good ideas for diversification by division A were squashed at corporate or group level because "that's another division's territory." Such ideas are seldom "donated" to the other division and, with no place to go, generally die at the point of origination. More on this issue of source and disposition of ideas is presented in Chapter 6.

8. *Size of the Idea Pool.* Closely associated with the previous factor is the question of how large an idea pool a small- or medium-size divisional lab is likely to generate—if indeed new ideas are welcome at all. The "hit ratio" for new ideas is very very small. Perhaps 1 in 10 or 1 in 100 ever makes it to successful commercialization. This depends on how radical they are—that is, how far they depart from the company's, the division's, or the industry's narrow scope (narrowing through time, if our perception is a fair one). A single small- or medium-size division is not likely to be the source of a large enough idea pool to beat the odds.

9. *Gaps and Overlaps Between Divisional Territories.* A gap in technology or "market mission" may seem to provide an opportunity for enterprising divisional technology people to plunge in with ideas for new products. Overlaps would seem either to be inhibiting factors or, to the contrary, to encourage cooperation between the "adjacent" divisions. In many firms both are rare events. "It's too much of a hassle to try to get a joint project going with that other division. They are into their own problems and so are we." Where a gap is clear, one of the divisions may run with the ball and try to move sideways to fill the gap, but they must always be watching sideways to be sure that their idea or project will not be taken away from them or made a multidivisional or corporate project. Some divisions do have highly aggressive and entrepreneurial technology people, as well as marketers and division managers. Of them, some will gladly cooperate with other divisions through cost and risk sharing and others will prefer to play it alone and take on the risk of success or failure themselves.

10. *Limited Role Models for "Far Out" Researchers, Projects, Ideas, and Approaches.* For a number of the reasons mentioned above, the younger and more adventuresome members of a divisional lab do not have senior colleagues immediately at hand whose path they can follow into new areas. On the contrary, the informal pressures are typically in the direction of "sticking to your knitting" and "let's get the jobs done that we're getting paid for. Leave the far out stuff to the CRL or people in larger divisions."

11. *Domination by Plants and/or Marketing.* Most product divisions, except for those in high tech areas, are motivated, dominated, and controlled by production and/or marketing. Those pressures emphasize cost reduction, customer and plant service, and other narrowly focused and short-term goals for ideas, projects, and technology fixes. By market dominated, I do not, in most cases, mean that divisional marketing is equivalent to the typical corporate marketing function, which is supposed to look beyond current and near-term products and markets.

The divisional marketing people may in fact be part of the sales function and may be very uninterested in radical plunges away from current product lines and markets. In a similar way, most plants have enough on their hands with current technology without worrying about radically improved or brand new manufacturing technology, which is sure to interrupt production, make life difficult for them, and "most likely fail anyway."

12. *Lack of Technical Audit, Peer Review, or Second Opinions on Technical Issues.* The divisional technologist is usually "the expert" on a given subject. He may be the *only* one with sufficient experience in a given area to make a decision or solve a problem, especially if it involves scientific or engineering principles. The lack of a critical mass or even another colleague in his area leaves him isolated in terms of technical judgment and approach. He seldom seeks outside opinions on such matters, unless he is really stuck, the project is of extreme importance, the risks of failure are high, and/or he is honest enough to admit that he cannot handle it alone.

13. *Not Invented Here (NIH), Rivalry, and Mistrust.* Strong team spirit within the division often builds a barrier against cooperation with "outsiders," especially other divisions that may be competing for funds, other resources, and top management attention. Perceived self-sufficiency and unwillingness to admit deficiencies or weaknesses further inhibit cooperation.

14. *Arena May Be Too Limited for "Hot Shot" R&D Group or Individuals.* Some divisions are fortunate enough (from the viewpoint of radical innovations) to have an individual or group that is willing to plunge into new areas, take on high-risk projects, and push for diversification. However, such an individual or group may soon find that the resources available and the constraints on such "venturing" may be too confining to allow them the leeway they need. As a consequence, most such people or groups will "subside" and adjust to the realities of the division's time horizon, scope, and risk aversion or "go away." Even under limited circumstances such as these, however, an entrepreneurial group can persist if given a minimum of encouragement by the divisional managers, the divisional lab director, and/or their peers in the lab. That is seldom forthcoming in most divisional labs, because such people are viewed as "not pulling their weight in terms of the division's current and obvious needs" and "trying for the glory road to make a name for themselves."

15. *Tendency Toward Obsolescence.* Given the narrow focus and modest size of the typical division lab, as well as the time and other pressures, there are little "slack" time and resources for technical renewal. Most engineers do not read much beyond their current technical literature, unless they are enrolled in an advanced degree program or are taking high-level advanced technical courses. Divisional technologists generally achieve a high level of competence and reputation for what they know. They are thereby encouraged to stick to what they do know and not to venture into new fields or the cutting edge of their own field, which may entail high personal risks. Groups operate the same way, through emulation, social pressure, and shared values. We have encountered groups of engineers and scientists in divi-

sional labs (as well as CRLs) who are very comfortable with using and reusing the scientific and technical knowledge they brought with them from school many years ago and learned on the job since. Many such people are not even very curious about the latest advances in their own field—"that's for the longhairs and ivory tower boys."

16. *Limited Slack Time for Renewal, Exploration, Multiple Approaches, and "Wild Cards."* All these excursions and diversions away from the divisional lab's assigned mission take time, resources, and forebearance by the divisional manager. These are in short supply and hard to come by in an organizational unit that is close to the firing line. Multiple approaches and playing wild cards in attempting to solve problems are not looked on with favor by any of the division's functional managers—production, marketing, sales, and general divisional management. "Do it right the first time" or "play it safe and go with what you know will work" are strong guidelines, permitting few deviations or redundant efforts.

17. *Operating Culture May Not Be Supportive of R&D.* Summing up many of the above factors, we can say that the culture and general atmosphere in a "typical" operating division is not conducive to contemplative, leisurely, exploratory activities connected with research, let alone development. The pressures are toward highly focused technical activities that will directly and quickly support operations—production, procurement, sales, and service. The diversified portfolio of the typical CRL is "not on" in smaller divisional labs and is even difficult to maintain in larger divisional labs. There are always certain exceptions—very large size, very diversified product lines and technology, fast-moving product changes, and high-technology products.

All of the above factors make it sound like there is not a very good case for divisional labs having the interest in or capabilities for spending time working with other divisions on "their" problems or problems of possible future interest to the corporation as a whole. On the other hand, many of these factors also indicate that the typical division strongly needs help in dealing with *its* technology problems, if the barriers to such cooperation can somehow be eased or eliminated.

Remember, I have said a number of times that such cooperation and networking of divisional technology is not "natural" and is not easily accomplished. However, it is also not impossible or beyond the reach of most firms with obvious needs or opportunities for such networking. All such efforts take time, patience, and resources away from other, more immediate, activities. So there has to be a good rationale in terms of cost/benefit ratio to each of the participants. These can include the following:

- Avoiding missed opportunities in the market because "we didn't have the technology to take advantage of them."
- Cost sharing for expensive and/or specialized equipment, facilities, and people.

- Improved problem solving by getting the best available technology in the company, not just what happens to be available in the division at the time.
- Backing up divisional technologists with others in their specialties who can provide enough muscle to meet the market's threats, needs, and opportunities to a greater extent than a single division is able to do.
- Joint venture opportunities for this division with others that will provide comparative advantage or new areas on a shared-cost and shared-risk basis.
- A source of new blood for divisional technology from within the company and an opening up of career paths for divisional technologists who have grown stale and/or dissatisfied.

Given the problems and limitations on the one hand and the potential advantages of cooperation and divisional technology networking on the other, what are some of the ways in which this can be accomplished or improved?

4.4   WAYS OF DEALING WITH THE ISSUES/PROBLEMS OF DECENTRALIZED TECHNOLOGY/R&D

Figure 4-2 lists 19 potential mechanisms for coordinating, networking, and gaining synergy from independent divisional technology programs. Some of them will be anathema to some divisional managers. They would involve abandonment or gross violation of their perceived need for autonomy and freedom from control or "coordination" by the corporate office. Some would be welcome means of "leveraging" their limited technology resources. And others might fly if presented properly and introduced carefully on an "experimental" basis, with the option to withdraw if too constraining, too costly, or not advantageous enough to the division. Many of these mechanisms are already in place in some companies, between some subsets of divisions, and in some special situations. Some may appear to be "expensive" in out-of-pocket costs and/or personnel. Others may be accomplished at only marginal cost or as spin-offs from other procedures or mechanisms. My intent here is not to prescribe a "fixed package" of such mechanisms and advocate it as a solution to the networking problem in all firms or in any given firm. Rather, it is to offer a set of insights into some mechanisms that singly or in combination can help an individual division or the overall firm get more from its technology dollars than it currently does.

1. *Technical Gatekeepers or Key Communicators (KCs).* This mechanism is promising enough so that a major discussion of it follows later in this chapter, including some illustrations of how it works and some of the advantages and limitations of it. It is a very attractive approach since it piggybacks on the natural tendency of some engineers and scientists to want to communicate with others in their field and in other fields. Many "informal" key communicators fill that role

1. Technical gatekeepers or key communicators (KCs)
2. Technical committees on specific topics or fields
3. Regular meetings of divisional R&D managers
4. R&D or technical advisory, coordinating, or steering committees
5. Formal/informal liaison/linkage agents between labs
6. Technical seminars
7. Exchange of "skill inventory" data
8. Cross-divisional project teams
9. Open across-division transfers and promotion opportunities for R&D people
10. Strong corporate R&D coordination staff to act as liaison agents
11. Systematic and effective (accessible, legible, timely, usable) report exchange program
12. Temporary cross-divisional transfers (for projects, training, renewal, communication)
13. Corporate R&D staff reviews and audits of divisional projects, programs, outputs, and personnel
14. Joint funding of projects and programs by two or more divisions
15. Strong incentives (positive and negative) to division managers to cooperate in R&D innovation area
16. Cross-divisional design reviews—"do unto others"
17. Joint idea generation efforts
18. Coordination of R&D/technology segments of divisional long-range or strategic plans
19. Mutual program reviews

Figure 4-2. Potential mechanisms for coordinating, networking, and gaining synergy from independent divisional R&D activities. Their individual and combined effectiveness depends heavily on those factors to which the issues/problems are subject—see Figure 4-1. The mechanisms in this list are not presented in any particular order.

naturally, since they like to talk, interact, show off, hear about new ideas, explore new fields, and otherwise use their intellect and communication skills outside the normal course of working on assigned projects. Some do it well and others not so well. The effective ones can create important bridges across organizational lines and directly bring or indirectly stimulate the flow of ideas, information, and pieces of technology to their divisions. Supporting the natural tendencies of such key communicators and/or appointing formal ones to liaison roles can be cheap or expensive, depending on how the bookkeeping is done. One major new idea can pay for a whole KC program for many years if it results in a new product, a major cost saving, or avoiding a technical disaster.

2. *Technical Committees on Specific Topics or Fields.* These committees are generally organized by the CTO, but some occur spontaneously at divisional level. Most of them are not as effective as they might be. A lot depends on the focus of and interest in the topic, the agenda, the membership, the meeting locales, and the perceived gains from them. We have observed a few very successful ones, each the product of careful planning, monitoring, and constant redesign to overcome barriers to effective functioning. We have also observed many very unsuccessful ones that were designed "from the hip" and that deteriorated rapidly from there. A

strong focus on specific topics of mutual interest, careful attention to the mechanics and location, close monitoring of the frequency and "energy level," and insistence on a high technical level of participation are some of the factors associated with success. Meetings that drag on for hours at a low-energy and low-interest level, regular scheduling without a real agenda or pressing need ("why are we here"), allowing substitutes for the lead technical people from the divisions and allowing a deterioration in the technical level of discussions, and lack of follow-up to decisions made and requests from the committee members are associated with failure of such committees or a very low level of effectiveness.

3. *Regular Meetings of Divisional R&D/Technology Managers*. Many of the same factors apply to such meetings as they do to the technical committees. Effective use of time (especially that involving travel) is even more important for this level of divisional technology person than for the technical people who are used to operating at a more leisurely pace. Substitutions for the divisional CTO can lead to deterioration, as can lack of an agenda perceived as important to each or most divisions. We have been involved in seminar series conducted by many such "technology management councils" or committees and have seen many of them disintegrate or go on "hold" when the participants became bored with them or found the cost/benefit ratio too high. The most exciting ones were those involving some real hand-wrestling and bargain-making between divisions on things like joint projects, mutual audits, and specific "design reviews" (see those items later in this section).

4. *R&D or Technical Advisory, Coordinating, or Steering Committees*. Some of these are more formally structured and empowered than the often informal meetings of the divisional technology managers. These are generally run by the CTO as part of his "coordinating and oversight" duties. They are often the forum for launching cooperative projects between CRL and one or more than one division, arguing about budget allocations or personnel allocations to the various labs, and major program reviews.

When the CTO has a strong personality and/or a strong backing from the CEO and perhaps some measure of control over at least the "advanced technology" portion of divisional technology budgets, these can be an effective control and coordination mechanism. More commonly, it is a "duty" call and divisional people attend because they do not want to miss something, rather than because they feel that such attendance is to their direct advantage. They keep trying to send substitutes if "nothing is really going on" and that leads to a self-fulfilling prophecy of ineffective operation.

In the more successful committees of this type that we have encountered, the future of the firm's technology is very much influenced by what these people do and say in the meetings and as a result of them.

5. *Formal/Informal Liaison/Linkage Agents Between Labs*. In general, this mechanism does not work as well as the key communicator approach which benefits from the inner drives of the KCs. The formal liaision agents are often reluctant agents who "pass paper" and sit in on meetings without contributing much. There

are exceptions, however, involving highly qualified and motivated people who go beyond the formal liaison role and get involved in actual idea, information, and technology transfers and who are well connected and well thought of in their own labs and divisions. Many of these people appointed reluctantly, however, to liaison roles, turn out to be "information sinks" or poor representatives of the lab in terms of quality, credibility, or enthusiasm.

6. *Technical Seminars*. These are usually "one off" versions of the technical committees discussed above. Except for very large divisional labs, they are most often organized and hosted by CRL or corporate technology staff people. This means that the divisional technology people attend as "guests" or "strangers" and are sometimes very uncomfortable doing so. Added to this is their often remote location, and attendance becomes a significant time and budget matter. They may have to make a strong case for travel to the CRL location on "other business" and that barrier may not be worth going to the mat over with their division technology manager or immediate supervisor. In addition, if such seminars are organized at a relatively high or deep technical level (not unusual for the CRL organizers to strive for), then the divisional researcher may be out of his depth or embarrassed by his level of contribution or understanding.

These barriers to the typical CRL-sponsored seminar argue for an alternate approach, in which the divisional researchers and their areas of concern dominate the choice of topics, the depth of treatment, the seminar agenda, and the participants.

Where several divisions have some common technical interests, and where there is a local college, university, or technical society willing to sponsor the seminar, this can be an excellent mechanism for networking of divisions. The presence of strangers from outside the company (faculty and students and other company personnel) may inhibit discussion of the company's specific technical problems. However, such meetings can and do present an opportunity for technology people from different divisions of the same company to get acquainted and set the stage for further contact or networking later on. Some CRLs and CTOs encourage such "grass roots" seminars and even provide a speaker's bureau from CRL or outstanding university groups. Others seem to get their noses out of joint at such divisional initiative. That, in my opinion, is a mistake. The out-of-pocket cost of such events is very small. Many of them take place after working hours and in the same city or within driving distance. The costs may be picked up by the professional society chapter or may be shared through a small fee to a local university group that organizes them. The payoff in terms of stimulation, specific ideas, and general keeping up with the state of the art can be great. Sometimes there is a sufficient technical critical mass within the firm, including local divisions or even an extended network, within easy reach of a set of the firm's divisions. Under these conditions, occasional internal and private seminars may be more useful than external, public ones. Internal seminars provide the opportunity, assuming the attendees are willing to take it, to get down to specific divisional technology needs, problems, and opportunities. We have also been involved, via such seminar series, in a lot of "dirty laundry" washing and "Monday morning quarterbacking" on failed or hung-up projects. If carefully monitored and guided, such potentially heated discussions can be very healthy for an individual division and the firm as a whole.

A balanced combination of inside and outside seminars, where feasible, may be the ideal use of this mechanism for divisional networking and general "renewal" of divisional technology people.

On looking back at this subsection, note the length of it relative to my discussion of some of the other mechanisms. Partially, it is due to my long association with the academic community and as an organizer and conductor of such seminars. But also, it is due to the tremendous technical leverage that such a mechanism can provide for relatively small, isolated divisional labs that have few other forms of technical stimulation and ideas for their professional people.

7. *Exchange of "Skill Inventory" Data.* This is a politically sensitive mechanism. In addition, it may be infeasible, where a division does not know its own skill groupings and individual skill profiles. This is a topic that is treated at greater length elsewhere in this book (in Section 8.3 on imbedded technology).

The political danger is that such an exchange of information about the special skills and capabilities of a division's technology group may be used by other divisions as a recruiting tool. And it sometimes is. However, there are other channels by which a division on the hunt for a particular specialist or hotshot can find out about him. In addition, having other people know about "our specialists" can lead to a constant stream of requests for assistance, visits, or "temporary loans to help us out of a hole."

Offsetting these disadvantages can be the ability to get help or an opinion quickly (the other side of the coin of getting requests) without having to go through the formal procedures and agreements needed to identify and retain outside consultants. Furthermore, such people in other divisions are more likely to be closer to the actual level of application required and may, through past encounters, already be somewhat familiar with the requesting division's operations, needs, and people.

The feasibility question is another matter. Few companies, even at the CRL level, go to the trouble to construct or maintain a detailed skills inventory. However, a "minimum skills inventory" can be set up and kept relatively current for a very modest outlay. It can consist of a loose-leaf notebook with a page for each kind of technical area or skill grouping and the names of people who might be contacted first in a search for assistance. A few phone calls can yield the person or group needed to help address, if not solve, a given technical problem.

8. *Cross-Divisional Project Teams.* Perhaps the best and most direct means of building technical networks between divisions is actually to work together. Of course, a specific project experience may backfire and lead to vows "never to work with those guys again." However, joint projects, undertaken for whatever motives, can provide an excellent way in which people can get to know each other, assess each others' capabilities, and lay the groundwork for future cooperation.

Rather than avoid joint projects, as many fiercely independent division managers do, they should try some out to see what the cost/effectiveness of such a mechanism really is. High levels of synergy can be achieved between divisions where the pattern of cooperation is carefully planned and the project team is carefully "designed" instead of "thrown together" in a crisis or cast-off mode: "Joe isn't doing anything useful; assign him to that cooperative project."

9. *Open Across-Division Transfers and Promotion Opportunities for R&D/Technology People*. This may be anathema to some division managers and will not happen as long as they are in office. Where there is an openness to the possibility, however, the advantages can be worthwhile. At the minimum, such movement of people can bring a breath of fresh air into a division's technology group(s), provide morale-building career path opportunities, bring in new ideas and skills, and open up channels for future cooperation.

We have encountered companies where this is variously encouraged, forbidden, frowned upon, overlooked, or treated in some other way as an acceptable but not necessarily desirable mechanism. In one large aerospace company a bulletin board notice threatened extreme reprisals if any engineer made advances to another division for a transfer, or even a job interview, without clearing with his boss first.

10. *Strong Corporate R&D/Technology Coordination Staff to Act as Liaison Agents*. In earlier days, when corporate R&D was riding higher in the corporate pecking order than it is in many companies in the 1980s, this was a principal mechanism for coordination and networking of divisional technology. Corporate technology staffs grew in the 1960s and early 1970s, apparently immune from cost cutting and the "leaning and meaning" of other corporate staffs. (Exceptions were data processing and information systems, which were galloping past R&D in terms of sheer number of people and percentage of the budget.)

In some firms, of course, technology was never very prominent at the corporate level, so there was little evidence of a significant corporate technology staff for coordination. In many companies in the chemical, petroleum, electrical/electronic, pharmaceutical, and aerospace industries, corporate technology staffs expanded to take on the coordinating role.

Most of those efforts have taken a beating, however, in the face of general cutbacks in corporate staff and in the cases of serious reduction or elimination of corporate R&D altogether.

Where such corporate technology staffs did exist and still exist, however, they have often been able to do a first-rate bridging job between corporate and divisional technology and between the divisions themselves. Such effectiveness has been contingent, as always, on a number of factors, such as the power and influence of the CTO, the skills and credibility of the staff liaison people, field of technology, affinity of divisions, and locations of labs. Even in adversity, we have observed single individuals in the corporate technology staff doing a remarkable job of oversight, liaison, technology transfer, and adding to the corporation's overall technology base.

11. *Systematic and Effective (Accessible, Legible, Timely, Usable) Report Exchange Program*. It is surprising how few companies insist that divisional technology reports be exchanged in a regular fashion between operating divisions. In many, there are such formal requirements (requests?) but not much compliance. In others, there is not even acknowledgement that such a policy or practice exists.

Despite this lack of general use of this mechanism, it can be a very inexpensive and cost-effective way of letting the other divisions know, in general, what one another's technology people are up to. Even if no one or only a few people in the receiving division actually "read" these reports, the covers or title pages themselves carry enough information to alert a scientist or engineer that someone in the company may be working in an area or on a problem in which he has an interest or need for assistance.

This minimal information, of course, can be more economically transmitted by abstracts of reports, rather than full copies, and that is indeed the practice in some organizations. Given that divisional technology people do scan such abstracts or lists, the purpose of minimal identification of mutual interests can be served. However, some researchers like to "touch and smell" the whole report—that is, actually to scan it for aspects of interest beyond the title or abstract, for example, curves or data tables. Those divisions that make it difficult for such first-hand contact with or ownership of another division's reports may be missing a good networking opportunity. Confidentiality, busy-ness, lack of funds for printing and distribution, or fear of exposure and criticism may prevent such exchanges. Care should be taken to see that the reasons for *not* exchanging technical reports and related documents—for example, program plans, program summaries, lists of problems in need of solutions—are sound and in the best interests of the company and the divisions.

12. *Temporary, Cross-Divisional Transfers (for Projects, Training, Renewal, Communication).* In a closed shop, such transfers, even (or especially) temporary ones, may not be allowed by the division manager and/or his technology manager. Many of the reasons cited earlier may prevent such seemingly harmless assignments. If the barriers to cross-divisional job posting and joint projects can be overcome in the first place, then such temporary exchanges of people can flow naturally. Some of my earlier comments paint a picture of many divisional labs facing obsolescence, stuck in a rut, and/or showing a low energy and entrepreneurship level. These kinds of transfers can provide a shot in the arm and stimulate the flow of ideas, information, technology, blood to the brain, and a general atmosphere of innovation and enterprise.

13. *Corporate R&D/Technology Staff Reviews and Audits of Divisional Projects, Programs, Outputs, and Personnel.* Such a role calls for even more power and influence on the part of the CTO and his staff than the "coordination" activities mentioned in item 10. This extreme degree of oversight and intervention does occur in some firms, with the good effect of keeping the company-wide technology program alive, vigorous, and forward looking.

It ranks along with "those damned bean counters from corporate headquarters" in the perception of many division managers and *their* staffs. If that attitude can be overcome, such strong guidance and "control" from corporate technology staff can pay off in higher-quality divisional programs and much greater cooperation between them than is likely to occur without strong corporate urging or pushing. Unfortunately, we have seen many cases of corporate technology staffs thinking they had the charter to perform this role and then finding themselves shot down in flames and barred from divisional premises when a showdown occurs between the CTO and certain strong division managers.

14. *Joint Funding of Projects and Programs by Two or More Divisions.* If this can be pulled off on other than an "exceptional" basis, many of the other coordinating and divisional networking mechanisms will follow naturally. Funding of important projects/programs at significant levels will assure the interest of divisional technology and management personnel. Their self-interest will lead them to make the coordination and networking efforts to assure that the joint venture has a good chance of success.

Initiating such joint ventures may take some doing. If an opportunity arises which has obvious advantages to two or more divisions (strangely, it may be easier to launch such a project when more than two divisions are involved), one of the DMs may take the initiative and propose a cooperative effort. Where the opportunities are longer range and less obvious, it may require the good offices and a lot of persuasion by the CTO and even the CEO.

Such ventures are easier for new *manufacturing* technology projects than for new *products*. Many companies have launched company-wide or multidivisional manufacturing projects/programs. Some clearly have the potential of benefiting all the participants without threatening an exclusive market position. A sweetener of corporate funds contributed "above the line" is often added so that divisions' operating profits are not as hard hit.

Corporate technology management and the CTO should be alert to such joint venture possibilities for their own sake as well as their value as triggers for interdivisional technology networking. Without being frantic about such projects succeeding, the CTO should provide whatever support he can to boost the odds that they *will* succeed. A lot may be at stake for the long-run health of the firm's technology, costs, and survival.

15. *Strong Incentives (Positive and Negative) to Division Mangers to Cooperate in R&D/Innovation Area.* This is easy to say but has turned out to be very, very difficult to accomplish. Given the self-interest and initiative to engage in such cooperation on joint projects of strong interest to his division, the DM may not need additional incentives. He may need some additional resources—people, money, and/or equipment. In the absence of such evident self-interest and initiative, however, the typical highly focused DM is unlikely to put his scarce resources into nebulous ventures that might have a payoff to his successor. They might, during the projects themselves, add to his problems and aggravation and divert the attention of his technology people from the division's immediate and direct technology needs.

Given the natural tendency of the DM to suboptimize and focus primarily on his own division's current, short-term concerns, some action by corporate management is needed to persuade him to engage in these "side excursions."

A number of large companies have tried a variety of compensation and incentive schemes that will weigh technology along with sales, profits, and growth in determining the base compensation and bonuses of the DM. Few have dealt with the essence of the problem. The message that the typical DM receives from corporate headquarters is that "it would be nice if you reinvest some of your operating funds or profits in renewal of all kinds—products, facilities, people—but it is essential that you maintain and increase current operating results." Often it comes down to a single number—return on assets or investment, cash flow, increase in sales or profits, or growth rate. Seldom is the technology factored in explicitly so that the DM can see the direct impact of enhancing or neglecting technology on his personal income. Formulas for accomplishing this have been developed, but the barrier to their use is the sharp conflict between maximizing current results and sacrificing some significant part of current results for the longer term. Given that conflict and the behavior it engenders by the DM, it is hard enough to get him to spend more and better for his own direct technology, without trying to get him to spend a significant amount on cooperation with other divisions in the hope of a more distant and indirect benefit to his division.

If the CEO really wants such interdivisional cooperation to improve the overall and long-term technology picture in the company, he has to communicate clearly and loudly to the DM that his personal income will suffer if it is not done and that he will benefit if it is done well and not via lip service.

Again, as in the case of joint projects/programs, a clear signal to the DM that cooperating and networking will be rewarded can stimulate many of the mechanisms presented here.

16. *Cross-Divisional Design Reviews—"Do unto Others."* In the few cases where we have observed an effective cross-divisional design review of products, processes, or even raw technology, cooperation has been enhanced. Sometimes this cooperation is accompanied by interpersonal conflict and below-the-surface resentment. This is because egos can be bruised, pet oxen can be gored, and technical reputations can be besmirched. A "real" design review is a no-holds-barred analysis of the flaws in a design concept or prototype which raises fairly basic issues about whether the underlying technology is sound and whether the item will work. It is much less common in commercial operations than in the military and aerospace communities, where it is required by contract and has been routine for decades.

Getting such a procedure started in the typical commercial firm is not easy. It requires a combination of good will and desperation to do one in a cross-divisional arena for a product/process that is in trouble. The "owner" of the item is on the spot and, if a design review is imposed on him and his colleagues, he is likely to be defensive. This means that he may or may not cooperate fully with the design review committee or task force and may even try to cover up or withhold information required for a good review.

The trick to overcoming this formidable barrier seems to be a routine program of design reviews in which *each* division gets a chance to critique and be critiqued for a sequence of projects that are not initially in a crisis mode.

An ideal way to start is with a program of design reviews for a set of projects contributed by the participating divisions in a routine way, where the divisions are genuinely seeking help in analyzing and improving their designs. Then, when a project gets into trouble, the mechanism is in place, people have gotten to know (and hopefully like) each other, and the full attention of the committee can be focused on dealing with the technical problems rather than each others' sensitivities.

It takes a strong hand and careful planning to achieve this state of affairs. Typically, it is organized and handled by a strong CTO and his staff, with the first participants picked carefully from among the divisional technology and management people.

Even though, as pointed out earlier, there is often a strong bond of loyalty between the DM and his technology team, an alert DM will want to be sure that the designs his technology people are coming up with are practical, safe, and cost effective. His own personal ox is not necessarily being gored if one of his division's designs is battered. He may actually gain in stature by showing that his division can gracefully accept criticism and benefit from outside help. He can be viewed as an antidote to the parochialism of his technology people.

Setting up and effectively operating design reviews takes patience, tact, wisdom, time, and good will. It is not easy, but the payoff in design mistakes avoided or fixed can be great.

17. *Joint Idea Generation Efforts*. This seldom happens naturally on its own. If it does occur, it is usually a spin-off or concomitant of one of the other mechanisms, such as design reviews, joint projects, and technical committees. However, as will be seen in Chapter 6 on idea sources and flow, anything that increases the "idea pool" for a division is worth considering. It was asserted earlier that for many reasons the idea pool—the actual and potential ways of improving a division's technology—is severely constricted in the typical modest-size and narrow-focus operating division. This means that opportunities for radical departures from current technology and major improvements, let alone new products, are also severely limited.

A cheap way of generating such ideas is to "steal" them from other divisions which may have generated relevant ones that they are not in a position to exploit. Casual idea generation as part of interdivisional networking can yield some potentially good ideas for participating divisions. Finding an internal champion and the resources to develop them further and to carry them through to commercialization may not be so cheap or easy. Thus, joint projects, interdivisional transfers (of the idea originator), and some of the other networking mechanisms may be needed to pick up the idea and make it happen. Much more will be said in Chapter 6 about how ideas are originated, what factors influence their flow, and how organizational and personal barriers can combine to kill them prematurely. In this section I merely point out that networking can be an excellent means of broadening the idea pool for a division and for helping to overcome the divisional time and scope limitations mentioned many times in this book.

18. *Coordination of R&D/Technology Segments of Divisional Long-Range or Strategic Plans*. This topic is also treated at length in a separate chapter (Chapter 5 on technology planning). In this discussion of divisional technology networking, the

main issue is to convince the DMs of the various divisions just to share their long-range technology plans with each other, let alone agree to the joint projects and mutual analysis/criticism mentioned earlier. Except for particular pairs or subsets of divisions, such as those within one product or technology group under a group executive, this is not likely to occur. In fact, sharing such plans, where there are any, with even the CTO and his staff is not easy to arrange. DMs very jealously guard their plans or intentions and defend them strongly when criticized or attacked by corporate staff people, or even the CEO. They plead the old argument about "how can I invest a lot in nebulous future technology when I have to meet your performance standards in the short term."

One of the caselets in Chapter 1 recounts the stonewalling that one CTO experienced in trying to respond to the CEO's request that he analyze and criticize the DMs' strategic plans from the viewpoint of the technology needed to support them.

Nevertheless, coordination of technology plans is another leverage point for initiating divisional networking. In fact, a DM is more likely to accept questions and criticisms from his peers—other DMs—than he is from a "bunch of desk-bound corporate staff people." More on this in Chapter 5.

19. *Mutual Program Reviews*. This may be the ultimate in cross-divisional cooperation in technology. It goes far beyond design reviews for individual projects and provides a forum for mutual help in formulating and improving the whole technology program in the operating divisions. This can either be a starting point or a culmination of cross-divisional cooperating and networking. I hasten to say, at this point, that a superficial level of this procedure is conducted in many companies through the various other mechanisms, such as meetings of the divisional technology managers and the many advisory, coordinating, or steering committees (see the extended discussion of these mechanisms earlier in this chapter).

However, many of the reviews are primarily "dog and pony" or "show and tell" events where most of the communication is one way—from the presenting division to its audience. It is often dominated by mutual forebearance to criticize another division. In some cases it has teeth and can be effective. Those cases involve DMs who are seriously concerned about the merits and structure of their overall technology programs—current and planned—and want an outside opinion from people they respect, that is, other DMs. When it happens one level down—between divisional technology managers—it can also be sincere. But the "absent members"—the DMs—carry a lot of weight in inhibiting free discussion and preventing effective action based on the group's recommendations: "That's all very well for those guys in the other divisions to pick away at our divisional technology program, but we have to live with the plants and the customers and we know what problems we have to tackle to keep out of trouble."

Like the design reviews, the joint projects, the sharing of ideas, and other mechanisms discussed in this chapter, mutual program reviews can reflect a true sharing of technology capabilities and technology between divisions. It is a goal well worth striving for in the interests of company-wide technology excellence.

4.5 SOME CASE STUDIES OF NETWORKING

As a small antidote to the many generalizations in the first part of this discussion on networking—"many companies," "in general," "most division managers"—this section presents two case studies of divisional networking or attempts at such. They illustrate many of the factors discussed earlier which can inhibit networking and many of the mechanisms for attempting to overcome these factors.

The two case studies are presented in outline form. The first—in Figure 4-3—is representative of many that we have encountered, where the corporate technology staff is attempting to network divisional technology groups with each other and with the corporate technology group.

All these attempts at networking failed, some because they were ineptly conceived and executed. For example, the first (and last) telephone conference (F6) was a bomb. It involved too many participants, no advanced planning, no agenda, no one to handle discussion priorities and keep the discussion on track, and other technical flaws—for example, the telephone connection was bad(!).

A. Background

A major "acquisition-grown" corporation, highly divisionalized, using a wide variety of technologies

B. Objective

Increase effectiveness of technical communication between divisions to obtain

(a) quicker technical problem-solving and

(b) quicker evaluation of new know-how and its incorporation into production

C. Organization

1. Corporate staff function created out of dissolved corporate R&D function (CRL) to

   (a) provide support and monitoring services for divisional technical activities and

   (b) search out new technology and products that the divisions can incorporate into their businesses

2. Maintain between six and ten "experts" for service to divisions

3. Maintain outside consultants, licensing agents, technology brokers, and "area" agents searching for new technologies

D. Major Problems Encountered

1. Low utilization of ideas generated by the corporate staff

2. Low utilization of the corporate staff "experts"

3. Low level of interdivisional cooperation

E. Barriers

1. Difficulty of stimulating genuine interest in the divisions for the new ideas generated

2. Establishing technical credibility that the corporate technology experts know their business

---

Figure 4-3. An example of interdivisional communication in company A.

The "brochure" (F7) in which one of our research teams was involved was an equal disaster. The corporate technical staff got cold feet about putting their biographies and skill descriptions, let alone their pictures, in the brochure for fear of seeming too aggressive and self-aggrandizing to their divisional colleagues. It sat at the printer for several months and was finally killed by the head of corporate technology.

The closest this set of mechanisms came to serving the purposes of technology sharing was a set of specialized relations, established between two or three of the corporate staff people with much needed technology *and equipment* and some of the more forthcoming divisions. The outcome was that, eventually, all the corporate staff people were "hired" by the divisions as part of a further corporate staff breakup.

This failed case does not mean that efforts to provide technical help to divisions and to stimulate mutual self-help are bound to fail and are not worth trying. In this "worst case," the attitude and actions of the CEO in the spirit of "complete autonomy" and "minimum corporate staff interference in divisional affairs" made

---

3. Establishing political credibility (trust) that they were not corporate spies
4. Overcoming NIH factor and the highly "autonomous" nature of the divisions (being continuously reenforced by the CEO—a former division manager)
5. Having to dispel the *notion* of "power" associated with their group which was resented by the directors of engineering in the divisions
6. The lack of *actual* "power" in their group to get actions taken
7. Overcoming the division managers' resentment of their "youngsters" talking to corporate people about their problems
8. Realizing that the corporate staff did not know enough about the precise needs of each division
9. Overcoming the perception by the corporate staff itself that they were not wanted in the divisions

F. **Attempts at solving the problems and overcoming the barriers**
1. Developing close relations between members of the corporate technical staff and selected high-potential divisions in the company
2. Identifying specific individuals within the divisions who would act as useful division contacts
3. Increasing dependence on them by the divisions via providing seed money for new product development within divisions and by centralized services needed by all divisions
4. Conducting technology audits to identify the state of a given technology in the whole company
5. Conducting technical seminars to acquaint key people in all the divisions of the state of the art in a given technological area
6. Conducting telephone conferences in specific technology areas
7. Circulating a brochure to increase general familiarity with the corporate staff on the part of the people in the divisions

---

Figure 4-3. *(Continued)*

any of these efforts hopeless from the start. His decentralization policy was so extreme, in fact, that he eventually fired or transferred all his corporate vice presidents with the exception of finance and one jack-of-all-trades brought in from outside the company. The latter handled manufacturing, technology, government relations, and several other functions and was warned not to interfere directly with divisional technology programs.

The second case was more promising and illustrates the potential value of the key communicator (KC) concept (see Figure 4-4).

This company had a formal KC program, as well as a network of informal KCs who emerged as additional points of contact and transfer mechanisms.

The quantitative economic benefits from this system of divisional networking through KCs were difficult to assess. Some KCs were acknowledged to have helped out in other divisions. Some were given credit for providing new ideas and new technology. Others helped out in crisis situations. Still others just helped

---

1. The goal of the formal key communicator (FKC—appointed by company) program was to promote technology within the company.

2. "Clients" who used FKCs in other divisions tended to contact them only after all intradivision resources were utilized.

3. Clients tended to value FKCs with the following characteristics:
   (a) They knew them.
   (b) They found them at least as effective as other information sources.
   (c) They were easy to use.
   (d) They provided useful information that could readily be applied.
   (e) They perceived the information to be current and of high quality.
   (f) They were challenged by application problems.
   (g) They did not insist on taking "credit" for the information provided.

4. Clients tended not to cross product lines, primarily contacting FKCs in other divisions who were within the same product group.

5. Contacts with FKCs in other divisions tended to be occasional.

6. Clients did not tend to call FKCs just to talk over an idea, or to "browse" for information.

7. FKC contacts with the most positive results were client-initiated.

8. The most *well-known* FKCs (the "marketers") tended not to be evaluated by clients as *most effective*.

9. Formal KCs were more likely to have clients in other *divisions* than informal key communicators (IKCs, who did it "naturally").

10. IKCs were more likely to have clients in other *organizational functions* than FKCs.

11. FKCs were more likely to know and be known by senior division management and corporate management than IKCs.

12. IKCs who were managers were likely to receive higher ratings for their role as KCs from their bosses than they received from their clients and colleagues.

13. IKCs who were nonmanagers were likely to receive higher ratings for their role as KCs from their clients and colleagues than they received from their bosses.

Figure 4-4. An example of the use of key communicators in a large decentralized company.
Source: Dissertation by Connie Knapp (1983).

to maintain and increase the "technology smarts" of other divisions and their own (through what they received in return). The company undertook a deep analysis of its KC program and concluded that it was beneficial enough to keep going but that it needed some modifications to make it more effective. Cost was really not a major consideration, since the KCs generally did their regular divisional jobs in addition to their KC duties.

Along with the kinds of contributions made through these individual contacts, other mechanisms mentioned in this chapter accompanied the KC program. These included technical seminars, technical committees, some joint projects, and a new mechanism that the corporate staff pushed—"centers of excellence." This was a way of describing a policy that identified divisions within which an individual or group had a leading technical edge beyond that of any other division in the company. These individuals or groups were then designated as lead technology centers in their areas of specialization and the attention of all technical groups in the company was directed to them for leading edge work, ideas, technical information, and technical assistance in their field of excellence.

It is a notion borrowed from or at least similar to that used increasingly by the federal government in funding research in a given area at one or a few places, so that the maximum in scientific and technical capability could be established and maintained. The Materials Research Centers program of the Department of Defense and the National Science Foundation are good examples. KC programs may be worthwhile in themselves, but they can also supply leverage for other effective networking mechanisms.

4.6 IMPACTS OF NETWORKING (OR LACK OF IT) ON GROUP TECHNOLOGY

There is a special area where the lack of divisional technology networking can hurt the firm's attempt to improve its manufacturing technology and cost structure. This is the area of *group technology* in which attempts are made to standardize parts, materials, or subassemblies across product lines. For a wildly diversified company, where the products and manufacturing technologies/methods of the divisions have nothing in common—for example, food processing, machinery, petroleum, service sectors—this is not an issue, although some combined work in supporting services, purchasing, and other areas might help. Other situations where divisional networking may not be a major factor are where all of the potentially common manufacturing or product technology is within one division. That is, all the products and product lines that might benefit from a program of partial common manufacturing are within the same operating division and under the same DM. A major opportunity lies in firms that have several divisions with products and product lines whose components and methods of manufacturing have enough commonality or potential commonality to make consideration of group technology worthwhile.

The incentives to try to bring about more cooperation in design of products and processes and their parts, materials, and manufacturing methods can be strong from the viewpoint of top management. They may see a major opportunity for cost savings and, as a consequence, price reductions, increased market share and profitability. Additional advantages might accrue in service and warranty costs, and the firm's image in its various markets. From the viewpoint of an individual division manager, the idea may seem attractive in concept but may have certain nonattractive features in practice. He may be doing "just fine" with his current manufacturing methods, design, materials, styling, cost structure, prices, and/or margins. Or he may be able to rationalize his possibly higher costs by the prices he can get for his products and his market share and operating margins. He may see long-term advantages to getting his costs down through improved design, stronger purchasing power for materials and parts, and even equipment. But he also sees several drawbacks to such potential cooperation with his fellow DMs. Some of them might be:

- The need to commit his scarce R&D/technology people to a long-term cross-divisional project when they are fully committed to important projects for the division currently.

- The probable long waiting time for the results of such a project before any possible benefits could be available to help his division.

- The uncertainty that such projects will indeed yield results or benefits to his division at all—it may help some other divisions and the corporation in general but not do him any direct good.

- The likelihood that such projects will result in the need for new equipment and facilities that are hard enough to get for routine replacement of his current equipment.

- The possibility that he will be expected (required?) to share manufacturing facilities, thus reducing his flexibility and autonomy.

- The basic idea of autonomy itself, which he has fought for and protected so vigorously, involving keeping corporate staff snoopers out of his plants.

- The possibility of having to give up some of his favorite long-term suppliers with whom he is very comfortable and take on new ones over whom he will not have as much leverage. Worse yet, he may have to go through a corporate staff middleman to deal with new suppliers.

For DMs in specific situations, the list can go on, and many reasons—some of them quite credible—can be presented for reluctance to participate in such cross-divisional technology projects. Some of these are from the lists earlier in this chapter on the general barriers to divisional networking. If such projects are to get off the ground or even be considered seriously, they have to be approached from the viewpoint of "what's in it for me" in terms of costs/benefits.

The pure logic of group technology or other means of bringing more standardization and purchasing power into the manufacturing and design process are not enough to bring many independent DMs on board, let alone generate high enthusiasm among them. Limited "dollars and cents" alone will also not convince many of them who have a keen sense of "opportunity cost." They may calculate accurately that the effort and working capital and investment capital required to develop, adopt, and benefit from group technology may far exceed the potential benefits when they factor in the other things that they and their divisional technology people might be doing with the same amount of time, effort, and resources. The potential benefit then has to be perceived as substantial by the DMs to get such projects started. Even when forced on them by a strong-minded corporate CEO and his staff, there are many ways in which the division can slow them down, kill them, or make them, ultimately, unattractive for the corporation to continue pursuing.

Some ways of influencing the DM to go along with, even demand, such cooperative projects include the following:

1. Make sure the corporate manufacturing or technology staff has done its homework and really knows what the current divisional costs are and what savings are likely to be made via group technology. This is a "carrot" approach and depends on the ability of the corporate staff or a multidivisional task force (a better mechanism, as mentioned below) to even get access to each division's actual cost information, detailed manufacturing and procurement procedures, and product designs.

2. Set up an interdivisional task force to carry out the front-end feasibility analysis so that the DM's people are participating and controlling the project from the start, rather than having it viewed as primarily a "corporate staff" project.

3. Tie the requirement to investigate and even implement group technology into the conditions for receiving capital investment funds for manufacturing facilities and major pieces of equipment—the "stick" approach. This can backfire but can force participation in self-interest.

4. Set up such cross-divisional projects on a subsidized basis by the corporate office, the CRL, the group executive, the CEO's own contingency funds, or some other device to show the DM that he is not bearing the full cost directly. Of course, the skeptical (cynical?) DM will say "eventually, it's all my money anyway, because that's where the cash comes from." However, such subsidies are frequently used to undertake corporate-desired projects that are unlikely to be funded by individual divisions. Indeed, the CRL itself is a major example of such channeling of divisional profits through the corporation to support company-wide technology activities. It's a game the DM is familiar with and, although he may grouse, he would rather see some of the money spent on projects of direct potential value to his division (e.g., through cost reduction or improved design) than see it "wasted" in the "usual kinds of blue sky efforts" in which CRLs engage.

5. Set up a corporate-level manufacturing engineering or manufacturing technology group, lab, or department. Wait a minute—another CRL? They seem to be falling out of fashion themselves; how could you get away with another one? The

difference is that a manufacturing-focused lab or group can (and should be) tied directly to divisional interests in the manufacturing and engineering areas. They should in fact, and in a few companies actually do, provide the technical backup that smaller and even larger divisions need to take advantage of the leading edge of bringing more standardization and purchasing power into the manufacturing, materials, process, and design technology. Unfortunately, there are not many of them yet, and of those that have been formed, many of *them* have been buffeted by the factors that have threatened and indeed taken the lives of an increasing number of CRLs (see Chapter 2 for a detailed discussion of that).

The successful ones, however, have proved their worth through being the principal or only means of bringing the latest manufacturing technology to the divisions. Many operating divisions are too small, too preoccupied, or too obsolete technically to search for, evaluate, adopt, and successfully implement leading-edge technology. Examples abound of strikeouts at the divisional level in computer-aided design, computer-aided manufacture, attempts to manufacture exotic new materials such as composites and crystals, completely automated materials handling systems, process control, and so on. There are many exceptions to these failures at the divisional level, especially when the division is large and sophisticated enough and has the backup of a good network of suppliers and consultants. For many divisions, these conditions do not prevail and a corporate-wide manufacturing technology lab or group can be very helpful, no matter who pays for it.

6. Another stronger and even riskier stick than tying such divisional networking and cooperation to capital investment proposals is to use such cooperation as a direct factor in determining the DM's personal compensation. "You did fine in your division, but you're not a corporate team player and that can hurt the company in the long run. I am therefore discounting your bonus to allow for future costs we could have saved if you had been more cooperative." Wow! What CEO would even try that, let alone make it a policy. However, it fits in with at least the verbal trend toward broadening the basis of compensation and incentives for the division manager (see Chapter 3 for further discussion of this).

If and when cross-divisional projects, such as group technology, can be launched and successfully concluded, many other forms of divisional technology networking can follow. Despite the kinds of reservations mentioned with respect to this kind of project, it can be much less threatening and autonomy challenging than other kinds of corporate office attempts to get divisions to cooperate.

For example, the battle for control of computer facilities, information systems, and related technologies is still raging in most decentralized firms, although there seem to be temporary agreements and solutions. The rapid pace of technology development in computers and information technology have made some of the outward struggles—for example, who "owns" the main frame(s)—moot. The flood of personal, mini, and microcomputers makes it feasible for divisions to computerize rapidly on their own without major capital investment requests and territorial hassles. The advent of local area networks (LANs) and other networking mechanisms can push the centralization–decentralization struggle either way.

Centralized versus decentralized (or cooperative) long-range planning is discussed in Chapter 5 and the issues of centralization–decentralization in other staff activities are mentioned in several places in this book as they relate, even if indirectly, to technology.

Cooperation in advanced manufacturing technology, through such "trial balloons" as group technology, can be eased into place if some of the concerns and barriers raised in this chapter can be dealt with effectively (the "carrots and sticks"). Once the divisional technology people have a sample of the benefits to them personally (prestige, collegial support, windows into new technology, someone's shoulder to cry on when things go wrong), as well as the ultimate costs and benefits to other divisions, they may be eager to try additional forms of divisional technology networking that appear riskier and more uncertain. Their enthusiasm can have a positive influence on their DMs if they do their homework and place their proposals in the proper perspective for approval.

__5
TECHNOLOGY PLANNING

5.1 INTRODUCTION AND CHAPTER OVERVIEW

There is commonly confusion about long-range planning in the technology area. Technology projects, especially those involving radically new products and manufacturing processes, often take a long time to come to fruition and to impact profits. Therefore, many nontechnical people, including top managers without direct experience in technology activities, often assume that some reasonable degree of planning must have gone into them. Certainly, on a project-by-project basis, R&D projects do in fact involve a lot of planning and scheduling. So do manufacturing projects, involving new manufacturing methods, new materials, retraining or recruiting new sets of skills, and lead time for tooling, equipment procurement, product introduction, and plant start-up. This is not the kind of long-range technology planning we are concerned with here. All of the above activities are taken for granted (although they often should not be) in a well-run or even a not-so-well-run firm. Technical people with many years of experience have learned, from that experience, that you have to anticipate problems and situations that might interfere with smooth transition from lab to factory and market. Some individuals and some firms do it better than others and that might make the difference between being competitive and profitable or not. Some of these "routine" planning aspects of technology are dealt with in Chapter 8 on commercialization and the technology life cycle.

This chapter addresses a system or frame of mind in which more general and less routine planning issues are raised concerning the general directions of the firm and its supporting technology. We look at technology planning with primary emphasis on the "upstream" technology activities such as R&D, product development, and new manufacturing process and materials development in the context of their relevance and value to the firm in the long run. Some of these issues are often considered under the heading of "technology strategy" or "technology policy" for the firm as a whole, as well as for individual operating divisions and product lines. Although it is not dwelt on in this book, *manufacturing* policy is a closely related issue and is discussed in the context of commercialization and the technology life cycle in Chapter 8.

5.1.1 Some Issues/Questions

- Who takes the lead on long-range technology planning—the planning people, the technology people, the financial people, or the marketing people?
- What is the starting point for long-range technology planning?
- How informed are the R&D people about company plans, aspirations, limitations, and the trends in markets and resources likely to affect the company significantly?
- Is there a "partnership" between the firm's business planners and technology planners?

5.1.2 Main Body of Chapter

This contains a discussion of the issues and problems in technology planning and how to address them.

5.1.3 Implications for Management

Effective long-range technology planning is a rare event in most firms. It requires careful organizing, close monitoring, persuasion to achieve cooperation, and patience in the usually long start-up period.

5.1.4 A Watch List

- Constant missing of time, cost, and performance targets for longer-term technology projects and whole programs.
- Mismatch between business plans, hopes and dreams, and the actual available technology to help achieve them.
- Apparent ignorance or indifference on the part of technical people about the directions and targets of the company, its markets, and its needs for new and improved technology.

5.2 ISSUES AND PROBLEMS IN TECHNOLOGY PLANNING AND HOW TO ADDRESS THEM

5.2.1 Assumptions and Issues Related to Long-Range Strategic Technology Planning (STP)

Initiation or improvement of the firm's ability to do realistic, effective long-term planning for technology requires, in itself, a great deal of planning and thoughtful analysis. The preconditions and continuing conditions for effective planning are not obvious, beyond a few platitudes (some of which are included in this section). In our examination of the planning processes and lack thereof in a large number of very large and not so large decentralized companies, a list of common issues arises. Particular ones are of more consequence in some firms than in others. However, all of them should be considered, even if only "scanned" to make sure they are not presenting barriers and difficulties to effective planning.

The list in Figure 5-1 contains some fairly critical issues for the effectiveness of technology planning. Others are more marginal and can help or hinder planning in nonmajor ways.

Since the list is long—25 items—I discuss them here only briefly. In later sections of this chapter more detailed discussions are presented of several sets of these factors, which, as a group, may be critical for the feasibility and success of technology planning in a given firm or situation. Like the technology/innovation (T/I) audit presented in Chapter 7, this list and the accompanying discussion about planning issues are primarily diagnostic and analytical. Removing or reducing the barriers posed by these issues or taking advantage of some of the factors facilitating technology planning has to be done on an individual basis—by each firm or operating division, according to its objectives, resources, constraints, and special circumstances. Unfortunately, the state of the art does not currently permit a universal "handbook" for technology planning. The discussion in this chapter should, however, provide adequate guidelines to support a serious effort at instituting or improving strategic technology planning in the firm or an individual division.

5.2.2 Brief Discussion of the Assumptions and Issues in Figure 5-1

1. How willing are the key people in the various business areas and functional areas of the company (including technology) to share information, ideas, and speculations about the future? Is there defensiveness or secretiveness that must be overcome in certain areas or across the company before effective technology planning can be carried on?

2. There should be a degree of control (including self-control) that would keep people from jumping onto or overreacting to a potential new product line, field, concept, problem, or other opportunity until it has been thoroughly (or adequately) checked out and the technology people are able to give some assurances of its viability as part of a plan.

3. A major weakness of strategic technology planning (STP) in large companies involves the lack of articulation of technology plans and planning into that of corporate business planning, including production, personnel, procurement, and logistics realities and constraints, and the general (even if vague) plans of marketing, finance, and the various "business" components of the company.

4. Realistically, there is no clear linear sequence of planning that will assure good and orderly planning in terms of *concepts* as opposed to *mechanics*. It is not clear whether top management, technology, or neither should lead off in the

1. Willingness to share information and speculate about the future.
2. Self-control to prevent overreaction to new product and market ideas.
3. Articulation between technology planning and business planning.
4. No clear *linear* sequence of planning.
5. General weakness of market research for new products and markets.
6. Frequent poor relations between technology and financial planning.
7. General dominance of the cash flow concept.
8. Need of top managers and "businessmen" in the firm for help from technology groups:
   (a) Early warning of technical threats, needs, and opportunities (TNOs).
   (b) General credible technological guidance and advice.
   (c) Scenarios (also credible) on future markets, technology trends, competitors' actions, materials and energy prices and supply.
9. Heavy involvement of R&D in manufacturing technology.
10. Potential conflict between company goals and that of individual managers.
11. The implementors of plans are line managers, not staff analysts.
12. Future actions, plans, and events must be tied to changes in company goals, criteria, characteristics, and capabilities.
13. Managers differ widely in their willingness to play the planning game.
14. Planning requires a marketing viewpoint (vis-a-vis internal "clients").
15. The regulation and economic/social environment are integral parts of the plan, not merely constraints.
16. The target periods (time horizons) for planning must fit with the character of the decisionmakers, the company, the industry, and the environment.
17. Strategic business units (SBUs) may or may not be logical elements for long-range technology planning.
18. How defensive is the posture of the company's technology policy in current products and product lines?
19. How strongly held is the "sunk cost" philosophy of the company?
20. How strongly is the company tied to existing operating units as vehicles for launching new products and technologies?
21. How conservative is the company's diversification strategy, if it has a coherent one?
22. Which functions in the firm dominate in planning and implementation of plans?
23. How closely is R&D involved in new ventures, technology purchasing and licensing, and mergers and acquisitions?
24. How close are the technology people to the hopes, dreams, and fears of their "clients"?
25. How enthusiastic are the technology people about doing long-range strategic technology planning?

Figure 5-1. Assumptions and issues underlying strategic technology planning (STP).

planning sequence. It is ideally a cyclical process that builds up momentum incrementally and hopefully converges on specific and reasonable plans and associated decisions and actions.

5. Long-range market planning is very weak in many large companies except for certain consumer and capital goods industries that are, of necessity, tied to long

market introduction and forecasting lead times. What passes for market research is often short term, tied to present products and business, and does not provide the scope and depth that R&D needs as inputs and as a basis for a tough, realistic dialogue on the "possible" and the "desirable" for the long run.

6. The relation of STP with financial planning is frequently poor, with finance usually setting the limits on funds availability and spending patterns often with little input from technology. The preferred method appears to be, in many companies (if not most), "tell me what you want to do and we'll tell you *when the time comes* if we can afford to do it or if it fits into our cash flow and investment planning."

7. How dominant is the cash flow concept in the firm? Does it dominate return on investment and other measures of performance in investment and planning? If so, how can STP effectively plan on the availability of cash or investment funds in a, currently, dimly perceived and uncertain future pattern.

8. In most companies, top corporate management and the "businessmen" of the company in divisions and groups would like to be able to depend on technology to be responsive to their pattern of operations. This is generally true whether they value technology greatly or merely tolerate it. They would like help from R&D (generally on request, and not unsolicited, except for alerting to problems and dangers) in these areas among others: (a) early warnings on threats or opportunities (they may be the same events or conditions but viewed differently by different observers), (b) advice and guidance on what is possible and better technologically that can make their lives easier and provide better costs, productivity, market share, and so on, and (c) scenarios or forecasts on what is likely to happen to their products, markets, materials, energy, competitors' actions and capabilities, environmental constraints and ways of dealing with them, and general state-of-the-art changes that could have a significant impact on their domains.

9. Part of any strategic technology plan should include consideration of the trends in manufacturing, including procurement and contracting out. In some industries and individual firms there is a growing doctrine that more money and more *certain* money are to be made in the factory and logistics areas (including materials and energy usage, quality control, and methods) than in the new products area. Part of that is based on increasing and increasingly visible costs of warranties, liability, repairs and maintenance, and replacement of products already in the field.

10. How closely do the personal objectives of key top management and businessmen (division managers) in the company mesh with formal or assumed company goals and objectives? There are several kinds of rationality that should be taken into account in attempting to formulate, sell, and implement long-range plans that can clash with nonobvious personal goals and frames of reference. For example, steady growth or even stability with low risk may be preferred over "go-for-broke" venturing into dangerous waters.

11. The businessmen or DMs of the company will carry out or not carry out the decisions and actions called for in the STPs, not the staff people, or perhaps not even top management.

12. A very valuable aid in the STP process is a set of scenarios or case studies of past attempts (fairly recent, to be sure the whole organization has not changed) to formulate and systematically carry out STP. Both success and failure cases might provide clues to mechanisms and concepts to push and to avoid.

13. Some potential clients in the company (CEOs or DMs) are more amenable to playing the planning game than others, especially as it involves technology. Those most amenable would be good clients to start with and try some of the ideas and mechanisms on as "demonstrations" or "experimental" exercises.

14. People involved in the STP process must take a marketing viewpoint— who are the clients (customers inside the firm) and what do they *really* want. This means getting to know them and pooling staff information on them.

15. The regulation, economic, and social environments are not merely constraints on STP but are integral parts of the conceptual and operational aspects of planning and can often provide new cost-saving and profit opportunities (e.g., energy, safety, product warranty).

16. Feasible and acceptable time periods for STP should be chosen as targets— for example, 5-, 10-, and/or 20-year plans.

17. Strategic business units (SBUs) appear on the surface to be logical units to build an STP on, but their limited scope, time horizons, and technical resources may make that infeasible, difficult, or inadvisable.

18. How defensive is the firm's technology policy in terms of *current* products, lines, and business areas relative to potential *new* ones? Even though it can vary over time, such a guideline is important for the technology planners.

19. How widely and deeply held in the company is a "sunk cost" philosophy— a willingness to walk away from a business or operation that is currently or is soon likely to become a loser? Most companies pay lip service to this but, when the crunch comes, there is generally someone to demand or plead for more time to make it work, turn it around, or revive it. Many "cash cows" should be put to pasture in time to avoid eventual cash drains.

20. What is the attitude in the company with respect to maintaining and building on or attempting to adapt existing organizational units to new products, lines, or technologies (e.g., the tube manufacturers who tried to make semiconductors in the tube divisions and the textile people who tried to introduce synthetics without revamping the production operation significantly)?

21. A good conservative diversification strategy, driven by STP, keeps one foot firmly planted in at least one familiar area—materials, manufacturing processes, or markets (see Chapter 7).

22. Which functional areas of the firm dominate in planning decisions and implementation?

23. How does the STP of R&D relate to the new venture areas of the company? In many companies the articulation is poor and the two groups operate at cross purposes or in a duplication or gap mode that causes confusion, delays, conflicts, and missed opportunities.

24. How familiar are technology people in the company with the problems, hopes, dreams, and fears of their clients? How much effort do they put into obtaining this kind of information? In some cases, outside intelligence is good, but internal intelligence is neglected.

25. How enthusiastic are technology people about doing STP in the company? In many companies they have a very low interest in and tolerance for STP and, although they complain about it, they actually use "firefighting" as an excuse for not "getting to" the STP and the kind of strategic explorations discussed in this chapter.

5.3 HOW CAN THESE ISSUES BE DEALT WITH? SOME FURTHER OBSERVATIONS AND SOME RECOMMENDATIONS

Following this presentation of issues and questions involved in strategic technology planning (STP), some additional observations and recommendations and suggestions for dealing with them are now presented. As mentioned earlier, individual solutions must be tailored to the specific circumstances of the firm or operating division that wants to initiate or improve its STP. Figures 5-3 through 5-5 in Section 5.4.1–5.4.3 group these key sub-issues for easier reference. In this section some of them are discussed in greater detail, in the same sequence as in Section 5.2. Recommendations and items for the watch list are given for each issue.

1. *Willingness to Share Information and Speculate About the Future*
    (a) A "safe" atmosphere for sharing information, ideas, hunches, and misgivings.
    (b) A mechanism for addressing the future—both inside the company and in the external environment: scenarios, data bases, projections of futurists.
    (c) Need for *responsibility* for speculations but not *punishment* for them.

In order to share information and speculate about and plan for the future, the people involved must feel secure. They must feel that the risk to their careers, reputations, and egos are not going to suffer significantly if they "stick their necks out" and put forth or support "far out" or unpopular views of the future.

Successful DMs and even CEOs who are quite happy with the P&L performance or cash flow of some of their mature businesses do not want to be badgered by "the possibility that these businesses will disappear" in the face of emerging or over-the-horizon technology. The plastics versus metals war, for example, has been raging in many industrial sectors and in many companies for decades. The advocates of current technology (typically metals) for containers, fasteners, parts, subassemblies, and whole machines or devices have lost ground in most sectors. But some of them still fight on, arguing the merits and profit capacity of metals for at least the time *they* expect to be around. Dozens of subsectors and market niches have been devastated by new plastics technology from both domestic and foreign sources. Yet the advocates still take a defensive posture in discussions about the future.

The planning atmosphere can be made safer by the behavior and attitudes of top managers, who are even willing to abandon their own pet positions or allow them to be challenged. Mechanisms for such planning are much less of a problem. Techniques and devices for planning have become very abundant and some are very sophisticated and even "user friendly." The key is establishing and maintaining planning atmospheres that allow for and encourage responsible speculations and forecasts. Some false starts and even planning fiascos may be required to establish such an atmosphere, but it is necessary to establish one.

2. *Self-Control to Prevent Overreaction to New Product and Market Ideas*
   (a) Clear differentiation between possibilities, likelihoods, and sure things.
   (b) Preparing the ground in the market but not getting customers overexpecting and then, possibly, disappointed.
   (c) Systematic and periodic evaluation and review of progress and likely introduction dates to alert sales, marketing, production, purchasing, and other functions.
   (d) A balance between excessive secrecy and control versus overpromising.
   (e) Contingency planning for schedule slippages and delays in R&D, production, or other areas.

Effective STP requires a balance between (1) willingness to go beyond what is known and what the participants feel comfortable with and (2) consideration of possible future events that may not happen or may not turn out as predicted or planned. Getting "downstream" groups (plants, customers) worked up to expect a radically new or improved product line or breakthrough in technology must be balanced against the effect of anger or disappointment if it does not happen. Planning for production and commercialization (see Chapter 8) does require that these other groups be alerted and even get directly involved early on in the planning of a major project or program, even a long-term one. But control of overreaction and unnecessary or prohibitively expensive contingency planning may be necessary to conserve resources and reduce the likelihood or intensity of disappointment.

A major protection against continued overpromising or overexpecting is systematic and periodic review of progress and forecasts involving reestimates of milestones for decision, action, and accomplishments. These reviews should not only be periodic, such as at budget time, but they should also be "event-driven," occurring at any time that a significant internal or external event or situation appears to threaten or put into doubt a plan or milestone.

Contingency planning does not mean a large number of possible paths that might be triggered by insignificant events. It does mean that the participants, including the downstream impactees of a major project or program, must be aware that conditions may change (after all, that is the nature of the technology game) and they should have some alternatives thought out that can be put into play as needed. An unintended consequence of some contingency planning is that it seems to provide *license* for slippage, delays, and missed targets. The atmosphere for contingency planning in connection with STP must make it clear that such usages are not legitimate—a promise is a promise, within the ability of the promiser to foresee and take account of future events.

Prompt signaling of changes and significant events can reduce a lot of the uncertainty and conflict that can accompany sloppy or "subverted" contingency planning (see Chapter 7 for discussion of this in relation to shorter-term projects and programs).

3. *Articulation Between Technology and Business Planning*
   (a) One of the most common weaknesses in technology planning.
   (b) Power relations affect the amount and timing of inputs and the influence of those inputs.
   (c) Shared experiences help—for example, R&D people with marketing and planning experience and planning people with R&D or technology experience (job rotation, career path, temporary assignment, mixed teams).
   (d) Planning in a continuing interactive and personal mode versus planning by documents.

Articulation between technology and business planning is the most critical factor in successful STP. The general recommendations given above—providing for shared experiences and planning in a continuing interactive and personal mode—will be worthless if the general atmosphere of trust and cooperation advocated earlier does not exist. If business planners, engaged in the firm's overall strategic planning, have no interest in including technology aspects except as afterthoughts or "add-ons," then little can be accomplished. This is especially true if the power of the business-oriented people is very dominant or if they have the "exclusive charter" for planning in the firm. "Don't you worry about where technology fits in. We'll develop the goals and plans for the company (division) and then show you where we need new technology, if any, to help meet them."

This lack of articulation and the attempts at "linear" planning (see item 4, below) are major causes of failure to develop and effectively use STP systems or even business planning itself. "We're planning on an x percent increase in sales, with a y percent margin, and an increase in z percent return on investment." "But where are those added volumes and cost reductions going to come from?" "Never mind, we'll all buckle down and cut out some more fat and come up with some good ideas for new products."

4. *No Clear Linear Sequence of Planning*
   (a) The start can be in R&D/technology or in planning—R&D as *pro*active rather than only *re*active.
   (b) The other parties must get into the cycle early.
   (c) The steps should be clear and understood, with a lot of feedback loops to account for changes in:
      • objectives (from top management),
      • external circumstances,

- technology opportunities,
- personnel,
- and dead ends and red herrings.

Effective STP has a sort of "linear envelope" within which the information-gathering, decisionmaking, estimating, and planning go on. But this is far from a linear, step-by-step process. Many aspects of a plan depend on other aspects that have not been recognized or analyzed yet. Others will be changed as more information is gathered, more analysis is done, and more decisions are made. Even the starting point is not rigorously defined. Although the formal meetings and documentation generally start somewhere—typically in the corporate or division headquarters—and "flow downward," the ideas for strategic technology planning can start anywhere.

Ideally, ideas for elements of the plan *will* originate in different places simultaneously or within the same time frame. The key is whether those ideas and the nascent plans they generate are based on the same set of planning and decision premises and can in fact be reconciled when the formal, documented parts of the plan are developed and promulgated.

If the business strategy, the technology policy, and the general decision premises used for looking ahead are well formulated and well recognized (not rigid and unchangeable), then multiple starting points may be very productive.

One continuous spin-off from a good planning process, for technology as well as general business planning, is this very opportunity to reexamine the policies and premises on which the firm is basing its future. Back to the first point: it is imperative that the people and groups involved in planning are willing to share information, speculate about the future, and question assumptions, decision premises, and policies as they go along. "I thought we had a policy against getting into that business." "Well, things have changed a little since last year and it has its attractions." Or: "Why are you wasting time developing a long-range technology plan for that division; we will probably be divesting it in the next year or two." (This situation arose during the writing of this section in a company with which we have been closely involved.) Whatever the starting point(s) of the technology planning process, the key players should be involved early on, so that their inputs can be examined in the context of ideas and potential programs. Marketing or production, for example, may not have the capability to commercialize and produce a new line of products within the planning horizon for that line. Provisions may have to be made in the plan for enhancing those capabilities or the program must be put off or dropped from the planning portfolio.

Allowing for feedback in the planning and plan-execution processes can provide the flexibility needed for changes in course, early abandonment where indicated, and avoidance of the "over-the-cliff" type of persistence (stubbornness?) often encountered in rigid planning situations.

5. *Weakness of Market Research*
   (a) Generally focused on current and definitely forthcoming products and product lines.

(b) Techniques developed from traditional consumer marketing—not much industrial market research taught in most business schools.

(c) Most techniques tied to time series data on same or similar product lines.

(d) Market research people typically from sales or marketing and are current-product focused.

(e) With exceptions, such people are generally not technical—little exposure to technology.

Most of the weaknesses listed above are fairly common throughout industry. They stem from the training received by many market research people, the low status of market research versus actual selling, and the lack of credibility of many long-range market predictions relative to new products, from "It won't fly, Orville" and "The telephone is just a toy" to "Plastics will never replace our products." The list above does not contain any specific recommendations, except the implied one: if you want market research to contribute significantly to technology planning beyond the next model year or the next minor improvement, then "fix it."

Informally, many technology groups have "made" their own market research for radical new products, rather than "buying" it from corporate or divisional staff groups. Many of them have also stumbled in trying to estimate the demand, price, appropriate niche, and features of products the firm or industry has never made before. However, there have been successes. Many of the radically new products in computers, communication, biotechnology, instrumentation, and others have been the result of careful and deep probing of the need and opportunities for new technology by technical people. Often this probing has been far afield from the customary markets and technologies of the innovating firms. Technical people, including members of CRLs and technoeconomic forecasting groups, have used all the technical and behavioral information they could find to scope out the market for a radically new product, material, or service. Frequently, they have been wrong. But they have also hit enough times to suggest that market research in support of radically new and long-term technology projects might be better off led and/or staffed by technical people who are able to take a fresh look at potential markets without the bias of past failures and successes. Combinations of traditional marketing people and technical people might even be better, if the open and cooperative atmosphere advocated here is attainable in the firm.

To keep from operating "blind" in potential future markets, the CEO, DMs, and their staffs would do well to place a lot of emphasis on building up a true market research capability for radically new products, in support of their technology planning.

6. *Frequent Poor Relation Between Technology and Financial Planning*

(a) Too often an adversary relation.

(b) Too much "bean counting" by financial people.

(c) Mutual distrust can distort the planning procedure—for example, mutual stereotypes of "bankers" and "mad scientists with no financial understanding or responsibility."

Whatever the level of discomfort and mistrust between marketing and technology people, it is seldom as high as between financial and technology people. Researchers look at accounting and financial people as "bean counters" with little imagination and financial people consider most researchers "wastrels" who are careless with the company's money and who are innocent of bottom-line considerations. Both stereotypes are probably correct to some degree, at least from each other's perspective.

If strategic technology planning and the fruits of that process—new products and processes of a radical nature—are to be effective, the stereotypes must be broken or surmounted. Unlike marketing and production, which share some functional activities with technology groups, financial and accounting people operate at a different level of abstraction (necessarily so perhaps) and are not notable for interfacing easily. Much of this difference in perspective results from differences in training, experience, personality, and decision premises. That is the hard core of the stereotypes and the sources of conflict. However, an example of bridging this gap in the arena of long-term technology planning suggests that it is not a hopeless situation.

In a large electrical/electronic company we have worked with closely over a number of years, for example, an accommodation has been reached. The usual method of trying to place engineers with financial training and experience in or as liaison with R&D people was not the solution, although there are some promising examples of that approach. Instead, a bona fide, full-fledged financial person was assigned to each major technology lab and became the "business manager" of that lab. This did not happen suddenly. It took time for language, viewpoints, values, and styles to be reconciled and for a close working relation to develop. After several years, the blending has reached a point where it is hard to tell which viewpoints and professions are being represented in strategic technology planning sessions. Like the situation between technology and market research (or marketing in general for that matter) the champions of the STP process must look closely at the finance-technology interface and, if it is causing difficulties in the planning process, fix it.

7. General Dominance of the Cash Flow Concept

   (a) Cash flow may not be an appropriate decision criterion for some high-tech products.

   (b) Cash flow, return on investment or assets used, and other measures of performance have to be compared for each project.

When making estimates of the total costs of a long-term technology project or program, many scientists and engineers without business experience will overlook the need for working capital to get the new business launched and sustain it until it begins to break even. They also tend to overlook the cash flow characteristics of a project or business and thus focus the "return" side of the evaluation equation (see Chapter 7) on a single measure such as return on investment. They are aided and abetted in that viewpoint by planners and top managers who tend to focus primarily on one such measure.

"Money men," on the other hand, tend to overemphasize cash flow in the spirit of keeping the business afloat and paying a return to its backers and, as a result, often tend to "waste or consume assets" by declining to feed enough cash back to replace and renew them. A narrow or one-sided view of the economics of a technology project can cause it to fall between the cracks in the planning process or endow it with a "figure of merit" it does not really deserve. One of the first tasks required for the kind of integrated technology and business planning discussed earlier is to get the financial terms of reference out on the table and agreed on. The definitions and implications of such terms as "cash flow" and "working capital" and "pay out" need to be disclosed and agreed on, or the various parties to the plan may carry, for the life of the project, different views of its financial merits. In this discussion, the possibility should be considered and planned for that some projects in the plan may involve very high risk and will turn into long-term programs. Some may have to be abandoned and a "sunk cost" approach to them agreed on.

A recent experience with the CRL director of a multi-billion-dollar company illustrates the need for clarifying the cash flow and ROI kinds of measures of success for a project. The CRL director, charged with "putting the firm into totally new technical and market areas," had developed his own financial targets for new programs. He wanted to reach a sales volume of a certain amount within a certain number of years and to attain a return on investment a bit above the firm's current ROI. Upon analysis, it appeared evident that even if his program had succeeded technically and commercially (which it finally did not) it would not be viewed as successful by the top management and by the financial people in the firm's corporate office. They were looking at new technology-based programs like this in terms of adding "cents per share" to corporate earnings. The volume levels he was planning for were "very nice" in terms of the new field and markets themselves, but the levels fell far short of what the firm wanted and needed in terms of significant increments to overall corporate revenue and earnings.

A further confusion in the choice of financial ratios or figures of merit concerns the nature of the technology itself. Many firms have traditionally made very heavy investments in plant and equipment where the rate of change of manufacturing technology has not been very great historically. However, when trying to break into new technologies, especially some sectors of computers, electronics, biotechnology, and communication, they find that the actual capital outlay for facilities and production equipment is not as high as they are used to, especially if they already have some product lines in those fields. As a result, and as a result of thin margins, long introduction and start-up times, and short product life cycles, they are finding that such product lines do not generate the kinds of cash flow they are used to. In the end, if successful, forays into such new high-tech fields may yield cash flow, primarily in terms of earnings. But the steady pattern of the depreciation component of cash flow is no longer available and the sporadic pattern of earnings in some of these fields further reduces their ability to forecast and count on the cash flow.

One further implication of this difference in cash flow pattern is the difficulty of "milking" such an investment by withholding a substantial part of the cash flow for other corporate uses. High-tech programs may not require as much "hard" capital investment, but they certainly chew up a huge amount of working capital at the front end (e.g., R&D, engineering, software, market introduction) as well as software maintenance and service downstream.

8. *Need of Top Managers and "Businessmen" in Firm for Help from Innovation/R&D*

   (a) Early warning of threats, needs, and opportunities (TNOs).

   (b) General credible technological guidance and advice.

   (c) Scenarios (also credible) on future markets, technology trends, competitors' actions, materials and energy prices and supply.

   (d) If internal R&D/technology cannot or will not supply this kind of information and advice, they will go outside for it.

   (e) R&D/technology should be prepared with this kind of information in case they are asked.

   (f) Credibility takes time to build. It is based on a good technical track record, high responsiveness to managers' questions, and willingness to take technical risks.

   (g) Information seekers may forego the *best* technical source for one that is readily *accessible* and *easy to deal with*.

Despite the jaundiced view and reluctance with which some top management and, especially, some division management support technology, it is generally critical to the survival and well-being of their operations. "Good" technology groups at corporate and divisional levels can serve in many ways to protect and enhance the firm's product lines, manufacturing methods, relations with its customers, and competitive posture. They can also keep the firm out of troubles arising from technology itself (pollution, energy consumption, safety, regulation) and arising from lack of adequate technology. The key to supporting top management with technology advice, information, and "fixes" is that they be delivered in an "appropriate" manner. In this case, appropriate means that the technology people have to be able to anticipate certain needs and have the necessary information or problem solutions "on the shelf" or close enough so that they can deliver it when needed and in a form that is understandable, credible, and usable by top managers.

The "elegant" technical solution to a problem or the "brilliant and complete" response to a request for information may backfire if it does not meet these criteria. One fast way for the technical staff to build credibility is to be ready and to deliver such solutions and information when and as needed. One way to lose credibility fast and perhaps permanently is to leave the CEO in a bind when he does need and ask for help. Perhaps some top managers will forgive one or two incidents of mismatch. But a perceived pattern that their technology people cannot handle simple or urgent requests for information and technical assistance can cause a permanent turning away from dependence on in-house technology and seeking for help outside. The wise CTO and his staff, as well as the managers of divisional labs, will be aware of this fact of corporate life and will develop and keep sharply honed a "quick-response capability" which can not only respond in linear sequence to questions and requests but can also anticipate and preempt such situations that require quick and simple inputs to top management.

"Well, what is energy costing us in our various operations?" "We don't have that figure at the moment, but we can do a study that...." "What's the cost structure of our competitor?" "We're not sure, but we can do a study...." "What's the impact of this new regulation likely to be?" "We haven't done that analysis yet, but we can...." As indicated earlier, in Chapter 3 on the viewpoints of the key players in the firm's technology game, top managers—corporate and divisional—tend to look on their R&D labs and other technology groups as part of their own support systems for addressing and dealing with technology issues. A lab or technology group that is not sensitive to concerns of the "boss" (even if he keeps them secret), not responsive to his needs, and apparently not really interested in the "business" side of his business is not really pulling its weight. This may be his perception even though they are "doing good things" in terms of developing and improving the firm's products and manufacturing processes.

Figure 5-2 presents the results of a quick "organizational design" exercise by a group of technology management people[*] aimed at developing an approach that would increase the credibility of R&D in the eyes of top management.

The CTO and his staff should spend some time on such exercises to diagnose whether in fact their credibility is low and to develop methods of increasing it and keeping it up. The CEO and his staff should give the technology people a fair opportunity to establish credibility as a critical significant part of the firm's policy-formulation and decisionmaking system. Giving enough time, where feasible, to respond to questions and spending some of their own time in increasing their

---

[*] In one of our seminars on R&D/technology management.

Design objective

Getting management and R&D
to play to the same music

Design features (manipulable)

Communication (periodic,
 multilevel, with feedback)
Mutual respect
Specific R&D planning
 with sensitivity analysis

Criteria (what we want to accomplish)

Establish and maintain consensus
between management and R&D on
goals and timing of R&D projects

Parameters (constraints)

Culture/Background
Organization/Hierarchy

Figure 5-2. An organizational design exercise.

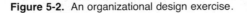

understanding of the responses can help in this direction. A mutual educational process can help make the technology planning process better by improving the channels for communication between managers and technical people. If credibility has been established by a sequence of cooperative acts (e.g., bailing the firm out of difficulties or responding quickly to opportunities), then the longer-term planning process has a firmer foundation than it might have had without such mutually reenforcing events.

9. *Heavy Involvement of R&D in Planning for Manufacturing Technology*

(a) All plans for new products should include feasibility and contingency plans for manufacturing to support them.

(b) Front-end R&D planning should not assume that somehow manufacturing will be able to produce the new products.

More will be said about the interfaces between R&D and production in Chapter 8 on commercialization and the flow of technology through its early life cycle on the way to plant and market. Procedures for specific projects and short-term technology transfers apply to longer-term technology planning as well. Surprises and uncertainty should be reduced as much as possible. Production people do not like surprises and the kind of uncertainty with which R&D people are used to dealing. They can become very unresponsive and unforthcoming when their workload and day-to-day problems are complicated by large and sudden demands on their time and attention, relative to new and unfamiliar technology. Many production managers will say that they prefer not to have to worry about long-term plans that could involve major changes in their manufacturing methods, equipment, materials, and people. They say they would rather let the R&D people worry about them and sort out the ones that *really* will make it all the way to market and *then* they will get involved. Do not believe it or accept that expressed attitude at face value!

Planning for transition to production involves many component activities, some with very long lead times and some involving long queues for attention and service. For example, a member of our research group recently completed a dissertation on the role of the pilot plant in transition from R&D to production for a large number of chemical projects. Not only does the pilot plant constitute a "critical path" in the life cycle of a project, but it can *add* uncertainties about time and cost of a project, as well as reducing them.

Production people or at least top production managers should be brought in and made part of the technology planning process early on to provide inputs about feasibility, costs, and alternative approaches to manufacturing as well as to prepare them for possible projects headed their way.

10. *Potential Conflict Between Company Goals and Those of Individual Managers*

(a) Clarification of the goals of individual managers is part of the planning process.

(b) Where conflicts exist or arise between corporate and divisional or other sets of goals, they must be dealt with early on, or effective planning can be difficult if not infeasible.

All or most managers in the firm will espouse the philosophical and general goals of the firm—growth, profitability, and enhancement of stockholders' value. When it comes to the next level of such goals—the actual *measures* of those concepts of growth and so on, there are frequently wide differences of viewpoint. Furthermore, when *means of achieving* these goals are being discussed and planned for, even wider differences appear.

An industrial firm is not a "democracy" in the sense of one person, one vote or cooperation by plebiscite. However, the very strength and resiliency of many successful firms is based on a diversity of views that are somehow reconciled into policies and courses of action. Where the differences in ends and means are not even recognized or allowed to surface with respect to technology and its impacts, STP can be a very empty and self-defeating enterprise.

We have been involved in a number of corporate technology planning exercises where it incorrectly was assumed that "everyone, including the division managers, were on board" with respect to the direction in which the CEO wanted to head. Elaborate plans for diversification, increased investment in technology, improvement of quality and reliability of products, and renewal of manufacturing facilities were developed and, apparently, "agreed to." Several years later, however, these plans have not been pursued, let alone fulfilled.

Analysis of what happened in several such cases yielded clear evidence that everyone was *not* "on board" and that significant differences of viewpoint on ends and/or means derailed the technology plans before they went very far.

Probing the individual goals and ambitions of the key players in the technology planning process can be very painful and frustrating for the individuals involved as well as for the planning group as a whole. Back to the first point in this section— "willingness to share information and speculate about the future" and the "safe" atmosphere that requires. Where the information to be shared involves individual hopes and dreams, it can be a difficult sharing process indeed, but an important one if it can be done within the context of the firm's structure. When sitting in on such meetings where there are obvious sharp differences below the surface, we are reminded of an exercise used by people in human relations training. It involves comic strips with the "balloons" overhead representing both the spoken words and unspoken thoughts of the cartoon characters. The training exercise requires the participants to fill in the other person's balloons!

If that technique or some like it were used and if the participants were "open and honest," differences in means and ends would soon become apparent. In some cases one individual, for example, the CEO, might attribute incorrect motives and perceptions to his colleagues in, say, marketing, finance, or production. But such a process can help to surface real, as well as incorrectly perceived, differences and so place the planning process on a firmer foundation. Of course, where there are irreconcilable differences, disclosing them may lead to more conflict than when they are just assumed—"if you don't say it, I don't have to react to it."

But somehow differences in perceptions of means and ends for technology planning have to be aired to a much greater extent than they are in the normal planning process in most firms. Procedures and the associated safeguards should be developed as integral parts of the planning process. Some progress has been made in this direction via simulations and games, which are used in training students for management. However, their use in "real" situations where much more is at stake—careers and power positions—need much further development and experimentation.

11. *The Implementors of Technology Plans Are Line Managers, Not Staff Analysts*

(a) Line managers must be involved in STP.

(b) Staff planners must know and be able to represent the goals and viewpoints of the line managers or plans will remain on the shelf or worse.

Ultimately, the policies, decisions, and actions of line managers such as the CEO and division managers (DMs) determine whether a particular technology plan or set of plans will be carried out.

Elaborate plans to enter new fields and new markets or exploit new fields of technology for manufacturing will depend on a series of "Simon says" interactions.

Having a plan to initiate or expand a new technology program does not mean that it will be funded, staffed, and otherwise encouraged or supported on a continuing basis. Even where a specific plan component was developed at the special request of a division manager, for example, he may not be ready or interested in pursuing it when the time for resource allocation and the definitive "go" decision needs to be made. This means that he should not be subjected to surprise requests for support on thrusts or programs that he has since forgotten about or changed his mind about, "even though his signature is there on the memo requesting the program." This means that implementation of the plan requires the *continual* involvement and, if possible, the reaffirming agreement of the executive responsible for providing the resources and the support for implementation of the plans.

This is not necessarily easy or straightforward. Some line executives, as mentioned earlier in several contexts, are not inclined to think about the long term too seriously and just about tolerate having their staff people do the necessary planning to satisfy corporate management requests or demands for formal plans. If staff planners, in spite of these possible barriers, do not have the support of the ultimate funders and implementors of the plans, the whole exercise is likely to be futile.

12. *Future Actions, Plans, and Events Must Be Tied to Changes in Company Goals, Criteria, Characteristics, and Capabilities*

(a) The planning premises must be open to continuing review.

(b) Planning should be an organic versus a mechanical process, accounting for the changes that occur in the organization—for example,

• market posture,

• market sectors,

- technology,
- labor force,
- management goals,
- new entries,
- financial position—long and short term.

It is very easy for staff planners, including those working on strategic technology plans, to be carried away with what they think is a mandate to make plans as they see fit. Of course, the sensible staff planner will try, at the outset, to determine the ground rules on which to base his plans and the resources and constraints that are likely to be present when the time comes for implementation of the plans. He may develop comprehensive plans for a thrust into a new technology or market, based on what he understands are the targets the firm has set in its overall business plans. He may, in good faith, assume that these initial conditions (goals, targets, resources to be committed, steadfastness of purpose) will persist at least through the planning period and ideally beyond that into the actual operating stages of the plan. He is often wrong in this assumption. External and internal conditions (profitability, cash flow, personnel) can change continuously or sporadically and the "initial conditions" are no longer valid. If the planning process is allowed to proceed under its own momentum and is insulated from such changes, their outcomes may be very inappropriate. Does this mean that long-term, strategic technology planning can and should be buffeted about by every change in the firm's environment or every change in attitude or perception that occurs in the minds of line managers? Definitely not. By definition, such planning should go beyond the momentary influences that make it difficult even to think about the long-term future, let alone deal with it in any detail. However, plans that are stubbornly adhered to when there is no longer in existence the threat, need, or opportunity which gave rise to them, or those to which they were originally addressed have changed significantly, may fall very flat indeed. We have observed a number of excellent plans for market penetration being pursued through advanced technology when the market itself has dried up or taken a turn that makes the plan obsolete even before it is put into operation.

The premises underlying STP must be continually open for review. Such premises or elements of "technology policy" as these must be continually subject to review and, where necessary, revision:

- A preference for "make" over "buy"—that is, doing in-house R&D versus contracting out or buying packaged technology (or the reverse).
- Being first in the market with a new product based on new technology.
- Being willing to try any new manufacturing approach (within reason) to support a new product line and, perhaps as a consequence, writing off substantial investments still functioning in existing plant and equipment.
- Granting early retirement to groups of skilled technologists in R&D, engineering, and production in order to reduce head count and staff costs. (See the

extended discussion of this in Section 8.2.5 on imbedded technology capability.)

- Contracting out all tooling and special production equipment, as well as "catalog" items.
- Going all out to support new products and technologies with customer service and warranties.
- The organizational structure and deployment of R&D in the firm, especially under conditions of merger, acquisition, or divestiture.

The economics of a new thrust in technology may turn out to be based on false or obsolete premises, such as some of the above might become as conditions change in the firm, the industry, and the economy at large. In that case, a whole program or the plan itself may lose credibility and be shelved or thrown out, along with some of the staff planners. Unfortunately, in some sectors and in industry as a whole, the pace of change in the 1980s and beyond makes it unlikely that a rigid and static set of planning premises will hold up for long or be a valid continuing basis for STP.[*]

STP, like other modes of long-range planning, requires an "organic" process that is open to and can accommodate changes even while planning documents are being typed and distributed, if necessary. This is not to be confused with drifting in the wind, when even a slight breeze of change comes up. Indeed, one of the main reasons for having a long-term strategic planning system for technology is to prevent the firm from persisting along a path to obsolescence or one of decreased competitiveness. If the plans are sufficiently well-founded and significant changes that are likely to occur in the period affected by the plan (and affecting it) are taken into account, then there is a better chance that the plan's component programs and projects will remain robust and not be thrown off course by minor shifts either within or outside the firm.

If specific technology plans and the planning process itself are too subject to direction change, then a desirable characteristic of planning—flexibility—may actually become a very undesirable one—flaccidity—where there is no constant direction or sustained thrust or underlying and defensible rationale.

13. *Managers Differ Widely in Their Willingness to "Play the Planning Game"*
    (a) Staff planners should choose their planning "partners" carefully to get the planning process started.
    (b) There is no point in forcing an operating (e.g., division or SBU) manager who is currently in trouble to do long-range strategic technology planning (STP).
    (c) Some managers are "natural" planners; others are nonplanners.

---

[*]While proofreading this section, I was shocked to learn that a large company I have been close to for many years suddenly shelved its exciting long-term technology and diversification plans, cut R&D by almost half, and accepted the resignation of its CTO—all in the face of a potential takeover threat.

One common mistake we have observed in the attempt to get STP started has been the CEO's and planning executives' insistence that "all managers must get into a planning mode." This is a laudable goal but may not be too practical, given the wide range of personalities and situations of the division managers in a typical company. Some of these DMs have a natural flair for planning. Indeed, even though they make disparaging remarks about the formal trappings of planning foisted on them by staff analysts, they are continually looking into the future and making their own plans for technology as well as marketing and other phases of the business. The style and time horizon with which some of these DMs do this may offend some orthodox planners and may cause them to believe that such DMs are recalcitrant and not willing to look to the future "in the proper way."

Other DMs and many CEOs and other top managers are inherently "short-termers" in their outlook and sincerely believe that if they do well today, the future will take care of itself. This perception is common among many entrepreneurs and businessmen who firmly believe in the "centipede" story—that if you probe too deeply into *what* you are doing, *how* you are doing it, and *why* you are doing it, you may lose your ability to walk. Preoccupation with the future, they hold, can paralyze you in the present. The optimism we have encountered among many such people, especially those whose executive "training" was in the school of hard knocks rather than in formal management programs, is born of a belief system that contains a lot of self-confidence. They believe that their strength is in dealing with uncertainty and problems *when they arise* and not in advance—that is too energy-draining. (See the section below about "methodical planner" or "anxiety-driven worrier.") Attempting to force such people into a formal planning mode too soon in the development of a company-wide STP process can be very unfruitful. If these are successful "businessmen" in the company and they have the backing of top management, they can easily stonewall and block the whole planning process. If they are *not* successful, in fact on the verge of failing, they are unlikely to have the time or the stomach for facing the long-term, especially for something as esoteric as technology.

So where should the STP process start? The most promising mode is to identify and approach a few or perhaps even one moderately successful DM who has exhibited signs of a "planning mentality," even if it is idiosyncratic and "sloppy." The staff planners should try to identify weak spots in his technology planning and come up with ideas to improve them. That is, they should do their homework on his likely future threats, needs, and opportunities (the TNO referred to in many places in this book). If this individual DM or group of DMs recognizes that STP can provide them with additional leverage on their markets, competitors, technology, costs, and position vis-à-vis other DMs in the company, they might buy in.

Going after one or more "natural planners" and forming a common cause can be the most effective way of penetrating general management resistance to STP. A head-on approach to the "hard cases" among DMs and other managers has very high failure potential. Pressure from the CEO may force nominal or superficial compliance with STP procedures, but the essence may be lost and the cause set back for some time when it fails. There is also the distinct possibility that staff planners can learn something from natural planners and incorporate those lessons in their planning formats and procedures. We have observed staff planners rejecting techniques, ideas, and mechanisms currently used by individual managers as "not fitting into the master plan," "too idiosyncratic," "not elegant enough," and "not according to the academic framework and catechisms for long-range planning."

Another variation among DMs and other executives which can influence their willingness to engage in the planning game is their chronological age or their "personal time horizon." On the one hand, a DM on the verge of automatic or forced retirement, with no prospects for further advancement in the company, is not likely to be a good candidate to plan for the long-term future of his division. Except, of course, if he was the founder of the division or the entrepreneur who entered the firm through sale of his independent company to the firm, with the arrangement that he continue as division manager. In such cases, his pride in the business may make him an excellent candidate for "chief planner" for the future of "his" division. Some perceptive CEOs have recognized this potential and, a couple of years before the DM's retirement, have relieved him of operating responsibilities and asked him to spend time on planning the future of his division.

The response of some of these people is astounding. They sometimes become more analytical and academic than the staff planners and their university consultants. They probe, plan, challenge assumptions, look to the side and over the horizon, and do all the things that a planner should. Such potential planners may be less rare than many people think.

After a career of working in the short term and keeping their noses to the grindstone, some entrepreneurs/businessmen welcome the chance to wax a bit philosophical and conceptual. We have worked with a number of such "changelings" with a great deal of pleasure and gratifying results. Of course, the thinking habits of a lifetime are not easily put aside and such preretirement managers may just not be able to overcome their constrained view of the world that has ruled or guided them throughout their careers. So be it. If there are some that *can* do it, identify them and try them out.

The other side of the argument is illustrated by the former line executives, weary from the day-to-day battles they have waged and cynical about the future, who are *forced* into corporate staff planning positions. Such people, tired in body and spirit, may actually inhibit the planning process and squelch the eager, young, formally trained staff analysts whose inputs are needed to make the planning process more than just a conceptual exercise.

The point is that there are wide differences among managers in most firms, with respect to vision, time horizon, flexibility of thinking, and taste for the planning game. Careful scouting and selection of those most compatible with STP can pay off handsomely.

14. *Planning Requires a Marketing Viewpoint vis-à-vis Internal "Clients"*
   (a) Who are R&D's clients?
      • They "own" the problem.
      • They can act—that is, take decisions or implement plans.

(b) What are their needs, constraints, demands, biases, style of operation, and "rhythm"?

(c) The (potential) internal client needs to be "market-researched" and marketed to on plans, ideas, suggestions, information, and proposals.

As the previous section suggests, staff planners have to get to know their clients—the managers who "own" the problems associated with planning and have the resources and power to do something about them. A market research approach to clients is essential, not just for the initial stage of "getting them involved" in STP but also for achieving significant results from the planning process itself.

For the planning group in a very large company with a very large number of operating divisions, this may seem like an impossible task. It probably is and should not be undertaken directly by the, typically small, corporate planning staff or the staff of the CTO. The way to a division manager's heart, or at least to his real commitment to the STP process, is through *his own* technology and planning people. If these divisional staff people cannot be brought on board the STP process, then it is less likely that their boss—the DM—can.

These "client surrogates"—people selected and maintained by the DM to help him—also have values, styles, ambitions, fears, and biases that can contribute to or detract from the planning process. Developers of a corporate-wide STP have to "market research" these divisional staff people also. They, in turn, have to sell planning to their DM and his other managers. Often, the cooperation or initiative of corporate staff planners, with the explicit or implicit backing of the CEO and top management—for example, group executives—can strengthen the resolve and efforts of divisional staff and help in that selling. The unfortunate adversary relation we have observed between many division and corporate planners can hurt both their efforts at preparing for the division's technological future. The history of long-range planning in large corporations suggests a very short "positional life" for professional staff planners. They either go "up or out" in periods like 2–3 years or even less. They have a brief window of opportunity in which to get STP started and to see that it is on a sound enough basis for its continuance. Good use of internal market research on their actual and potential clients can help provide that sound basis. Ignoring the differences in needs, constraints, demands, biases, style of operation, and "rhythm" of their different clients can spell failure from the start.

15. *The Regulation and Economic/Social Environments Are Integral Parts of the Strategic Technology Plan, Not Merely Constraints on It*

(a) One planner's constraints are another's opportunities.

(b) Cost savings have been stimulated by environmental constraints and other regulations.

One of the wonderful qualities of a good company technology function is its ability to respond to crises and apparent adversity with new opportunities. The energy crises, pollution regulations, foreign competition, and product liability were initially viewed as unmitigated disasters by some firms. Others, however, have turned them into opportunities for improving their products, reducing costs, identifying new product opportunities and generally revitalizing their technology efforts. Some of these advantages have been accomplished in the short term as the result of knee-jerk responses to outside influences. Other longer-term and sustainable benefits have accrued in firms where such factors are incorporated into the strategic technology planning and "made to happen" in a systematic, predictable way. The flurry of early responses to rising energy costs and shortages yielded some quick-fix savings—"turn out lights you are not using." But more substantial savings and opportunities were soon perceived in some companies whose technology planners could look beyond the immediate crisis and see how this stimulus could lead to basic rethinking of product and process design, facilities planning, and the general acquisition and use of energy.

Pollution control had less obvious benefits in some industries, where requirements for installation of control equipment resulted in a major capital outlay with no apparent payback. Again, however, some firms used that opportunity to identify places in their manufacturing systems where pollution could be curbed with net savings in costs through more careful handling of pollutants, recycling of water and materials, and other means.

Product liability, that huge monster that haunts many companies and their lawyers, can also have a bright side. Accidents are frequently caused or exacerbated by poor design, manufacture, and/or maintenance. Careful analysis and long-term technology planning that goes beyond the present production run or model year can lead to redesign or totally new designs that will not only reduce liability exposure but also provide new competitive advantages to the company's products.

Is this looking on the bright side of a mostly negative situation? Sure it is, and effective STP will incorporate such optimism and an open-minded approach to finding longer-term and cost-effective solutions, rather than merely sticking fingers in the dyke. Every STP for a division, a product line, and SBU or the company as a whole should have to withstand external reviews and criticisms that raise "what if" type questions about the market and the environment in general. "What if the clean water or toxic wastes laws are strengthened?" "What if product liability suits and awards escalate?" "What if the costs and availability of energy once again hit you, your suppliers, and customers?" Long-range strategic technology plans that are not able to address such questions with well-thought-out contingency plans are incomplete and likely to be faulty in other areas too.

16. *The Target Periods (Time Horizons) for Planning Must Fit with the Character of the Decisionmakers, the Company, the Industry, and the Environment*
    (a) Planning horizons should not violate the natural time constants and rhythm of the organization.
    (b) STP targets must be consonant with corporate financial planning periods (e.g., related to stock offerings and long-term debt flotation).

Planning targets that violate the natural rhythm of the organization—division or the whole company—can lead to failure. In the 1960s we encountered extremes of planning horizons. Some firms in the auto and other industries would not approve of capital investment, technology projects, or other types of project whose time horizon—the time to payoff—exceeded 18–24 months. One appliance company had a 25-year plan for revolutionizing the very concept of the roles of appliances in the home. And a major computer company had a 40-year view of the potential opportunities in the general field of "information transfer."

Certainly, the corporate planning horizon is or should be longer than that of most of its SBUs or product divisions. The methods of control and evaluation and the many other pressures on the division manager assure that. However, the time horizon for a given component of the strategic technology plan also has to be consistent with the rate of advance in the technical fields on which it depends. The switch from mechanical and hydraulic controls to electronic caught some firms by surprise. They were not watching out for the rapid advances in cost reduction and operating performance by electronic components and systems. Many metal-bending companies have been overtaken by the advances they failed to take account of in the nonmetals material fields—for example, plastics and composites.

Above all, however, feasible plans must be consonant with the traditional existing, and likely to persist, financial time periods for which management is continually planning—cash flow, borrowing, debt retirement, stock issues, capital investment portfolio, and so on. We have frequently encountered "completed" technology projects that die before reaching plant or market because their timing was wrong in terms of the availability and cost of funds. "If you had had that new product (or new plant design) ready 2 years ago, there would have been plenty of money to commercialize or build it." "Can you keep those prototypes on ice for a while until the money situation gets better and our backlog of capital projects thins out a little?" Although early stages of technology programs that are parts of the STP are not very amenable to precise cash flow and investment milestones, flexibility and contingency time estimates are needed to meet the changing character of company fortunes and the investment queue.

"Do now or die" arguments in favor of commercializing or exploiting the results of a technology project "now" sometimes work, in the presence of convincing cost–benefit calculations and a hard sell. However, a succession of such forced decisions by top managers may lead to future unwillingness to be panicked into precipitate action: "Why don't you plan better, so we could fit this project into our financial and business plans."

This again is an argument for "living" technology plans that are constantly under review and that do not sit on the shelf becoming obsolete. Milestones, time horizons, and contingency plans need to be continually reviewed and updated as internal and external conditions change, including progress on the projects themselves.

17. *Strategic Business Units (SBUs) May or May Not Be Logical Elements for Long-Range R&D/Innovation Planning*
    (a) By definition, most are constrained in scope—little "side" vision.
    (b) Many are also constrained in time horizon by P&L requirements in the short run.

(c) They seldom have a long-range technology outlook.

(d) They may be the best planning unit for product line maintenance and minor improvements.

Here is where technology planning requirements clash with conventional wisdom and the inexorable tide of corporate organization change over the past two to three decades. There is no question that strategic business units (SBUs) have become basic building blocks of many modern decentralized companies and have served them well in the marketplace. SBUs have succeeded in focusing the attention of small groups of managers on very specific, in many cases, "niche" markets and enabled some giant firms to compete with smaller independent firms in terms of flexibility, service, pricing, and other aspects of competition.

Coupled with the large firm's financial resources and technical power (sometimes through its CRLs and divisional labs) they have helped in product differentiation, product improvement, and cost reduction to the dismay of the large firm's smaller competitors who lack such corporate resources.

An SBU may be the best planning unit for the short and, perhaps, intermediate term for product line maintenance and minor improvements in products, manufacturing methods, and cost structure. However, the very characteristics that have made many SBUs howling successes in their markets have sown the seeds for ultimate decline in their competitive position. Many of them are run by the closest things some large corporations have to "real businessmen" or "entrepreneurs"— even more so than some of their division management bosses. As part of that "package," however, many of them are overfocused on a very narrow product line or market niche and are overly constrained in their ability to diversify and move "sideways" in products or markets. This is especially true if the firm has many SBUs in closely similar fields and market niches.

The familiar adherence to or forced restriction to "big green ones" and not "medium-size blue ones" keeps their planning arena very small and very narrow. We have seen many potentially great ideas for new and radically improved products squelched by the directive that "that market/product line belongs to another SBU— stick to your own knitting."

Add to this the short tenure of many SBU managers—fast up or out—and it is clear that the SBU is not the best unit for strategic—long-term—technology planning. Another major factor is that few SBUs, unless they are very large— in effect, operating divisions—have their own technology capabilities and are dependent on the division or the corporate staff for their technology support. Often the queue in a divisional or corporate lab is very long and the future needs of specific SBUs may be very low down on the priority list.

18. *How Defensive Is the Posture of the Company's Technology Policy on Current Products and Business Areas?*

   (a) Technology planners need a clear statement of technology policy with which to work.

(b) Even current profitable products and product lines need to be included in the technology policy.

(c) Technology policies should indicate the defensive/aggressive nature of the technology program backing up products and business areas.

In Chapter 7 I present a set of categories that can be used to describe and analyze the strategy behind the firm's overall technology policy or technology policy related to particular products and business areas. A range of strategies is illustrated, from very conservative and defensive—for example, service to customers and minor improvements—to very radical and aggressive—the *future market mission*. I also present below a strategy for enhancing and capitalizing on even "mature" products and product lines through incremental technology projects that could prolong their life in the market and/or add to their profit potential even in the latter phases of their life cycles.

In order to include current operations—products, manufacturing processes, market niches, and business areas—in the STP, planners need clear statements of the intentions of the "owners" of these operations with respect to technology.

Sometimes these "statements" are indeed explicit policy formulations that guide the technology projects and programs supporting these operations. More often, at the divisional level, they are policies that must be inferred from the behavior of the division manager and his staff. The DM may be willing or even insistent on treating a particular product line or business area as a "pure cash cow," with no further investment devoted to it than absolutely necessary to meet current competitive threats and keep the business alive in the short term—for example, responding to customer complaints or meeting competitors' new features with superficial "dress-up" or "rechrome" features.

In such a case, the planners have no option, if they cannot change the DM's position, but to relegate the particular product line or business area to a defensive strategy that omits any new major technical initiatives. Butting their heads against a dug-in position like this will not win them any points from the DM and in fact can "contaminate" the rest of the plan in his eyes.

However, if the DM, in consonance with company-wide technology policy, is willing or even eager to include current products and product lines in his strategic technology plan, many things are possible. The STP process may open up the way for a new fresh look at the technology underlying a particular product, product line, or business area. Considering changes in technology which have occurred since the last examination, it may be possible to identify major new strategies for enhancing the ability of the product or product line to survive and thrive—extending its life and improving its cost–price structure. It may be possible to "rethink" a business area and see where new or available technology might strengthen the division's and firm's position.

Careful consideration of the potential impacts of further investment in technology for a particular item (e.g., product) may make it clear that it is indeed not cost effective to pump more funds in or that the technology does not allow for anything short of abandoning the product and starting over. Then, at least, clear technology strategy for the item will have been formulated.

If the DM wants to launch an aggressive technology program for an item that cannot sustain it, based on the current or likely future technological state of the art, then it is the duty of the planners to make that clear and to recommend a defensive or "harvesting" strategy.

Whichever way the technology planning for an item goes—defensive, aggressive, or some gradation between—STP should make these pieces of technology strategy part of the division's and, eventually, the firm's overall technology policy and plans.

For example, if the firm espouses a "first in the market" strategy for new, technology-based items, then a particular division's policies and practices with respect to that should be made clear to the planners as well as the executors of a technology plan. If "quality over output" is a *real* policy and not just a euphemism, then each division's policies and practices on quality should be made explicit in STP to identify potential conflicts and synergies.

19. *How Strongly Held Is the "Sunk Cost" Philosophy of the Company?*

(a) If there is a strong sunk cost philosophy, then it should be reflected in specific technology plans and decisions about projects and programs.

(b) There should be no recriminations from abandoned projects if the decisions to stop them were taken within a sunk cost philosophy.

It takes a lot of guts to advocate walking away from a project or folding a tent in which other people are urging continued investment. If the particular project/program involved big bucks and was championed by someone high up, there is likely to be a reluctance to say that "the emperor has no clothes and let's fold our tent and walk away from it."

A truly entrepreneurial viewpoint about technology projects will include the very real and very common possibility that a significant percentage of technology projects never will pay off or will not pay off in anything like the pattern originally anticipated. If such a viewpoint is embedded in the firm's technology policy(ies), then the planners—line and staff—have permission to think the unthinkable and recommend cutting the losses on "bad" projects well before they reach crisis proportions and make abandonment a necessary and obvious choice. In order to avoid wasteful drains of cash, energy, and other resources, such contingency abandonment plans should be built into the strategic plan for each major program and even for individual projects other than minor or routine ones—for example, the upper categories in Figure 7-28. These contingencies should be part of the milestones that should be part of planning for any level of technology—projects, project phases, programs, or the whole technology enterprise.

This is strong medicine in a field that is not known for its sophisticated and tough management style. Technology people and programs tend to drift on in search of the breakthrough or the unexpected success that will pull a failing or lagging project out of the hat. Effective STP, backed by a realistic and sunk cost philosophy that is honored in deed and not just in rhetoric, can strengthen the whole technology enterprise and keep its components alive—the promising ones, that is.

20. *How Strongly Is the Company Tied to Existing Operating Units as Vehicles for Launching Radically New Products and Technologies?*
    (a) What is the track record?
    (b) Existing operating units have many strengths, including stability and people experienced in working together.
    (c) They can ease the "reentry" problem for internal entrepreneurs and venture project people.

Some of our arguments in the sections of this book dealing with technical entrepreneurship in the firm and venture/entrepreneurial (V/E) projects in general (Chapters 6 and 8) may seem to imply that the existent divisional environment in most companies is not adequate to the task of launching and commercializing such projects. In many decentralized firms this is indeed the case. The philosophy and practice of decentralization may have been carried so far that division managers are unwilling or unable to do much about such "maverick" projects within their existing framework. The many constraints on such V/E projects, discussed in Section 8.4 on technical entrepreneurship and in Section 6.5 on technology acquisition via purchase of high-tech small companies, may be difficult if not impossible to overcome within specific divisions or even within the entire divisional structure of some firms.

The other side of this issue is that *some* operating divisions in *some* firms have been able to originate and successfully exploit *some* V/E projects. Obvious cases are very large divisions that are themselves equivalent to large diversified companies. Others are divisions in high-tech businesses with very large concentrations of scientists and engineers. Other cases may be associated with an entrepreneurial DM, a highly volatile market, competitive pressures, or a windfall technology opportunity.

The task for strategic technology planners is to identify these special cases and to differentiate them from the divisional situations that are not suitable for launching radically new technologies and products. As part of the planning process, the track record for originating and commercializing radically new technologies and products should be examined carefully as well as the likely future conditions that might change the division's willingness or ability to do so in the future.

Many firms that charged blindly ahead with setting up new divisions, task forces, V/E departments, and subsidiaries for new technology projects have found themselves with a lot of nascent, immature organizational units that do not have the resources and experience to pull off a complete life cycle of the project.

As argued elsewhere in the book, the environments of most operating divisions are not conducive to the origination and early nurturing—the front-end or research phase of V/E projects. However, when it comes to the downstream or commercialization stages, the situation may be entirely different. The free-wheeling, creative, entrepreneurial environment needed for the front end of such projects may be entirely unsuitable for the downstream stages of getting the products or processes to plant or market. Existing operating units typically have the stability, resources, and experience needed for commercializing and implementing technology projects. They may have strong teams of people who have been through the cycle many times and know how to deal with the uncertainties involved and the frustrations encountered. Many V/E teams are lacking fundamental knowledge of or experience with the painstaking steps required in getting a piece of new technology ready for production or market introduction. They often tend to cave in when things do not go right in what they had thought were the "mindless, routine, unchallenging" aspects of the project. They may find that the supporting services—for example, engineering, tooling, production planning, procurement, facilities planning, and training—are not willing to support these "new boys" who have been swashbuckling around. This choice between the nascent V/E project structure and the existing, stable divisional structure as a launching platform for radical new technology projects is a critical part of an STP and the component plans that make it up. (See Chapter 8 for more on this.)

Such deployment decisions for technology projects are often made on very ad hoc grounds—personality of the DM, political power of the various players, superficial similarities between a V/E project and a division's interests, top management's frustration or boredom with a V/E project and desire to get it off their plate, organizational slack in a division that is looking for another mission, and so on. These are critical decisions and should be part of the strategic technology plan—both for specific V/E projects coming up for deployment in the plan's time period and for such ventures in general. Assigning V/E projects to an operating division—perhaps with the potentially relevant divisions "bidding" for it—can also ease the re-entry problems for members of the V/E team by providing a more stable career path and the support they need to reintegrate with the main stream of the organization.

In Chapter 8, there is a discussion of alternatives to feeding new V/E projects into existing operating units where the project and unit just do not "fit" together in a cost-effective manner.

21. *How Conservative Is the Company's Diversification Strategy, If It Has One?*
    (a) A conservative diversification strategy keeps the firm's or division's one foot firmly planted in areas it knows well.
    (b) Radical or wild diversification strategies leave the firm exposed in areas with multidimensional threats and uncertainties.

One of the important roles for an STP is to guide and interpret the firm's diversification strategy, as it concerns technology. Among effective technology planners, their guidance and planning go beyond the "pure" technology aspects of a diversification strategy and look at the general exposure particular kinds of diversifications entail in terms of experience and knowledge base in the company.

Some current management dogma is moving back toward the "stick to your knitting" philosphy of diversification—a very conservative approach indeed. It argues that a firm should stick to its main lines of business and expertise and not go off into entirely different, unanchored areas. Had this advice been followed by many major corporations over the past few decades, many of the debacles that accompanied radical diversification might have been avoided and the customers, stockholders, and employees might have fared better. This is guesswork, however, or hindsight at best. Some radical leaps into new fields have greatly strengthened some firms and have contributed to their very survival when their traditional markets have shrunk or collapsed. The argument, in principle, will continue until there is a sufficient body of evidence favoring one kind of strategy over another. Even then, bold or desperate managers will defy conventional wisdom (whatever it happens to be at that moment) and go off in search of better returns on their own efforts and their stockholders' investment.

General business judgement is not the province of technology planners and many have come to grief overstepping their bounds. However, it *is* part of their territory to examine the impacts of technology on possible forays into new fields for the firm and to sound clearly the tocsin when particular diversification efforts— internal or external—are likely to fail or go off the track due to changes in or unforeseen aspects of relevant technology. We have been involved with a number of attempts to diversify into fields where the technical barriers stood squarely in the way of meeting the financial objectives of the diversification—sales, costs, margins, growth, and return on investment. Failure to recognize and act on such barriers has led to disappointment and eventual decisions to dump or disinvest in the venture later on.

The STP process should, as argued throughout this chapter, be able to identify pitfalls as well as opportunities related to technology in all aspects of the firm's operations—products, processes, business area, markets, specific market niches. It should be able to give early warning of technical and related pitfalls (e.g., the threat from liability, the cost of quality, the changing cost structure) and should contain recommendations for and against general patterns or specific examples of diversification. Of course, there are other major factors than technology which affect the merits of a diversification and some decisions may be overwhelmed by dominant financial or other considerations, even though technology remains a potential barrier or threat.

However, the strategic technology planners who succumb to rhetoric or power plays without at least notifying their clients of technical pitfalls of a diversification plan or program are not doing their job adequately. It is not easy to be a Cassandra and to dwell on all the risks and bad things that can happen. It is even worse to have been right and to be in a position to say "I told you so." But such is the short and risky life of a planner and the bolder ones are always on the edge, in the position of displeasing some constituencies and goring someone's ox or pet project.

When setting up an STP group and getting the process started, the CEO and his advisors have to decide whether they really want such a group to range broadly and concern themselves with (stick their noses into) the firm's diversification policy or to observe strict limits and boundaries on their activities. It is the CEO's choice on what benefits he wants from the STP process and the people who perform that important function.

22. *Which Functions in the Firm Dominate in Planning Decisions and Implementation of Plans?*

(a) Is R&D/technology a leader, follower, or go-along in planning—how proactive?

(b) Implementation is a line function: staffs can help but not "do" implementation independent of the owner of the resources (people, dollars, facilities, "charters to act").

One of the themes which runs through this book is that technology people—especially the CTO, his staff, and the key R&D people at the corporate and divisional levels—have to take a *proactive* stance toward technology planning. If they passively accept other groups' "drop-ins" or "add-ons" concerning technology in the business plans, they are not doing their job effectively. Exceptions are, of course, situations where the corporate or divisional planning groups include individuals who are knowledgeable and farsighted about technology or where the business planning process *begins* with a number of premises and assumptions, including those about future technology.

In some industry sectors these "exceptions" are the rule and the long-range business planning group can build technology considerations in as integral parts of the plan. In most sectors, however, this does not happen, and technology, if included at all, consists of the very add-ons, drop-ins, and "oops-type" modifications implied above.

What do I mean by the "technology component" of plans? Most business plans of any quality do mention the need for and hope for new and improved products to meet changing market conditions or to open new market opportunities. We have worked with many companies whose "hockey stick" projections of sales and growth of profits are footnoted by such comments as: "product line *x* will have to be completely modernized," "40% of the incremental growth in sales and profits will come from new products over the next 5 years," or "manufacturing costs will be greatly reduced over the next 2 years." The missing dimensions are the feasibility (in terms of the state of technology) and cost effectiveness of product/process development that will or is likely to allow these fine results to be achieved.

In the "high-tech" industries, technology can, does, and should play a dominant role in long-range planning, and the strategic technology plan should be an integral part of the business plan. The STP should, where feasible politically and temporally, *precede* the business plans. This is so that the planners and the *implementors* of the plan can know what is possible, feasible, likely to be achieved, and cost effective in the way of new technology-based products, product lines, materials, manufacturing processes and services. In no case is it a purely linear process, with step A always preceding step B. STP and long-range business planning involve an incremental, iterative, integrated, self-challenging process, where the "final" plan may bear little resemblance to the outline prepared at the start of the cycle.

If the technology people are doing their job, for example, they should be able constantly to challenge the assumptions, targets, and many specifics of the evolving plan from a technology perspective. This does not mean their contribution is primarily negative or critical. Over-the-horizon technical possibilities, due to the evolving state of the art, may make it possible and desirable to enhance or enlarge the planning scope by presenting options about new products, markets, or cost structures. CEOs, DMs, and other top managers should look carefully at the role of their technology people in the planning process. If it is passive, serving primarily as a tacit endorsement of the targets and premises of the nontechnical planners, then they will be ill-served by the planning process.

This brings us back to another continuing theme: staffs can help with strategic planning or do much of the formal work involved. But *implementing* plans is the responsibility of the line executives—corporate and divisional. This does not neatly divide the planning process into "planning" and "doing." If the "doers" are not intimately involved in the planning itself, especially the underlying framework and premises, they are likely to be reluctant to actually implement plans that entail risks and bold thrusts into areas about which they do not feel comfortable. This "lack of implementation" is not necessarily an up-front statement that they do not accept certain components of the plan—for example, a foray into a new technology or market. They may indeed go along with the early stages of such forays and then, at a later and critical time, decide that "they really didn't want to be in that market, business area, or technology anyway and let's sell it or fold it up." While writing this section I had a long session with the former CTO of a large firm who saw exactly this happen—moves into several new technologies and markets in accordance with some long-range plans of several years before, and subsequent bailing out of *all of them* when things got tight and the mood changed to "lean and mean" and "focus on our main businesses."

There is no guarantee that top managers will persist with new thrusts arising from a long-range plan even if they did indeed participate in its formulation initially. Conditions change, CEOs and their close advisors change, and what appeared to be a good move 3 years ago may seem less attractive now. So participation of the implementors (line executives) in the planning process is not a *sufficient* condition for success of a plan. But it certainly approaches a *necessary* condition.

23. *How Closely Are the Technology People Involved in New Ventures, Technology Purchasing and Licensing, and Mergers and Acquisitions*

   (a) There are critical phases of these activities for such involvement.

   (b) Such involvement is a reflection of management's trust in and dependence on technology and its proponents.

   (c) The technology people may not be asked but should be monitoring and doing their homework.

(d) Bad news may not be welcome, but it may urgently be needed at management levels.

This topic is also discussed in the context of technology acquisition from outside, in Chapter 6. It is mentioned here because many long-range plans involve acquisitions, mergers, joint ventures, partnerships, and other arrangements which involve technology considerations. Merger with a company in the same field may involve decisions (tentative until the merger partner is actually identified and the deal approaches consummation) about combining, eliminating, or dividing up technology groups such as CRLs, divisional labs, or technical service groups. A paradoxical situation may arise in this connection, with the CTO being asked to analyze these possible options and to make recommendations. The paradox comes from his position as perhaps the most knowledgeable person in the firm with respect to technical capabilities and consequences of different options. At the same time (and we have observed many such cases), his technology organization (CRL) is generally one of the pieces "on the table" in the deal, with the possibility of its being eliminated or losing some of its current position and power.

Despite this, there are many reasons for including the CTO and his people in the analysis and planning for such major moves. If such planning is not done early on, even before the merger or acquisition is a certainty, key people from both companies can be lost, programs may bog down, and the overall technology venture may be damaged while the rumor mills and personal anxieties take their toll.

Analogous roles can be played by technology in other "outside" moves, as well as formulating new enterprises within the firm. The timely phasing of technology inputs into the planning process for such moves may be critical to their success. Whether the CEO and his M&A* people bring technology people into such planning depends very heavily on how much they trust such people in matters of knowledge, judgment, representing the best interests of the company, objectivity, and style of participation. If, for example, the CEO feels that dealing with technology people "gives me a headache," then he is unlikely to depend on them for this kind of planning and decisionmaking.

Asked or not, the CTO and his people should be proactive here also, even if it entails some political risk. They should be doing their homework on potential partners or targets for takeover, acquisition or joint ventures. They should be assessing the technical capabilities of such firms or groups. Chapter 7 on evaluation contains a procedure—the technology/innovation (T/I) audit—which can be used externally as well as internally to assess technical capabilities and limitations of firms, divisions, labs, or programs. Working from the outside is more difficult than internal use of the T/I audit but can give a first cut appraisal of the technology in a potential partner's organization and can signal potential pitfalls and surprises. If bad news comes from such a "preemptive" analysis, especially if it was not asked for, then the CTO and his people may be in *political* trouble. But

---

* Merger and acquisitions.

presenting their bad news—poor products, out-of-date production facilities, little R&D capability, indefensible patent positions, bad reputation for quality—can save their top management and the firm a lot of *economic* trouble later on.

M&A people can muster plausible arguments for keeping their activities separate from the firm's technology functions. They can claim that secrecy in analysis and negotiations is critical, that technical people take too long in making recommendations and tend to waffle and add too many caveats and "yes · · · buts" to their advice, and that they are not really "buying technology" so it's none of their business. Such attitudes, common among the M&A community, have led to some tremendous blunders, where lack of technology or misinterpreted technical promises and capabilities have undermined the value of the deal. Discussions with M&A specialists at several major investment banking firms discloses that they rely heavily and pay heavily for advice from lawyers, accountants, and financial people in their deal making. But few of them employ more than superficial measures to assure that the M&A candidates are *technically* as represented or implied. These attitudes are also reflected in "internal M&A" activity and can strongly affect the probability of failure or of disappointment with the deal. Obsolete product lines, unmanageable quality and cost problems, and liability-prone products are only a few of the weak spots that a thorough (or even not-so-thorough) T/I audit can identify, when performed by competent, knowledgeable technology people with the interests of their company in mind.

24. *How Close Are the Technology People to the Hopes, Dreams, and Fears of Their "Clients"*
    (a) A part of the "internal market research."
    (b) Can technology people serve as "technical therapists."

"Arm's-length" insertion of "technology components" into long-range plans will not do the trick of protecting the firm against technical surprises and helping it to take advantage of technical opportunities. As part of the internal market research on their clients, which was advocated earlier, the technology planners have to really get to know their clients. Given that they are indeed trusted by top management or that they are not actively *dis*trusted, they can go much beyond formal technical inputs. Most of their clients, except in some high-tech industries or sectors, have little or no direct knowledge of or feelings of comfort about technology.

Managers have been alerted by journalists, consultants, management authors, conference speakers, and government officials about "the importance of technology" to the firm and the country. Frequently, they are able to articulate clearly this importance and in turn write papers and give speeches about it, including making it an occasional lead feature in the firm's annual reports. This is a far cry, however, from really understanding or feeling comfortable about technology issues as they are likely to affect the firm in the future.

He seldom openly asks for it, or may have given up asking for it after a series of disappointments, but a CEO may strongly need and want advice on technology from someone whom he trusts. What do I mean by trust? Not just the literal definition, relating to the honesty or integrity of another person. The kind of trust that will cause him to seek and take advice also relates to his perception of how competent the adviser is, how much he understands and accepts "my position and requirements," and how much the advisor really has, in his heart, the interests of "me and my firm" (if they are the same or compatible). Also, trust may imply the perception that the advisor will not embarrass, betray, abandon, or do other bad things to the client as conditions change or if mistakes are made. This does not (necessarily) mean he is looking for "yes" men. He also wants/needs an independent viewpoint. Given such conditions of trust, some CTOs have indeed been able to act as the prime technical advisor and even "therapist" to their clients, in a manner that many lawyers and financial advisors are able to do. This is not a common situation, but it is something that CTOs and their staffs should be striving for, if it is within their power to achieve.

25. *How Enthusiastic Are the Technology People in the Firm About Actually Doing Long-Range Strategic Technology Planning (STP)?*
   (a) Half-hearted, insincere planning efforts can and do backfire.
   (b) Technologists must be carefully vetted for their ability and propensity to do STP.
   (c) Bad plans can come from bad planners as well as flawed planning procedures.

Appearances to the contrary, most research people do not have a long-term strategic view of technology. The appearances may come from the long time it takes for them to get organized, to perform projects, and to get the results out into the plant and market. The kind of planning required for the STP process described in this chapter is *pro*spective—forward-looking rather than merely taking a long time to do things. There are also wide personality differences within the technology community. Some people love to look ahead, speculate, take risks or advocate risks, and deal with uncertainty and ambiguity. Others run as fast as they can away from such conditions.

In forming the planning team and calling on technology people to participate in it, care should be taken to sort out these people and to select "real" planners for the strenuous and often hazardous job of helping to chart the course of the company in the dim and distant future (even 5 years out).

Technical competence and the right personality should also be accompanied by enthusiasm and a sense of excitement. Lacking those, the technology inputs into planning may be "ho hum" and reflect playing it safe and protecting one's rear.

Many technology people complain bitterly about the burden of short-term and firefighting projects that "don't let them do any long-term research or planning." Some of them, carefully selected, may indeed throw off the shackles and sparkle when given the time and charter to do STP. Others may find that they never really meant it, may have lost the knack and stamina they once had, or are not able to recognize that they are not really doing the job and fairly representing technology in the planning process. Top management and the planning staff must be careful, in selecting technology representatives, to avoid taking on dead wood, talkers instead of doers, and people who are technically obsolete and too narrowly focused to help in STP.

5.4 A SUMMARY AND REGROUPING OF THE 25 ISSUES/PROBLEMS

At the risk of too much redundancy (some is inevitable in dealing with a subject such as strategic technology planning where the field is just emerging), this section recaps some of the items in the long list of issues discussed in the previous sections. Figures 5-3 through 5-5 regroup 15 of the individual issues into three major clusters and discuss them in terms of these three macroissues.

5.4.1 Relations with the Businessmen and Top Managers in the Firm

The key to the status of technology and how it fits into the near-term and long-term strategies of the firm lies in how it is perceived by the people who run the company and influence its decisions and policies. Although the specific people who fulfill this role vary from company to company and are subject to particular circumstances and personalities, this key group is generally drawn from the top corporate management and the managers of the operating divisions and/or groups in the company. Their experience with, attitudes toward, and evaluation of the specific technology or R&D/innovation activities in their companies, and in general, are key factors affecting how such activities fit into the company's strategic planning as well as its tactical moves. In the latter context, for example, companies can clearly be differentiated according to how much they rely on their own R&D to meet near-term external market or other threats to their profitability or survival. A major indicator of the role of technology in planning and decisionmaking is the degree to which a set of explicit or implicit "technology policies" have been formulated or have developed in the firm, which reflect management's perception of the importance and the role of technology in the business.

   If technology is to have an important role in the decisions and plans of the businessmen and top managers, it must be able to meet their needs for help on early warning, technical guidance, and scenarios for the future (item 1 in Figure 5-3). In order to make such formal contributions successfully, the people in technology must have a marketing point of view toward their internal "clients" (item 3) and be close to the hopes, dreams, and fears of their clients as they look to the future. Part of the marketing approach is the ability to identify their "real" clients — people who "own" the problem and who can act to address or solve the problem through making decisions or implementing plans. Familiarity with clients requires an intimate and accurate perception of their needs, constraints, demands, biases, and style of operation, including the "rhythm" of their decisionmaking and action-

1. Need of top managers and businessmen in firm for help from R&D/innovation on:
   (a) Early warning of threats, needs, and opportunities (TNOs).
   (b) General credible technological guidance and advice.
   (c) Scenarios (also credible) on future markets, technology trends, competitors' actions, materials and energy prices and supply.
2. How close are R&D people to the hopes, dreams and fears of their "clients"?
3. Planning requires a marketing viewpoint (vis-à-vis internal "clients").
4. Strategic business units (SBUs) may or may not be logical elements for long-range R&D/innovation planning.
5. How strongly is the company tied to existing operating units as vehicles for launching new products and technologies?

Figure 5-3. Relations with the businessmen and top management.

taking. The potential internal client needs to be market-researched and marketed to, in terms of plans, ideas, suggestions, information, and proposals.

In the absence of these two conditions—closeness and a marketing approach—there is little likelihood that their clients will perceive technology people as an integral part of their planning process and their specific plans for the firm or call on them for help or advice on important matters. If the internal technology people cannot or will not (through reluctance or poor marketing skills or lack of responsiveness) supply the kinds of information and advice in the firm and at the time that their clients need it, it is very likely that these people will go outside for help—to consultants, suppliers, friends, and other sources of technology. CTOs and their staffs should be prepared with this kind of information in case they are asked, since questions often are posed as part of a short-term crisis, where technology may or may not be the main problem or the solution. Such preparedness and actually "coming through" in crisis situations can help build credibility for technology. It may also take a long time and require a number of response incidents to build an adequate level of credibility so that top management will perceive technology people as part of their team and as the logical place to go when issues of or related to technology arise. In the absence of this credibility and the responsiveness it is built on, such information seekers are likely to forego internal sources, which might be highly competent, for sources that are more readily accessible and easier to deal with—that is, where the delay and aggravation factors are less than they perceive they would encounter in dealing with their own people.

Depending on the circumstances in the company, the newly emerging strategic business units (SBUs)—a further escalation of the decentralization philosophy—may or may not be the logical elements for long-range planning in the technology area. Depending on their size, diversity, maturity, scope of activities, and time horizon, they may suffer from the same kind of shortcomings that whole operating divisions in many companies do, with respect to long-range planning. They may be too narrowly focused and too short-range oriented to do an adequate job of strategic technology planning. This ability to plan long range is also heavily influenced by how strongly the company is tied to existing operating units (operating divisions,

SBUs) for launching new products and technologies. In many of the companies we have worked with, such organization components are not willing or able to transcend the current product lines, technology, and marketing to do an effective job of long-range planning. With respect to technology, they may do a splendid job of planning for product line maintenance and incremental improvements but are not in a position to undertake and implement long-range plans for radical innovations and significant diversifications from their existing product lines and manufacturing technology.

Existing operating units (divisions and SBUs) have many strengths, including a successful track record, stability, and people experienced in working together. However, when considering the possibility of radical departure from existing lines of business and technology, these strengths may turn out to be barriers to innovation and "moving away in different directions," based on emerging or over-the-horizon technology.

5.4.2 Ideas, Information, and Planning

Figure 5-4 lists a number of factors affecting the success of the planning activity in the area of ideas and information. Future actions, plans, and planning-related events must be tied to changes in company goals, criteria, characteristics, and capabilities. Many technology-related plans are static, with no provision for adjusting easily to changes in company circumstances. The response time of technology-related plans may be very long—in some cases on the order of a decade or more. Plans to explore new areas, develop new technical capabilities, or shift emphasis of major research programs can quickly diverge from relevant paths as company objectives and plans change. Many "new starts" in R&D turn out to be off-target after a year or two because of diversification or divestiture moves, because the market has changed, or because the company's niche in a market has changed or is changing. The planning process must be open to continuing review, despite the difficulty and long "turnaround time" for R&D programs. Once underway, certain commitments may be hard to reverse without major upset or even organic damage to the technology program. However, "successful" research that is no longer *relevant* to the company's goals and needs can help hasten the demise of the R&D/innovation activity and weaken its credibility among its important clients and sponsors.

---

1. Future actions, plans, and events must be tied to changes in company goals, criteria, characteristics, and capabilities.
2. The target periods (time horizons) for planning must fit the character of the decisionmakers, the company, the industry, and the environment.
3. The general weakness of market research for radically new products and markets.
4. Self-control to prevent overreaction to new product and market ideas.
5. Willingness to share information and speculate, safely, about the future.

Figure 5-4. Ideas, information, and planning.

An organic process of planning is needed versus a mechanical one that identifies goals and then selects paths to follow blindly without continuing inputs from the environment. Among the specific types of change that can affect the validity and usefulness of technology plans are changes in:

- Market posture of the firm
- Market sectors and niches
- Technology itself—that is, the direction of the leading edge
- The labor force
- Management goals
- New entries into the market or the field of technology
- Financial position—long and short term
- Government actions such as regulation/deregulation, incentives, and tax laws
- Costs and prices of products sold, materials and services purchased

Despite the frequent inertia of research projects and programs themselves, the planning process for these projects and programs and the whole technology activity must track such events and trends and be able to influence the general direction of specific projects and programs, including terminating those that no longer fit.

The target periods (time horizons) for technology planning must fit with the character of the decisionmakers, the company, the industry, and the environment. There is no point in developing elegant long-range plans that are not going to be compatible with the time horizons adopted or preferred by the top management and their business and financial planners. Certainly, the technology aspects of company plans should realistically inform management and the other planners about the time required to fulfill certain technology-related objectives—for example, to develop a proprietary position, to build an in-house capability, or to design and build a new plant. But plans that clearly go beyond the ability of the management and business planners to predict or to make realistic commitments are not likely to be well received. At minimum, technology plans must explicitly spell out the time horizons they involve, in terms that business and financial planners use for making capital commitments—payout period, timing of capital expenditures, needs for working capital, and likely patterns of returns. Much planning involves negotiating, bargaining, second-guessing, and gaming. But a track record of unrealistic milestones and capital requirement estimates can mortally damage the technology activity's credibility and reception as part of the corporate planning team. Milestones should be flexible, with contingencies identified and multiple estimates given, based on alternative scenarios and contingent events.

It is important for technology to fit the "rhythm" of the company and to avoid showing either impatience or an undue "drag" effect on company planning. All this caution about fitting in with the company's time horizons and planning periods does not suggest that technology should take a back seat and fall into only a *response* mode of planning. It should, by its very nature, be *leading* the company in terms of technologically feasible and desirable courses of action. However, if it outruns company planning or hangs back unduly, it can fail to be an important partner in company-wide planning.

One constraint that many technology activities and their long-range plans must contend with is a widespread weakness of market research for new products, markets, and technology. For a number of reasons, market research in many firms is focused on current and "definitely upcoming" products and product lines. Among the reasons for this tendency are these:

- Many market research techniques have been developed primarily from the tradition of consumer marketing; this in turn reflects the experience of marketing faculty in most business schools. Industrial marketing courses have been rare but are increasing slowly.
- Most techniques in use in market research groups are tied to time series data on the same or similar products and product lines to those currently being sold.
- Market research people in most companies typically come from sales or marketing functions that are current-product focused.
- With some exceptions for high-tech industries, most market research people are not technically trained and have had little or no exposure to the process of R&D/innovation and the inherent uncertainties and time lags involved.

Many companies that have a good reputation for market penetration or introducing new products quickly and effectively have accomplished this inadvertently, through the natural process of entrepreneurship growing out of the R&D/innovation process itself and, in effect, have grown some do-it-yourself market researchers in the technical areas of the firm. In turn, a weakness of some of those groups is that they lack formal training in marketing theory and experience in market research methodologies and often make mistakes that more conventional marketers would be likely to avoid. Some combination of the free-wheeling creativity of R&D and the experience of marketing people could provide an ideal market research capability, if the usual organizational boundaries and territorial jealousies can be overcome or avoided.

Although enthusiasm and confidence in the potential of longer-range technology projects and programs are essential in selling them to management, it is important for technology planners to impose enough self-control to prevent overreaction to new product and market ideas in terms of timing and commitments. It is important that plans clearly differentiate between "possibilities," "likelihoods," and "sure things" that only require enough time and money to make them happen. The general odds are high against a new product being fully successful in the market — that is, returning an acceptable multiple of its total investment, including cost of capital sunk into it over a period of years. So, a necessary balance should be struck between generating acceptance for the idea and having people downstream in the R&D/innovation process move too soon in terms of stimulating customers to expect the new development that may never come to pass or that may take much longer in transition than expected. Careful preparation of the market is needed, based on systematic and periodic evaluation of progress and the most likely introduction dates, with optimistic and pessimistic limits. Realistic introduction dates should be generated continuously and changed as the project moves closer to manufacturing and market introduction, so that sales, marketing, purchasing, production, and other involved functions have enough *lead* time but not too much "anxiety" time.

A balance is also needed between excessive secrecy and this control of "over-promising." An effective STP will incorporate contingency planning for schedule slippages, delays in R&D (which almost seem inevitable), production delays, and other foreseeable delays. Finally, the planning process must reflect a willingness to share information—even though some of it is speculative. This requires a "safe" atmosphere for sharing information, ideas, hunches, and misgivings without fear of punishment or ridicule, although there should be a willingness to accept responsibility. An effective planning process should have mechanisms for addressing the future in a cooperative mode among the parties who will be involved in the overall R&D/innovation process. These mechanisms should be capable of addressing the future both inside and outside the company in terms of data bases, scenarios, trend projections, assessment of likelihood of specific events, and general environmental conditions.

5.4.3 Relations Between Technology and Other Groups

The ability of the people in technology activities to participate equally and effectively in company long-range planning depends on a number of factors, such as those shown in Figure 5-5. Some of these might be described as issues of communication or organizational structure, others as "organizational politics." A highly significant influence on the role of technology in planning relates to the issues of which functions—for example, finance, marketing, corporate planning, or manufacturing—dominate in the initiation, drawing up, and implementation of plans. Certainly, as planning doctrine suggests, top management should and must be in charge of both plans and their implementation. However, plans are generally prepared by staff people and their implementation is heavily influenced by dominant levels and groups in the company, such as the heads of operating divisions. If technology does not have a leading role in decisionmaking and policy formulation in general, chances are that it will not be prominent in the strategic planning process, even for technology itself. The issue is whether technology is a leader, a follower, or a go-along activity: How proactive is its role in the firm?

---

1. Which functions in the firm dominate in planning and implementation of plans?
2. Articulation between R&D/innovation and business planning.
3. No clear *linear* sequence of planning.
4. Frequent poor relations between R&D/innovation and financial planning.
5. How closely is R&D involved in new ventures, technology purchasing and licensing, and mergers and acquisitions?

Figure 5-5. Relations between technology and other groups.

One interesting example is the secondary role played by technology people in many firms in purchasing of technology via licensing, joint ventures, acquisitions, and other means. There are critical phases for their involvement in the overall and sometimes lengthy technology acquisition process. The actual degree of involvement is a reflection of management's trust in and dependence on technology itself and its advocates. For example, are they perceived as necessary evils or nuisances encountered in business? Even if not directly asked and included, technology people should be monitoring such acquisition activities and doing its homework in preparation for having to "catch the baby" and make the technology work once it has been acquired. It may stretch their risk-taking propensity or perceived role for them to advise management, unsolicited, against a bad technology deal; but such inputs may urgently be needed before the acquisition process has gone too far.

One manifestation of the status of technology, as it affects their participation in the planning process, is the level of articulation between the people in the R&D/innovation process and the business and strategic planners. Many business plans are drafted with a "technology component" or important technology implications (e.g., the ability to develop or acquire new products) without reference to or inputs from the people in R&D. This is a common weakness in planning. In other cases, R&D makes inputs, but only as "add-ons" to the business plan, when it has been essentially completed. Some of the lack of articulation arises from the relative power relations and/or from a poor historical relation between technology and financial managers. Each may view the other as unrealistic in its approach to the others' areas of concern—that is, the uncertainties of the innovation process or the need for credible financial estimates of future capital requirements. Too often, the ongoing relation is an adversarial one, where mutual stereotypes influence perceptions and willingness to consult and listen—for example, viewing financial people as "cold-eyed bankers" and R&D people as "unrealistic scientists" with no financial understanding or responsibility. Where such stereotypes persist and carry over into the planning process, there is little hope for a good articulation between the functions.

The ability of technology people to intervene in or even lead the strategic technology planning process depends heavily on such relations as those mentioned above. Without credibility, their inputs may be foregone or treated in an offhand manner. Shared experiences between technology and financial or business planners may help—such as engineers or scientists with marketing, financial, or planning experience and planning people with technology experience. Such cross-fertilization can be accomplished in a number of ways, including deliberate recruiting of "hybrids," job rotation, temporary assignments, task forces, or combined planning teams.

Establishing good working relations among the groups that should be contributing to the STP process requires a continuing interactive and personal mode of planning versus planning by documents alone. Overall, successful technology planning requires careful and continuing attention to integrating efforts with other corporate functions, especially those having dominant roles in corporate planning.

5.5   STRATEGIC TECHNOLOGY PLANNING
AT THE DIVISIONAL OR SBU LEVELS

The bulk of the discussion in this chapter involves issues in strategic technology planning across the corporation, no matter how it is organized (centralized, decentralized, or combination) as far as operations and technology are concerned.

In this section, however, I want to dwell on the particular problems of long-range and strategic technology planning for individual operating divisions and their subdivisions or strategic business units (SBUs), where they exist.

At the outset, it can be said that some division managers and their technology staffs do engage in respectable strategic technology planning, equivalent to that done by many leading corporate technology groups. This is especially true for divisions in high-technology fields and rapidly changing markets and for large divisions with a diversified portfolio of projects and technical fields.

For most operating divisions, however, this is rare. The business and strategic plans that many corporate financial and business planning staffs force the divisions to construct seldom contain more than a passing reference to technology, let alone detailed and credible plans for advancing it within the division in the long term.

This is not surprising. The picture presented up to now is of the division manager and his technology staff concentrated on short time horizon and narrowly focused projects and programs, which are the product of the constraints under which the typical division operates. Time and resource limitations as well as constrained market segments do not make it attractive or compelling for divisions to spend a lot of their time and effort on longer-range technology efforts that may never come to fruition during the tenure of the DM or even his immediate successor. This may appear to be a selfish attitude, and it is. However, it is naturally fostered under the conditions existing in most highly decentralized companies.

With his reward system and his energies focused on short-term improvements, there is little incentive for a DM to engage in fanciful "if · · · then" exercises about the great benefits his division *might* reap if certain long-term, risky, and off-the-mainstream projects are launched and nourished with scarce divisional resources. So why bother planning for them or making promises to top management that they will engage in such projects if there is little chance that they can be launched and sustained in the face of continuing firefighting and crises?

Digging below the rhetoric of many divisional strategic business plans (where they are required to construct any), we find that references to technology plans are primarily "more of the same"—continued minor and an occasional major improvement in the division's products, manufacturing processes, or materials. Digging further, even such plans do not have much credibility without the accompanying provisions for new resources and for staff and facilities renewal that would permit advanced technology projects to be pursued successfully.

In other words, with the exceptions noted above, most divisional technology planning is fairly short term, fairly extrapolative, and not likely to lead to radical departures or breakthroughs, except by happy accident.

If the advantages of strategic technology planning discussed in this chapter are to accrue to the operating division, and thereby to the firm as a whole, then what can be done to stimulate and improve it? Many of the actions and decisions described here for general strategic technology planning in the firm can be applied at the divisional level as well, provided that the division manager and his CEO or group executive agree on the needs for it, the means of doing it, and the direct and visible benefits to the *incumbent* DM to do it in his division at that time.

This entails many of the factors discussed in Chapter 4 for stimulating divisional technology networking, but especially resources and personal incentives (positive and negative) for the DM personally.

Given this understanding and the accompanying resources (including the patience of top management to wait for results), what about the feasibility of division-level strategic technology planning? Again, leaving out the large and/or highly technical divisions that *must* plan to survive and to meet their rapidly changing markets, what limitations must the division overcome?

First, the divisional technology staff, if one exists, is seldom staffed with the level of people needed for strategic technology planning. They often consist of "old soldiers" who have served well in divisional technology roles for many years and know the current operations and needs of the division intimately.

With few exceptions, they are not out exploring new technical fields. They are close to home, helping out with plant and customer problems and, perhaps, they are in liaison roles, monitoring ongoing projects or outside contracts. There is seldom a deep scientific or economic or professional marketing capability available in these staffs. Where the division lab is of modest size (several dozen professionals), the group of section or department managers may serve as a sort of planning staff for the divisional technology manager or even the division manager (although that is rare). However, these individuals are typically preoccupied with the immediate and near-term problems of their own specialty areas or product lines. They are not generally up to speed on over-the-horizon technology and market opportunities in other fields.

If the pace and resource limitations of a division reinforce the DM's bias against or lack of interest in strategic technology planning, it is a lost cause—until perhaps a new DM takes office and selects "technology" as a platform for improving things or making a name for himself.

Where there is an openness to serious strategic technology planning by a division manager, stemming from whatever motives, the recommendations made in this chapter can be applied at the divisional as well as the corporate level with perhaps some help from specialists (economists, technical forecasters, market analysts) from corporate staff, other divisions, or outside sources.

Strategic technology planning at levels below the operating division—specifically the typical strategic business unit (SBU)—is almost a contradiction in terms. Except again for very large SBUs in rapidly moving and high-technology-based fields, the typical SBU is not in a position to do much about its technology base in the long run.

The original concept of the SBU and the way it is indeed practiced in some progressive companies allowed for and even *required* the SBU manager to engage in longer-term planning for his unit. In most cases, however, the SBU manager is even more frequently evaluated, and less flexible than the division manager. He is often boxed into a narrow market segment that is often contiguous with the market segments of his fellow SBU managers in the firm or the division. If he wanders too far afield in terms of new products or new markets for his existing products, he is liable to lose his investment when his excursion is discovered. This is especially likely when the successful results look like they belong in someone else's SBU segment.

In terms of time horizon and resources to apply to long-term planning, let alone *carry on* such planned projects, he is again worse off than the typical division manager. In particular, many SBU managers do not even "own" the manufacturing and technology facilities that support their product lines. Attempts in a number of companies to split up such divisional resources and piece them out to a large number of SBUs have been disastrous from a technical and cost point of view, as well as from the viewpoint of efficient management.

So all the odds are against most SBU managers engaging in strategic technology planning, unless *their* managers and top corporate management really believe in long-range strategic planning, including technology planning. In this case, the recommendations in this chapter can also be applied at the SBU level, in a more focused manner than for the operating division or the firm as a whole.

Figure 7-28 contains a set of general categories for describing the "project portfolio" or allocation of resources to projects and programs within a firm, a division, a specific lab, or another technology-performing unit (including some large SBUs).

These categories form a rough continuum from very short-term, narrowly focused efforts (e.g., technical service and minor improvements) to exploratory research and technology-base-building efforts. The latter represent long-term, often high-risk investments in the future of the firm or operating unit. We have collected data on project portfolios according to these categories from hundreds of firms, including corporate and divisional technology programs.

As indicated in Chapter 2 in the discussion of "basic research," few industrial firms have much of an effort in this category comparable to their university equivalents. Many CRLs have parts of their effort (5–25% for large, technology-reliant firms) devoted to categories 7–9, including "exploratory research in fields of potential interest."

The dominant categories for divisional R&D projects are in the first three or four categories, with some efforts in category 5 and an occasional thrust in category 6. The percentages of their technology budgets and personnel and the actual amounts devoted to such forays and advanced technology efforts are typically low—under 20% for most. The overwhelming majority have no discernible efforts in advanced and nonspecific areas (categories 8 and 9). Given this situation, is it fruitless to argue for long-range strategic technology planning in such divisions? The answer is no, if we can decouple the ideas of strategic technology planning from long-range university-type or "breakthrough" research.

Strategic technology planning can be focused on a long-term program of service, minor improvements, and occasional major improvements to the division's underlying technology base.

Figures 5-6 through 5-8 illustrate an approach to strategic technology planning in a general product line (SBU product managers note this) or even for a specific product that has potential for cost-effective technological improvement.

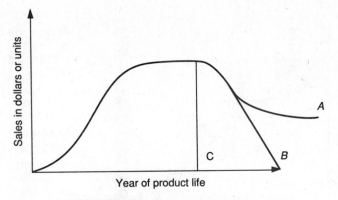

Figure 5-6. Product life cycles.

Figure 5-6 portrays classical product life cycles that end due to technical obsolescence of the product, substitution by other products, cost pressure from competitors, market share pressure due to competitive features, and other factors. Some top managers, division managers, and SBU managers are willing to accept the "harvest" approach to such end-of-life products or product lines and withhold all but necessary capital expenditures in their support.

Figure 5-7, however, notes the potential effects of *planned* investments at different stages of the "mature" phase of the life cycle, when many managers would pull the plug and let the product cycle ride out on a cash flow cushion.

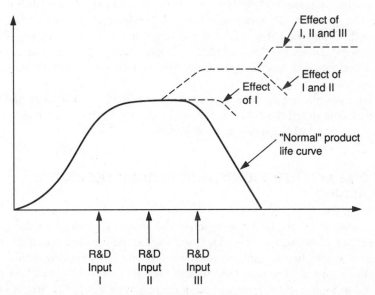

Figure 5-7. Possible effects of incremental R&D on product life cycle (subject to analysis of *costs*, including opportunity costs as well as benefits).

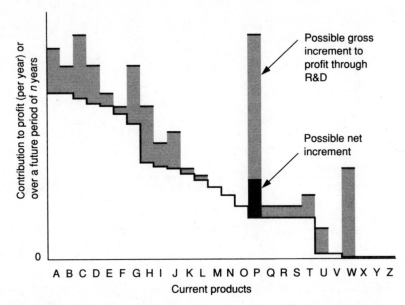

Contribution to profit (per year) or over a future period of *n* years

Possible gross
increment to
profit through
R&D

Possible net
increment

0

A B C D E F G H I J K L M N O P Q R S T U V W X Y Z
Current products

Figure 5-8. Contribution to profit as a guide to investing in R&D for an "old" product.

The potential cumulative effects of a deliberate strategy of investment in product (and underlying manufacturing method) improvement, especially with new technology approaches, can be tremendous. Figure 5-8 indicates the potential from such a strategy.

Of course, pumping technology dollars into a "dead horse" can turn out to be a bad strategy, especially when it is done "a dollar at a time with your eyes on the ground." Without a strategic plan for doing this in a programmatic way, lots of technology dollars can be wasted in a product line or product that is going to die soon and that will not likely be helped by an infusion of new technology just before the end.

We have observed and participated in many of both kinds of strategy and believe that the wisdom to tell the difference is one of the great gifts with which a division or SBU manager and his staff can be endowed.

5.6 SOME AFTERTHOUGHTS (NOT SECOND THOUGHTS) ON PLANNING

After looking over my many arguments and recommendations in the direction of long-range planning for technology and related matters and for more strategic thinking, I would like to suggest a bit of caution. Although the profit-oriented businessman knows very well that "in the long run we're all dead," he may be persuaded to look further ahead and think deeper about where his firm is or might be going and the role of technology in its future. All to the good! If, however, he lets such planning and futurizing distract him too much from the shorter-term goals of keeping the business going, he may not have the resources, or even the company, to use for the long run. Current operations generate cash flow and profits and, without those, long-range technology plans are not very feasible. So, we are arguing for a balance of short- and long-term thinking with respect to technology and the businesses it supports. Perhaps the reason for the heavy emphasis in this book on the long term is that in most decentralized firms, the balance is very much in favor of the short term. In the next section, the distinction between "planning" and "worrying" is made.

5.6.1 Methodical Planner or Anxiety-Driven Worrier

Plant Manager: If it ain't broke don't fix it.

Planner: But it's liable to break soon and if we let it happen, it can cause damage and make us lose production, and....

Plant Manager: Never mind, as long as it doesn't happen on my watch. Why should I pay for fixing things that might break on someone else's watch?

Sounds crass and selfish? With perhaps a bit of exaggeration, this attitude is not rare in the production area and many other areas of human activity including medicine, public works, and industrial technology.

The technology analog of this attitude can be illustrated by a recent case we were involved with in an operating division of a very large company that had a commanding share of the market:

Product Planner: One of our European salespeople has been telling us that several of his customers have been starting to unbundle some of our technology (which consisted of equipment, materials, and service) and making or procuring some parts of it separately.

Executive V.P.: They wouldn't do that; they know ours is the best technology around and they have been using and depending on it for years. Tell him they'll come back to us after they find out they can't do it on their own.

Planner: Well, actually, they *are* getting some advanced designs and new material from other suppliers.

Executive V.P.: Never mind, tell him to sell harder.

Planner: Oh, by the way, you know that order we lost from one of our largest customers in the Midwest?

Executive V.P.: You mean they're trying out some new materials to substitute for our products? It won't work. When they find they can't meet the cost and performance specs they'll be back.

Planner: Well, that may not be the only case, the field is changing from metals to plastics slowly but steadily and we ought to move in that direction also, so we can....

Executive V.P.: Don't worry, we are following that trend and when the time is right, we'll make our move.

We could have selected half a dozen companies in a range of fields with whom we have worked and the dialogue would be essentially the same, with the specific words and circumstances changed. Why is one person's "planning" viewed as "worrying" by others?

For the division manager, who operates within a tight scope and time horizon, long-range planning, speculative gaming, and even broad contingency planning may annoy him or make him ill. If, by nature, he is a worrier and has repressed his desire to look ahead and get ready for the unknown, the CEO's demand that he engage in long-range business planning and even strategic technology planning might be "liberating." He might even find it fun and approach it as he does his hobbies, which may take some careful contingency planning (skiing, sailing, golf, tennis, scuba, surfing, skydiving, or other sports).

For the DM who truly does not like, in fact hates, trying to go beyond the immediate present, forced planning may be torture and may in fact be faked. Our experience with "what if ..." and "what then..." exercises with top managers, including DMs, clearly differentiates the two general attitudes toward planning. The "planning-philes" readily fall into a frame of mind that allows them to bypass their current preoccupations and really look at future possibilities and what has to be done to realize them. The "planning-phobes" may go through the motions, but do not really "get into it." The "yes but" syndrome illustrated above keeps them from benefiting by, or contributing to, bold and realistic planning.

Before the CEO tries to force his division managers to play the strategic planning game, he should calibrate them carefully and decide which ones will be wasting his and their time in doing exercises without really meaning them. He may then try subtly to identify one or more members of the DMs' staffs and try to persuade them that they can do worthwhile planning without the direct participation or even approval of their DM (a potentially dangerous game for the integrity of the decentralization system). Or he can assign or allow his CTO and his staff to do the technology planning for those recalcitrant divisions and hope that the plans they make will be acceptable to the DM and/or be implementable even without his full-fledged cooperation.

An interesting and valuable spin-off often comes from strategic technology planning. This is a deeper and broader understanding of the division's (or firm's) current products, processes, materials, and their underlying technology. In order to engage in the "future market mission" exercises described in Chapter 7, the DM and his staff have to know or learn more about their current technology posture and its limitations:

DM: But why haven't you guys told me in the past few years that we were reaching the performance limits of that machine (or material, or technology)?

Divisional Technology Manager: It never came up; we were so busy squeezing cost and marginal performance out of the current technology.

DM: Well, what are the actual limits and how can they be lifted?

DTM: One way is by looking into this new technology that's just coming onstream and which two of our competitors have been experimenting with. We could....

Both the planning procedures in this chapter and the technology/innovation (T/I) audit described in Chapter 7 can contribute heavily to making the DM and his management group more aware of both the limitations and the untapped potential left in the technology on which his products and lines of business are based.

Another analytical technique, described above in Section 5.5, is the product life cycle approach to investing in R&D for existing, mature products and product lines.

All these management approaches emphasize looking carefully at and thinking carefully about the technology on which the business is based—both the current technology and the potential future technology. The next section of this chapter presents some characteristics of an ideal STP.

5.7 AN IDEAL STP SYSTEM AND ITS SUPPORTING INFRASTRUCTURE

Figure 5-9 lists some of the characteristics of an ideal strategic planning system for technology. It is "ideal" because some of the conditions mentioned are not fully attainable in the real world of organizational politics. However, progress toward establishing these conditions can provide an environment more conducive to effective technology planning than currently exists in most companies. Some of them are easier said than done, and others may not be entirely necessary but "would be nice" in establishing a supportive planning environment. In addition to the technical conditions (e.g., points 5, 7, 8, 10, 11), a major thrust of this list is in the area of interpersonal relations and the atmosphere in which planning occurs.

---

1. A true partnership between R&D, top management, operating management, and other functional area management.
2. Challenging, "mind-stretching" sessions involving all key participants.
3. A nonthreatening organizational climate for risktaking.
4. Instant access between all parties when necessary.
5. A secure but accessible information system.
6. An open, honest, nonthreatening method of generating probabilities.
7. An effective external technical and economic/social/political intelligence system.
8. Effective analysis and synthesizing techniques.
9. An atmosphere of enthusiasm for planning.
10. A monitoring and control system to follow planning decisions and results.
11. Action and decision plans for following up and changing plans.

Figure 5-9. Some characteristics of an ideal system for strategic technology planning.

Further along the technical direction, however, there is a strong need for a supporting infrastructure of information and decision support activities which generally come under the heading of Project Selection and Resource Allocation (PS/RA). Without an effective PS/RA system, plans for technology programs will remain paper plans and have little chance of being put into practice or coming to fruition. Figure 7-8 lists some desirable characteristics of an effective PS/RA system. Figure 7-9 lists the components of a PS/RA system to support strategic technology planning. Again, as with the ideal planning system itself, not all these components may be feasible or even desirable in a given company situation. However, they constitute the building blocks for an effective PS/RA system that can support and enhance the STP process.

5.7.1   Some Comments on the Points in Figure 5-9

1. A true partnership in STP between top, functional, and business management in the corporation with technology being a full partner instead of a "corresponding" or silent partner. This means open, frequent, and frank discussions of ideas, hopes, dreams, limitations, and the harsh realities that can affect the technological future and the business future of the company.

2. Challenging, mind-stretching sessions in which people from all relevant groups are stretched in their knowledge and judgment to initiate or react to ideas and possibilities (threats, needs, and opportunities).

3. A nonthreatening organizational and personal environment that does not punish for wild or wrong ideas or ideas that challenge accepted doctrine.

4. Mechanisms for "instant" access to all levels involved in STP when an idea is hot or a planner is hot and wants a reaction without waiting for the next committee meeting or the next planning cycle.

5. An information system that protects information that is sensitive from the company viewpoint but does not withhold it from planners who need it.

6. An open, honest, and analytically powerful and nonthreatening method of generating probabilities and their associated states of nature for planning purposes. (Some aspects of Delphi have this capability, but it is sometimes abused.)

7. An effective external technical and social/economic/political intelligence system that provides credible and continually updated assessments and early warnings.

8. A set of analytical and synthesizing techniques for data analysis, including qualitative, ill-structured "judgment" data.

9. An atmosphere of enthusiasm among top and all levels of management and technology people for the STP process and for the benefits it can achieve, including the side benefits of improving communication within the company and raising the general level of company "smarts."

10. A monitoring and control system that continually corrects and incorporates information and decisions without waiting for formal requests, time periods, or crises.

11. Action and decision plans (meta plans) for following up, verifying, stimulating, and elaborating outputs in a form that enhances proper implementation.

5.8 WHAT ABOUT FOCUS?

Much of current management literature advocates more "focus" than has been customary in recent years in many decentralized companies. Perhaps as a reaction to the sometimes wild diversification programs that many companies plunged into during the 1960s and 1970s, observers of the management scene are counseling, in the 1980s, that firms should "return to what they know"—markets, products, technology—and to "disinvest" in many of the businesses they entered while diversification was the dominant trend. For some companies in some industry sectors—such as steel, machinery, oil—the firm's original or earlier focus is no longer a viable option, unless they want to "shrink down," as a few of them have opted to do.

Focusing seems like good general advice, where feasible, because many top managements and their staffs have gotten in over their heads in terms of trying to guide large diversified companies in many different fields with little synergy between them. The philosophy of extreme decentralization, which appeared to allow for almost unlimited diversification, was counted on to keep the entire diversified corporation "going and growing" through heavy dependence on businessmen who knew their sectors—the division managers. These DMs were allowed high autonomy, except for financial and legal matters of great significance, and it was expected that they would maximize the firm's interests as well as those of their own divisions. This worked very well for a while and is still working for many individual firms. For many others, however, extreme diversification away from businesses the top management knew and were comfortable with, combined with extreme decentralization where the locus of most decisions was at the division level, has not served them well. Thus, the 1980s have become, for some at least, the era of de-diversification and redeployment of assests. The sell-off or close-down of businesses that "don't fit" has accelerated.

But this is not a book about general *business* strategy, although it is hard to deal with *technology* strategy and management of technology without considering business strategy as the context.

This *is* a book about *technology* strategy, however, and the issue of "focus" or "refocus" is a key one that must be addressed. Along with the advice that corporations refocus their lines of business, comes advice that they should also refocus their technology programs. In general, that sounds reasonable, but let us take a closer look. One "soft" statistic that needs to be kept in mind is that most (in some fields up to 95%) technology products do not succeed in meeting their original goals or the expectations of the people who finance them. The higher success levels occur for minor or incremental variations in products and processes (see Figure 7-28). The higher chances of failure occur for attempts at radical innovations—"really new" products, processes, and materials.

In some industries and technologies, the *actual* success rate (as compared to the predicted or hoped for success rate) is about 1 in 100. In particular industries, in fact, radical products come along only once or twice in a decade, if at all. While writing this section I had lunch with a research manager from a leading food company who said his company had not developed a major new food product in over 15 years. Given the staggering odds against a hugely successful new product or revolutionary production process that will have significant impacts on the firm's sales, cost structure, and profits, there is a strong need for a "portfolio" approach to technology projects. This means that the technology program of the corporation and even of a particular operating division should be diversified enough to "play the odds."

A company that continues, in its internal R&D, licensing, or technology acquisition program, to "flog" a dead or dying technology may appear to be playing it safe in terms of disastrous short-term results. There is *always* some more cost or value to squeeze out of an existing product line or technology (e.g., a material or a production method), if enough skilled personnel are devoted to the squeeze. However, while the firm's funds and technical personnel are being devoted to this "incremental" technology effort, major changes may be going on around them and leaving them behind.

It is hard to count the number of companies we have worked with which persisted in focusing their entire technology effort on the products, processes, and materials they were familiar with and had been successful with for many years. This is not a bad *partial* technology strategy. Certainly, people who know a lot about a certain field of technology have the possibility of further milking it for economic benefits. Indeed, earlier in this chapter I attempted to make a strong argument for doing just that—continuing to invest in technology focused on current product lines. This is in the expectation of prolonging their lives and/or reaping more profits out of them while they are still viable in the market. However, this effort should only be part of a *mixed* technology strategy that also includes risk taking and attempts to play the odds of coming up with a radical or "discontinuous" development.

Such a mixed strategy involves distributing the technology or R&D funds in the company over the categories listed in Figure 7-28 of Chapter 7. The strategy includes investment in categories in the lower part of that figure, including subprograms of developing and pursuing a "future market mission" and exploring extensions of the current fields of interest. It can provide the basis for growth from within and survival in a market that is highly competitive due to technological differentiation of products and the processes that produce them.

In this sense then, I strongly urge management *not* to focus too strongly or too narrowly if that means concentrating (almost) entirely on current product lines, production technology, and other traditional technical approaches (e.g., materials, methods of servicing products). One of the biases in this book is that somewhere in the firm there should be a capability or set of capabilities for probing the technological future, planning on how to exploit it or avoid being crushed by it, and actually doing something about it in tangible form, such as developing radically new products and technology. If my picture of the severe limits on this broad technology strategy at the divisional and division manager level are accepted, then the alternative has to be some form of corporate-level effort.

Of course, in the 1980s, this is a countertrend. Many more firms are closing down their corporate technology efforts (more than 10 major cases occurred during the writing of this part of the book in 1986–1987) than are starting new ones. Firms that want to stay alive technically and gain the competitive advantages of internally generated new technology must seriously consider bucking that trend. Along with going against this trend—which by now appears to be developing into the conventional wisdom that corporate-level technology or R&D is not cost effective—they must also broaden the focus of their technology programs to play the odds in an era of rapidly advancing technology and increasingly ruthless competitive "crossing-over" into "other peoples' markets."

A policy of diversifying the technology strategy, both in terms of the portfolio of internal projects and the sources used or considered for obtaining technology from outside, has certain consequences—some positive and some potentially negative. The obvious positive consequence is the likelihood of increasing the chances of making a hit—a radically new technology. Up to a reasonable limit, the more "balls in the air" in a given program of development, the more chances of achieving a good outcome, as long as each line of development is sufficiently staffed and funded at a high quality level. And that, of course, is the obvious disadvantage: such a strategy can be very costly. The cost of apparent duplication of effort can be offset by the spin-offs that can come from saturating a field with talent competent enough to be alert for other uses of the technology in the firm's lines of business. This is why it is important that such an effort be cross-divisional, as well as at the corporate level (see Chapter 4 on divisional networking). However, there is no escaping the fact that a diversified, broad-range technology program aimed at radical developments is not cheap. As has been said elsewhere in this book (perhaps too often) such an approach requires an *investment* view of strategy for technology rather than an expense view or a "damned expense" view. Too little focus in technology programs can be costly and can fail due to inadequate concentration on its component lines of investigation. But *too narrow* a focus, which has become an increasing characteristic of many diversified firms in their component operating divisions, is very unlikely to yield the kind of payoff that good in-house technology programs are capable of producing.

____6

SOURCES AND FLOW OF IDEAS AND TECHNOLOGY ACQUISITION: FROM WHERE DO IDEAS AND NEW TECHNOLOGY COME?

6.1 INTRODUCTION AND CHAPTER OVERVIEW

On the one hand, many managers say that "ideas for new products and manufacturing methods are our life blood and everything must be done to encourage their
generation and use in the organization." On the other hand, however, some of the
same and many other managers believe that internal sources of good ideas have
dried up, never were much good, and are not likely to meet the firm's needs in the
future. This latter attitude can be a self-fulfilling prophecy. If managers believe
that internal ideas are no good or not forthcoming, then their attitudes and their
turning away from internal sources will certainly not encourage their scientists and
engineers to go out of their way to generate, communicate, and push their ideas.
In this chapter we look at the perceptions and behaviors that affect the process
whereby ideas are generated, developed, communicated, screened, and placed in
the portfolio of projects to be worked on. We also look at the barriers and pitfalls
involved in this delicate process.

This internal idea flow process is also examined in the context of the general
issue of "make or buy"—to what extent the firm does and can rely on internal
ideas and to what extent it has to and does go outside for ideas and more-or-less
fully developed technology. A particular source of outside ideas—the university—
is discussed. It has not yet been exploited very much by most commercial firms,
although many have tried with less than satisfactory results.

A third source that has met with a lot of interest but mixed results is the buying
or buying into of small "high-tech" firms—many of them recent start-ups. Issues
in using this source of technology are also discussed.

If indeed ideas are the life blood of a firm's technology program and the source of support for its technology base, then a better understanding is needed by managers of the sources and uses of technical ideas and the technology that arises from them.

6.1.1   Some Issues/Questions

- From where do internal ideas for new and improved products and manufacturing processes come?
- What encourages their generation, emergence, and flow?
- What discourages idea generation, emergence, and flow?
- How can more and better ideas be encouraged?
- How does the issue of "make or buy" of technology arise?
- What are the advantages and disadvantages of make versus buy?
- What mechanisms can be used to establish and carry out make or buy policies?
- What role does and should R&D play in acquiring external technology?
- Are universities good sources of technology for new and improved products and processes?
- What are the issues involved in dealing with universities?
- What are the issues involved in "high-tech" acquisitions?

6.1.2   Main Body of Chapter

This contains a discussion of the idea, flow process, make or buy decisions, relations with universities, and high-tech acquisitions.

6.1.3   Implications for Management

In general, technical groups in the company (e.g., R&D, product and process development, production, engineering, technical service and customer service, market research) are potentially good sources for product and process ideas. Sometimes, however, ideas from such groups are only partially developed, withheld, bogged down in a screening procedure, or kept in the minds and notebooks of the technical people. Specific, direct means have to be used to identify, screen, and develop such internal ideas, including those gleaned from customers in terms of "needs or notions"—ideas that may not yet be fully developed in terms of a clear need and a means of realizing them in concrete and cost-effective form.

Acquisition of technology (products, know-how, materials, processes, equipment) from outside the firm should be a company-wide effort with all the relevant technical groups participating, from R&D to production and technical service.

Where feasible, new technology should be a combination of the "most appropriate" technology, from whatever source—internal or external. There should be a significant contribution by internal technical people to assure their motivation, continued interest, "handles" for continued improvement and innovation, a residual of imbedded technology capabilities related to the particular field of technology (see Chapter 8), and the competitive advantage that goes with having the capability to innovate continuously in a particular technology or set of technologies.

6.1.4 A Watch List

- From where are the ideas coming?
- Are there "enough" in quantity, quality, and timing?
- Are internal ideas getting bogged down in generation, flow, and review procedures?
- Is externally acquired technology being effectively adopted and adapted to company needs and operations?
- How do the costs and overall effectiveness of "make" versus "buy" compare?
- Does external technology acquisition lead to a "one off" event, or are there continuous spin-offs and follow-ons to the firm's know-how and technology base?
- Are the incentives and barriers to effective idea generation, flow, review, and adoption understood and under control?
- Are relations with universities yielding useful technology?
- Is the high-tech acquisition being integrated in a cost-effective manner?

6.2 IDEA SOURCES AND FLOW INSIDE THE FIRM

Why focus on *ideas* in a discussion of technology management and sources of technology for the firm? My reasons for including this discussion in the book stem from three sources:

1. The importance of the idea as a starting point or vehicle for generating and developing technology inside the firm or as a vehicle for searching externally for technology.
2. My own involvement and that of a large number of my graduate students in basic research on idea flow over two decades through the Program of Research on the Management of Research, Development and Innovation (POMRAD), first at M.I.T. and then at Northwestern University.
3. The opportunity provided by the discussion of idea flow to suggest to the practitioner of technology management some modes of analysis which can help improve the generation and flow of ideas in his own organization.

6.2.1 The Importance of Idea Flow and Project Selection in the Total R&D/Innovation Process

The early phases in the life of an R&D program—the generation and communication of ideas, their initial examination for feasibility, and the choice of which ones to support—are critical to the overall effectiveness of the program. The decisions resulting from these phases involve the allocation of resources and assignment of personnel in a pattern that is costly to modify or reverse. Where large efforts are involved and choices are mutually exclusive, these decisions may, in the short run, be irreversible.

Particularly in those areas that entail specific objectives, time constraints, and limited funds, the point of greatest flexibility in resource allocation is during the idea proposal and initial project selection phase. In later phases, when work is actually in progress, it becomes increasingly difficult to change direction or reallocate resources economically.

Effective project selection from among "good" idea proposals is critical to both small and large organizations. Several thousand operating divisions of industrial companies are currently supporting R&D at what may be a minimal level or a level that may be less than the minimal effort needed for accomplishing anything more than routine product or process improvement or technical service work. When such an organization does decide to fund one or more projects aimed at a radically improved or brand-new product or process, it may be stretching all its available technical resources to do it. Under those circumstances, the choice of which project or projects to support is critical.

Even for the large company with massive technology resources, these are critical choices. If the technical personnel are competent and in the forefront of their fields, there may be many more promising project opportunities than can be supported at any given time. The search then ensues for an optimal "project portfolio" that will provide the highest expected returns on the funds and other resources committed to it.

Managers of all sizes of companies have become increasingly insistent in the past few years that their technical personnel devote more time and more careful analysis to this critical problem of idea generation and project selection.

In many technologies—both civilian and military—the total allowable time from conception of an idea for a project to introduction of its results into the factory, the field, or the marketplace has been drastically shortened due to competitive pressures. In many instances, total allowable time or "lead time" has been cut to the point where an orderly sequence of follow-on activities is not possible. That is, functions such as plant and equipment design, plant and equipment construction or purchasing, product and process engineering, test and evaluation, market research, initial promotion and advertising, and training of workers must be started before there is assurance that the new product or process will be successful technically and economically. Consequently, additional resources are allocated without the assurance that the project will ultimately be completed and be economically successful. This acceleration of a previously "leisurely" process places additional importance on the idea flow and project selection stages of R&D/innovation.

Since we know that *good* ideas are very infrequent occurrences, even in R&D, we have not attempted, in our research, to use these units as the only or major "proximal output" of R&D. Other lines of our research have concentrated on information output and, in some cases, on material output such as prototypes and samples of material. The "idea," however, seemed to be a potentially powerful basis for getting at essential aspects of the R&D/innovation process in a systematic way and assessing its ongoing performance, short of evaluating its ultimate— economic—outputs.

Unlike the exact time at which the idea of "ideas" as important units of analysis occurred to us, I can fix the exact day when the idea or potential proposal for mounting a study of idea flow on a "stand-alone," systematic basis arose. It was during a session of our annual Seminar on Organization of R&D in 1960–1961 when a member of the seminar (whose members were local managers of R&D) said: "It's O.K. to talk about selecting the best projects out of those proposed in an R&D lab, because you can't fund all the good ones. But our big problem is that we don't have enough good ideas to even use up the money we have available."

We were to remember that comment vividly over the next few years as we carried on one of our major field studies in his laboratory and uncovered scores of "excellent" and "good" ideas which had existed in the minds, notebooks, and memoranda of people in his laboratory during the period when he complained of a lack of enough good ideas to fund.

Our methodological reason then for selecting the idea as a unit of consideration for research on R&D/technology was in addition to its substantive importance in the total R&D/innovation process. It was the hope that it would lead us into some of the key aspects of the R&D/innovation process in a series of discrete events that could be identified and crudely measured in time and place.

6.2.2 What Is an Idea?

Our focus was on the *idea*, which we define as "an actual or potential *proposal* for undertaking new technical work which will require the commitment of significant organizational resources, such as time, money, personnel, and energy." Typically, such a proposal, if accepted by the organization, will result in the establishment of a new project. Thus, we set about studying the origins and adventures of the following:

1. Proposals that eventually are accepted and supported as projects by the organization.
2. Proposals not accepted or supported.
3. Potential proposals that never arrive at a decision point for the organization but that are disposed of in some other way than outright acceptance or rejection.
4. "Nonproposals" that relate to work being carried on without having been formally proposed to the organization. This category entailed some difficult conceptual as well as empirical problems. For example, could we properly say that an idea or a proposal "exists" in the organization prior to the time that it is communicated to someone either formally (such as in seeking official approval and funding for it) or informally (such as mentioning it

to a colleague). Furthermore, if we could properly say that it *does* exist in this precommunication state, how could we gain access to it for purposes of studying its evolution?

Managers of technology continuously seek to avoid two general kinds of error: (1) failure to undertake "good" projects and (2) undertaking "bad" projects. The reasons for the difficulty most organizations have in avoiding these two types of error are inherent in the R&D process itself:

1. The outcomes of individual projects, programs, and the whole R&D process are highly unpredictable. That is, for other than technologically trivial projects, project selection involves decisionmaking under (at best) risk where probability distributions can be associated with outcomes—or (at worst) uncertainty—where such probability distributions are not available or feasible to construct.

2. The outcomes of individual projects occur with time lags of months or years, during which period some of the factors entering into the initial project selection decision—for example, market demand, material prices, competition, internal skill levels in R&D, available supporting technology— may change significantly.

The original study design (involving half a dozen dissertations, M.S. theses, and staff studies) was aimed at a better understanding of the following aspects of the R&D process as they affect idea flow and vice versa:

1. Factors associated with the *generation* and *communication* of ideas for new projects.

2. The *criteria* used in *selecting* among proposals, for example, time horizons, relation to goals of the organization, the originating individual or organizational component, estimated economic outcomes.

3. The *estimation process* for the significant aspects of proposed projects, for example, time patterns of costs and returns and probability of success.

The general classes of factors which were believed to influence idea generation, idea communication, estimation, and project selection behavior were technical, economic, individual, and organizational.

1. *Technical.* These include the state of the particular arts involved in the idea and related arts at the time an idea is proposed or a project is selected. Some fields are newer than others and may be more attractive or easier to exploit than others which have been under study for some time. Some fields require a high concentration of talent from various disciplines. Some are amenable to pencil and paper analysis; others require massive experimental or test equipment and facilities.

2. *Economic.* In some lines of business, almost any major improvement can lead to high economic gains for the company. In some of the older industries such as

primary metals processing, food processing, and metal working, however, technological breakthroughs are rare and many of the basic products and processes, which involve massive capital investment, have resisted drastic change for decades. Some companies and operating divisions view this as a highly attractive situation, where first-rate R&D can bring about revolutionary changes and great economic gains, as have occurred in more science-based industries such as chemicals, pharmaceuticals, and electronics. Others view it as an absence of opportunity for great economic gains and rely on R&D primarily for modest improvements in the basic manufacturing process and in the products produced by that process.

Many of the economic computations used in project selection are highly simplified and fall far short of representing the actual economic relations between the variables of cost and return (see Chapter 7). Many companies have adopted or modified the discounted cash flow technique from capital budgeting, for use in project selection. In this technique, the time patterns of costs and returns are calculated and, by using a proper discount or interest rate, the present value of the project is calculated. This has tended to focus some of the behavior of the various people involved in project selection on a common decision model and holds promise for improving the actual estimates that go into the computation. In the past decade, many companies have programmed the cash flow model on computers and continuously simulate the cash flow from particular projects as progress is made on the projects and as new information about the relevant economic variables becomes available.

3. *Individual*. Despite the formal terms in which the project selection procedure is often clothed, emphasizing the economic rationality that underlies it, the raw material for these decisions is heavily influenced by personal factors. These include risk propensity, career aspirations, scientific level, type of orientation (organizational versus professional), time in the company, personal reputation, historical "batting average" (proportion of good guesses or proposals to bad), and other factors that contribute individual bias to the decision process.

4. *Organizational*. Some of these factors are the distribution of power and influence in the laboratory and the company, relations between functional areas (e.g., marketing versus research), organizational location and status of the estimators and the decisionmakers, criteria for evaluating R&D performance, reward and penalty system (degree of asymmetry and response time for risk taking), general reputation of R&D in the company, analytical sophistication of the estimators and decisionmakers, availability and use of records on past performance, and organizational climate for cooperation, communication, and joint decisionmaking.

In the series of idea flow studies we undertook in the 1960s and 1970s, primary emphasis was placed on individual and organizational factors. This does not mean that technical and economic factors were ignored—a technically or economically bad idea is still bad, no matter how open, trusting, warm, and friendly the people in the organization are. However, the basic motivation for this study was to help redress the balance in perspective of the research on project selection which had been done up until the early 1960s when our formal studies began. The overwhelming majority of academic research and in-house company research on the subject appeared to concentrate on two aspects of the process: (1) technical and economic and (2) creativity and the creative process. Our efforts were aimed at contributing to knowledge about organizational factors and other aspects of individual behavior and perception in addition to those primarily associated with creativity or "rational" decisionmaking.

Figure 6-1 suggests a sequential process in the generation, communication, evaluation, and disposition of ideas for R&D projects (with, of course, many feedbacks, recycling, and short-circuiting of steps). In general, our idea flow studies involved the first six or seven steps in Figure 6-1, although many of the hundreds of ideas identified and examined not only went all the way through the formal project selection process during the course of our field studies but ended up as successful (or unsuccessful) business ventures or product and process improvements. Later studies in our own "project portfolio" followed ideas all the way to their end-of-life (see Appendix B for citations to such studies with terms

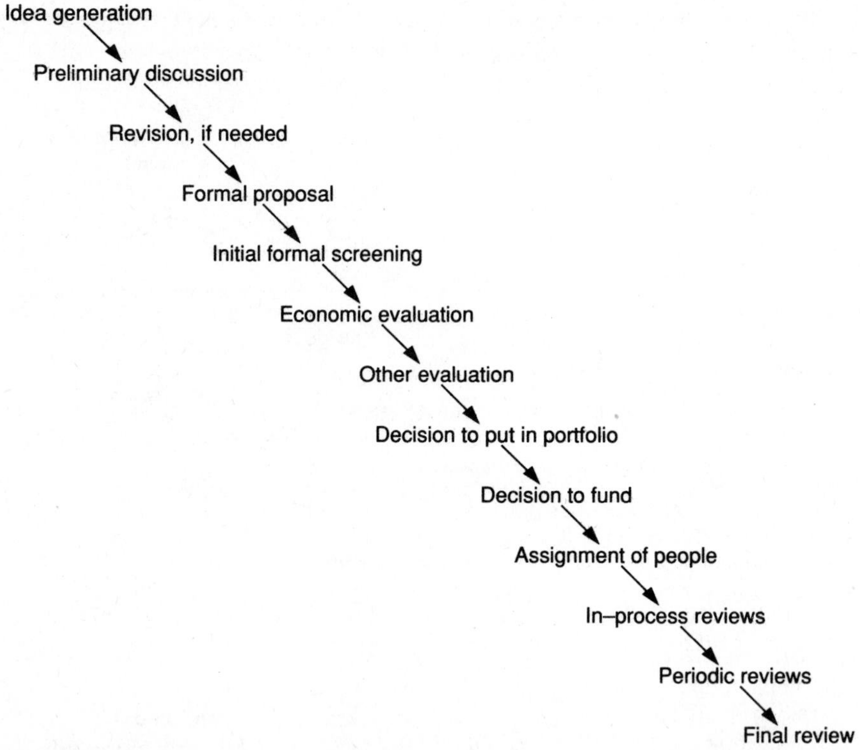

Idea generation
　Preliminary discussion
　　Revision, if needed
　　　Formal proposal
　　　　Initial formal screening
　　　　　Economic evaluation
　　　　　　Other evaluation
　　　　　　　Decision to put in portfolio
　　　　　　　　Decision to fund
　　　　　　　　　Assignment of people
　　　　　　　　　　In–process reviews
　　　　　　　　　　　Periodic reviews
　　　　　　　　　　　　Final review

Figure 6-1. General phases in the project selection process for a given proposal. Not all steps occur for all proposals, and not all steps occur in the same sequence. Also, recycling to an earlier phase is not uncommon; see Figure 6-3, for example.

such as "project selection," "barriers to innovation," "technology transfer," and "evaluation" in their titles).

In addition to a focus on the individual and organizational aspects of the idea flow process, the main study was also confined primarily to the R&D laboratory itself and the "subjects" to R&D professionals. Although the idea flow and project selection process in a firm typically involves many people in addition to R&D personnel, our particular interest in the idea flow studies was in the perceptions and behavior of the professional researcher (scientist or engineer) whose duties include, in most laboratories, the initiation and proposing of ideas for new projects.

Certainly, ideas for R&D come from many sources, both inside and outside the laboratory. In some instances, where the company is highly dominated by marketing, a significant percentage of ideas come from salespeople, market planners, and customers. Likewise, in a production-dominated company, many of the ideas come from the factory or production staffs.

6.2.3 A Flow Model for Ideas

Figure 6-2 starts with a first approximation to the total list of possible ideas that might be proposed for project status in a particular organization. This, of course, is sheer speculation without a thorough knowledge of the technoeconomic environment and capabilities of the organization. That is, in order for an observer to make a reasonably comprehensive list of all the ideas that *might* be proposed in a given organization at a given time, he would have to know a great deal about the business the company was in, its economic resources, the current state of its technological sophistication, the technical capabilities of its personnel, the states of the various arts that were involved in its underlying technical fields, and so on.

Even with all this information, however, it still might not be possible for any two experts to agree on a common list. Fortunately, or perhaps as a consequence of this difficulty, the design of the idea flow study did not require that such an actual list be drawn up. It did, however, require the *concept* of a feasible list of "technoeconomic opportunities" for R&D work by the organization. This should be a "feasible" list in the sense that even a casual, if not expert, observer can distinguish between the kind of realistic opportunities that are available to an organization with great technical and economic resources as compared with one that has modest technical and economic resources.

One clue to this feasible list is the behavior of other organizations engaged in the same fields as the organization being studied. This notion was more closely examined in a separate set of studies (McCarthy, 1965; McColly, 1967) where we attempted actually to establish a rough feasible list of potential projects for the firms in two narrowly designated areas of R&D.

The use to which we wanted to put the concept of such a feasible list was as a starting point in attempting to define an *actual* list of potential ideas that might be proposed by the individuals in the organization. That is, the interaction between this total feasible list of technoeconomic opportunities for R&D projects

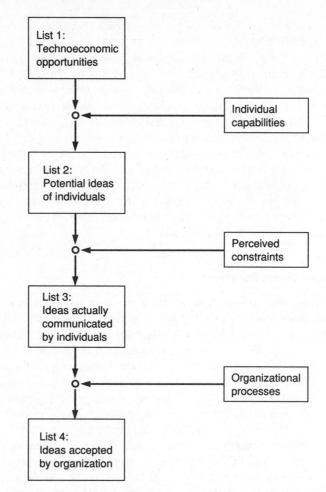

Figure 6-2. A flow model of the sources of ideas or project proposals and the factors affecting them.

and the abilities and other characteristics of the individual members of the R&D organization generates a second list. This second list might be called the list of all *potential* ideas that actually might be proposed by the individuals in the organization at a given time.

As an illustration of the source of a portion of such a list of potential ideas, consider a new researcher entering an organization. He brings with him certain abilities, formal training, skills, knowledge, interests, and experience with certain classes of problems. Upon arrival in the laboratory of his new employer, he learns about the businesses the company is in; the nature of its products, production processes, and services; what is currently going on in the laboratory and elsewhere in the company that relates to technology; what has been tried in the past; and what people are saying and thinking about future possibilities. Through some mysterious psychological process—which we variously call creativity, inventiveness, problem solving, or serendipity—he combines some of his abilities and the information he has collected into what might be called an *idea* for a potential project. He might, depending on his capabilities and the amount of information he has absorbed, have a number of such ideas over a period of time or at any one time. The sum of these individual lists of ideas may be conceived of as the total list of "potential individual ideas" that "exist" in the organization over a period of time or at a particular time.

In several of the more than a dozen field sites involved in the study, we did indeed make an attempt to take a total inventory of such *potential* individual ideas as well as ideas that had already been communicated to others in the organization. We attempted a complete inventory among all the professionals in four smaller laboratories and made partial inventories in several larger laboratories. Allowing for the many possible errors in obtaining these kinds of data directly from subjects, we believe that we did obtain a fair picture, in the four smaller laboratories, of the kinds of ideas that were in existence at the time we took the inventory.

Once we have this second list of potential individual ideas, we consider a third, reduced list. The ideas on this list are the ones that individuals in the organization actually *do communicate* to others in the laboratory and, in some cases, formally propose for project status.

The major reduction in size from the second list to the third list occurs through another mysterious psychological process within the mind of the potential proposer of an idea. This process has to do with his perceptions of the possible consequences for him—as an individual, as an employee, as a professional in his field, and in other possible roles—if he *does* actually communicate his ideas to other people in the organization. Although we did not intend to and were not equipped to probe very deeply into all the motivations involved at this stage, we attempted to get at some important aspects. For example, we attempted to learn how idea generators perceived the constraints placed on R&D work by the various levels of supervision and management. We also attempted to develop ways of evaluating individual risk propensity as a clue to how far an individual will go in testing the limits of such constraints. When list number two—potential individual ideas—is exposed to this set of factors, we can then expect to find a reduced list, number three, which consists of the ideas that the individual *actually does communicate* to others.

The factors that tend (1) to reduce the third list to the size of a fourth list—ideas *actually accepted by the organization* as projects—and (2) to modify specific ideas on the third list were the major foci of our study. We called these "organizational processes." These included, among other things, the communication systems, the power and authority systems, the systems of rewards and penalties, and the decisionmaking systems.

6.2.4 A Flow Model for Communication and Decisionmaking About Ideas

Our early concepts about the nature of communication and decisionmaking in the idea flow process are embodied in the schematic model of Figure 6-3 (Rubenstein and Hannenberg, 1962). The model was based on initial observations of the idea

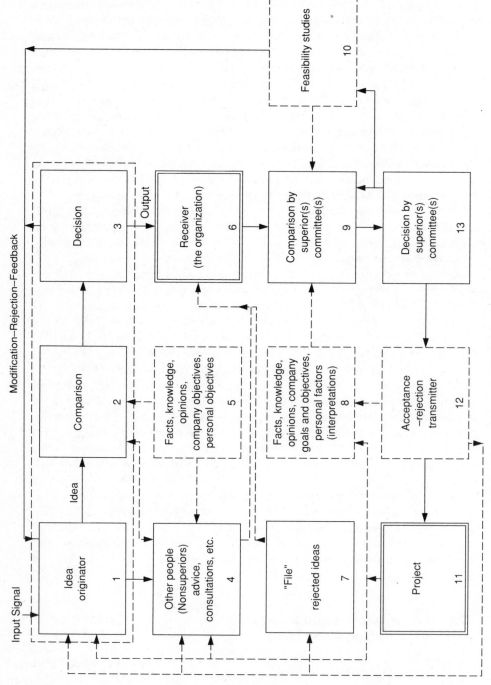

Figure 6-3. A flow model for communication and decisionmaking about ideas.

flow process and was helpful in indicating the factors that should be considered and the places at which they might influence the flow process. It suggests, as did the preceding flow model, the role played by accuracy of perception and individual enculturation and organizational learning in the idea flow process. An especially important aspect of the model is that it shows the researcher in the multiple roles of idea originator, evaluator, and decisionmaker, respectively.

For the purpose of discussing the model, consider a single individual in an R&D department, remembering that several individuals might be involved in one idea. The general flow of an idea that will be accepted as a project for technical work will be via the boxes numbered 1, 2, 3, 6, 9, 13, 12, and 11. Boxes 1, 2, and 3 represent the individual acting in three separate capacities: originator, evaluator, and decisionmaker, all relative to his own idea. At 6, he passes the idea on to the organization; this would be to any person or persons organizationally responsible for doing something about the idea. Boxes 9 and 13 repeat the evaluation and decisionmaking stages, this time at the level of the organization; at 12, the idea is being communicated and translated into a project and other information.

An idea originator may do a number of things with his own idea: for example, talk about it to others, consult others, try to convince them that it is a good idea, modify it, or forget it. In doing this, he may or may not reveal the complete idea. This interaction on the idea is shown by connections from boxes 1 to 4, 2 to 4, and 4 to 2. As the originator, he may just inform others about the whole idea or parts of it (1–4). Evaluation of the idea might be expressed by others even though advice is not requested (1–4–2), or advice may be obtained as the direct result of a request (2–4–2). On the other hand, the individual may "consult" only the body of facts, knowledge, personal perceptions, and so on that he himself has built up (5–2). Also (as shown for convenience by the same number 5 box), the colleagues consulted have bodies of facts, knowledge, and opinions that *they* consult (5–4).

Parallel relations are indicated as occurring in other parts of the organization. The company evaluator (evaluators) calls on his body of facts, knowledge, and especially on knowledge about company goals and objectives to see if the idea is relevant (8–9). In certain instances, the decision might be made that not enough is known about the idea, and, therefore, a feasibility study is undertaken (9–10). The information supplied by such a feasibility study enters at 9 from 10 and is really a special case of the relation between 8 and 9; thus, 10 could be included in 8. On the other hand, an idea may be directly modified at the decision stage, then evaluated before the final decision—the cycle being 13 to 9 via the right-hand path, then back to 13.

The scheme is conceptual, and nothing shown is indicative of implied time lags in the process. For example, modification coupled with decisionmaking and a new evaluation might be instantaneous, the validity of the modification having been determined almost directly through the process of comparison prior to modification.

The remaining lines represent feedback relations and an alternate path for a given idea to enter into the organization. When the organization processes the idea, the results become available for reinforcing or altering its own body of facts, knowledge, and so on; it is available also for influencing the processes and behavior of the idea originator and other participants, as shown by the connections from 12 to 8, 12 to 1, and 12 to 4. Since similar relations would be expected to exist for feedback from idea originations to the general store of "facts, knowledge, . . . etc.," a connection from 3 to 5 would be appropriate. If details of the decision process at 4 were shown, then explicit connections from 4 to 5 might also be indicated. These possibilities are relevant, but their inclusion would tend to confuse the other relations in the model.

Once project status is reached, there is further opportunity for the results to influence the processes (11–8, 11–1, 11–4). Some of the possible results are that a project is dropped before it is started, is held up indefinitely—being displaced by more important projects, is a failure in some or many aspects, or is a success. The results may have a more profound influence on the body of beliefs (boxes 5 and 8) than other occurrences because, for example, it is easy to refer to a past failure as sufficient reason for not attempting a similar project. At another stage, 7, there exists a file of rejected ideas, potential or actual, that might be called on to supply "new" ideas to the process.

While the idea is still internal to the individual, he might modify it or reject it; there is the possibility that he may start it back through the system as shown by the connection from 3 to 1. Also, the feasibility study at 10 may influence the scope of the idea so that the originator may wish to make changes (10–1). An alternate path from 4 to 6 is meant to suggest that other people—for example, senior researchers or "gatekeepers"—have an opportunity to transmit the idea or influence its reception. This might occur with or without the permission of the originator and before or after he has made a decision about it. It is even possible that if the originator told some others about a part of the idea (1–4), or about something general that should be done, these others might subsequently come up with the same idea. Thus, in this manner or by coincidence, two or more people might originate the "same" idea.

Previously, the organization was mentioned in a way that suggested it is capable of making comparisons and decisions as a single unit. Actually, we view the organizational level first via the interaction of individuals in positions of organizational responsibility and then as a composite unit. In a given case, one or several managerial people may make a decision about an idea. They may have identical or very different perceptions of company objectives and criteria for approving projects. Each individual also has perceptions about what is expected of him in his organizational role. Thus, the study ultimately concerned itself with the influence of these perceptions at the different levels—researcher, group leader, department head, and so on—on the flow process.

6.2.5   A Conceptual Model of Constraints on Idea Generation and Flow

Figure 6-4 is a conceptualization of the process whereby the intentions, beliefs, attitudes, and criteria of higher levels of management are transmitted, perceived, attenuated, distorted, and otherwise changed as they "flow" through intermediate levels of management to the potential idea originators in the laboratories and elsewhere in the organization.

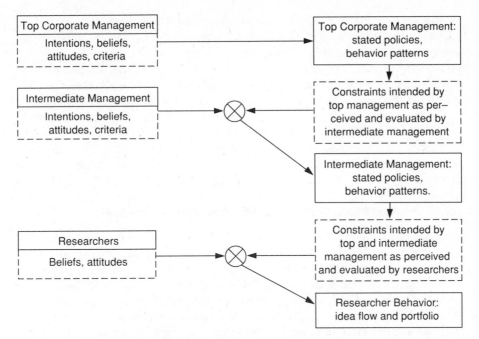

Figure 6-4. The chain of perceived constraints on R&D decisionmaking.

The main point of this diagram, originally developed in connection with our earlier studies of R&D in decentralized firms, is that many changes do occur in the transmission of such information from one level to another. Not all of these changes are, as perhaps implied above, in the direction of distortion or loss of information. Many of them involve clarification, elaboration, operationalization, and other processes that help make the higher-level intentions, beliefs, attitudes, and criteria more meaningful and more feasible to act on than they otherwise might be. Indeed, this is one of the main tasks of management in any organizational activity. In the case of R&D, however, we have observed more than the actual amount of dysfunctional change and loss in information as it is transmitted through the hierarchical levels. In several of the organizations we have studied systematically and the many more we have observed and been otherwise involved with, such information had to pass through as many as half a dozen or more individuals and organizational layers between the minds of top management and those of researchers who were generating ideas. Distortion and loss, under those circumstances, are very frequent occurences. In addition to each level (and intermediate individual) contributing change due to his own beliefs, attitudes, intentions and criteria, several of the transitions involved basically different organizational value systems and special languages. Objectives for R&D, when transmitted by division managers, for example, are often drastically changed en route *to* them from top management and en route *from* them to divisional laboratory researchers.

Several of our early exploratory studies of idea flow probed into this process of value and criteria transfers. Several of the main studies also focused on aspects of this same problem.

The very simple flow diagram of Figure 6-1 can help keep a perspective on the role of idea flow in the overall project selection process. It indicates that the generation of ideas and their formal disposition by the organization constitute only preliminary stages of the project selection process and an overall R&D/innovation process that is supported for the primary purpose of achieving economic or other socially useful results. It further indicates that mere formal approval or acceptance of an idea as a project is no guarantee that it will indeed be initiated and actively pursued as an R&D project. The simple linear flow portrayed in Figure 6-1 omits, of course, all the feedback, recycling, and reordering of steps which occur in a real, operating environment. But it can be helpful in keeping the idea flow process in perspective.

6.2.6 Researchable Questions, Propositions, and Models

The idea flow study (or series of studies) was the first project in our Program of Research on the Management of Research, Development and Innovation (POM-RAD), in which we tried a full-blown application of our evolving approach to proposition development and testing. Although individual theses and dissertations had previously used our evolving "research paradigm" (Figure 6-5), the idea flow study was the first instance in which we attempted to use it outside the classroom to structure a major piece of research, consisting of multiple theses and staff studies. Some of the theses and one or two of the staff studies followed the paradigm fairly rigorously. Others merely used it as a checklist or occasional recycling mechanism. This presented a dilemma in the structuring of this discussion of idea flow. Some of the "potentially researchable questions" and their associated "potentially testable propositions" (Figure 6-5) were indeed generated at the same time (we too have problems of tracing the genesis of some of our research ideas). Others were developed somewhat out of sequence, with propositions leading to further questions or to other propositions, without cycling through prior questions.

In an attempt to present the actual chronology of question-raising, proposition statement, and model development, I have followed a convention similar to that

---

Research area of interest
Potentially researchable questions
Potentially testable propositions and models
Variables and definitions
Indicators for the variables
Research instruments
Field study design
Field study
Proposition testing
Recycle*

Figure 6-5. The POMRAD research paradigm: our approach to proposition-based field research. (*Of course, there is recycling at each stage and the paradigm is not followed in rigid sequence but in a continuously developing and checking mode, involving continuous reexamination of previous steps.)

used in R&D itself. The basis for attribution of question, proposition, or model to a member of our "idea flow group" was its reduction to a written form of some kind—for example, memo, working paper, thesis, dissertation, or publication. In some cases, the only basis for attribution is a "back of the envelope" notation. Where feasible, this is also used as a "documented" source. In addition to the name(s) of the POMRAD originator(s) or transmitter(s) of the item (question, proposition, or model), the citations in this chapter also indicate the date of the written source of the item.

This, unfortunately, will not completely resolve the citation problem (a perennial one in group research). In any research group, certainly in ours, individual members vary widely in their propensity to reduce ideas to written form at all and in the speed with which they do this. Certainly, particular members of the group will recognize, in the cited work of a fellow group member, an idea or suggestion that he remembers giving to the cited author. Unless the cited author acknowledges this original (even if partial) source, I had no choice but to attribute the item to the author of a written document. I hope that my former and present students and colleagues who worked on the idea flow project will accept this attribution procedure, recognizing its inherent weakness in terms of our own theoretical and empirical work on idea flow (Barber, shear thyself!). One further complication, of course, is the relation between student and thesis advisor with respect to ideas—the classical problem of how much of a thesis is a product of the student alone versus the student–advisor team.

I have attempted, in this section, to group research questions and propositions under a few major headings. This task has proved quite difficult, due to the multiple-category nature of many of the questions and propositions that we generated. However, I have decided to place a proposition in only one place. The criterion for categorization used is the distinction between independent and dependent variables, with respect to the major focus of the research. That is, the propositions are grouped according to independent or explanatory or apparent "causal" variables with respect to the behavior we are attempting to predict or explain. This is a convenient scheme, since we were attempting, in the overall study, to explain or provide a basis for predicting a small set of behaviors related to the idea flow process:

1. The origination of an idea.
2. The decision to communicate it and the act of communicating it.
3. The decision to submit it to the organization and the act of submitting it.
4. The disposal decisions and behavior of those organizational members with the power to act on proposals (submitted ideas).

Therefore, the categories under which I have grouped the items in this chapter have to do with sets of independent or explanatory variables which might shed some light on the four sets of dependent variables listed above. In addition, of course, there are some propositions and questions related to parameters (e.g., environmental or other constraints) which influence the behavior of the dependent variables but were not in themselves major foci of the empirical research we undertook, except as "controls."

Finally, there is the unclear distinction between independent and dependent variables in organizational (as well as other) research. In most of our conceptual models or "box and line" diagrams of the idea flow and related processes, the convention used seems to imply that causality and influence of one variable or another is generally unidirectional. This is far from the case. The incidence of mutual causality and mutual influence of a pair of variables on each other is very high in organizational behavior. To keep the flow diagrams relatively uncluttered and to keep our attention focused on the hypothesized *main directions* of influence and causality, we generally do not explicitly include the "feedback" relations between variables, except when they themselves are a major focus of the research or conceptualization.

For this and other conceptual and methodological reasons, the distinction between independent and dependent variables (and in many cases, parameters) is not always clear in our models and those of other researchers. As an operational rule, we generally take the *dependent* variable(s) in a proposition or model to be that (those) whose behavior or variability we are attempting to explain or predict. The *independent* variables are those which appear to have or which we postulate to have explanatory or predictive power with respect to the, arbitrarily chosen, dependent variables.

As an illustration of the problem this attempt at distinction leads to, consider the relation between "submission behavior" (submission or nonsubmission of an idea for formal evaluation) and "disposition decisions" (decisions to accept or reject the idea for the organization's project portfolio). Certainly, they each have an effect on the other. Submission behavior directly affects disposition behavior, since the rate of disposition (among other characteristics) depends (among other things) on the rate of submission of ideas for consideration and disposition by the decisionmakers in the project selection process. However, it is also clear from the results of our idea flow studies that disposition decisions, when fed back to potential idea submitters, do influence their future submission behavior.

The key then to the way in which we originally formulated and presented the propositions generated in connection with the idea flow study lies in the focus of the particular investigator—the behavior he is attempting to explain or predict. However, a particular independent variable (e.g., "organizational age" of idea originators or their "attitudes toward organizational objectives") may have effects on more than one of the four dependent variables listed above. Furthermore, the literature is much more lavish in its treatment of what we are considering in the idea flow research to be independent variables than it is of our small set of dependent variables. This then posed the dilemma of how to present our combined sets of propositions in this chapter. The compromise was to present them in groups according to the dependent variables we were attempting to explain, facilitating later re-grouping into a matrix according to the independent variables and parameters which, according to the propositions, were believed to have some influence on the idea flow behavior with which we were concerned.

This elaborate discussion of the ordering of our propositional statements has been presented in support of our research methodology, whose pivotal point is the *a priori*, potentially testable proposition. That is, most of the empirical work carried on in this study, subsequent to the "early empirical explorations" of the study, were based on propositions generated and clearly stated in operational terms *before* the data were gathered. Although this may appear to be only a natural step in the scientific process, it is not common in much of the empirical literature of behavioral sciences, organization theory, and management. Few of the studies in the open literature contain *a priori* propositions derived independently of the results of the empirical work itself.

6.2.7 Which Came First—the Proposition or the Model?

One of the questions that frequently arises in connection with our research paradigm is whether models generate propositions or propositions generate models. The answer clearly is: both. The research process is far from a step-by-step, never-look-back venture. We would therefore expect that at various times we find ourselves (1) generating specific propositions prior to having a general scheme to fit them into and (2) constructing conceptual flow models which we expect will, in turn, generate a set of related propositions. Even more frequently, as we go through our stereotyped blackboard version of the research paradigm, we find ourselves generating both simultaneously. For example, as we write the first variable on the board and wonder how its value is determined, we visualize (and write to the left of it) the names of other variables which we believe may influence the first variable. The very act of writing the names of this second set of variables implies a set of linking propositions between each of them (or the entire set) and the first (dependent) variable. So that we must say that both of the following statements are true descriptions of our research behavior:

1. We generate models from our propositions.
2. We generate propositions from our models.

The reason for introducing this issue at this point is to clarify the fact that even though this discussion presents propositions first and models second, there is no necessary implication that that is the order in which specific items (models or propositions) were actually generated in the course of our research.

6.2.8 Potentially Researchable Questions

The overall research question with which we were concerned in this series of studies was: What factors influence the generation, communication, and disposition of ideas in an industrial R&D organization? Breaking this overall question down into stages, we were attempting to shed light on the factors influencing these dependent variables or stages in the idea flow process:

Time devoted to idea generation and development

Actual generation of ideas and their characteristics

Decision to communicate ideas or about ideas

Actual communication or consultation about ideas

Choice of a receiver

Decision to submit

Actual submission behavior

Disposition of ideas

Regrouping these questions into a smaller number of categories under which the propositions will be presented, we have:

I. What factors influence the idea generation process and the characteristics of ideas?

II. What factors influence communication of and consultation about ideas?

III. What factors influence submission behavior?

IV. What factors influence disposition of ideas?

V. What other factors and relations (interrelations of dependent or independent variables and parameters) have significant effects on the idea flow process?

6.2.9 Propositions About Idea Flow

Since the basic unit of work in our research on this subject was the proposition, we spent a good deal of effort throughout the life of the study (and since) in developing propositions that could be tested by our field studies and by others who were interested in similar or different methodological approaches.

Our aim was to generate the propositions through our evolving research strategy, using the literature, our own past and current research findings, and intuitive notions gained from first-hand observations of or involvement in the idea flow process.

Although we made several efforts during the course of the study to pull together propositions being developed by the various members of the project team, we did not achieve a full "proposition inventory" during the course of the formal project. This section then is the first attempt to do so for the idea flow study. Based on our experience in that, however, we established proposition inventories for several of our other continuing projects—Organization of R&D in Developing Countries ("Developing Countries"); Liaison, Interface, Coupling, Technology Transfer ("LINCOTT"); Technical Entrepreneurship in the Firm; and Information Searching Behavior of Researchers ("Infosearch"). The gathering and recording of propositions on these projects have been systematized and a computer storage and retrieval system has been established as a working tool of our overall research program.

The idea flow propositions in this section represent a relatively nonsystematic listing of propositions generated at various stages in the project by members of the group. As a concomitant of the open exchange of ideas in our research group, we are very conscious of authorship of ideas and make every effort to identify the individual who originated, or at least first communicated, the idea in the group. This is not always possible, since many of the ideas for propositions arose in group meetings or conversations between two or more members of the project team. Some of them, as in any academic setting, reflect discussions between thesis student and advisor or between fellow graduate students. The principal source used for attribution is therefore the existence of a written record of a proposition in a thesis, a memorandum, meeting minutes, or other written form. Where the source document identified a prior source of the proposition, such as a colleague or the author of an article in the open literature, double attribution is used. The notation used below is:

> Single source   Statement of proposition (name of author if he was a member of our research group)
>
> Multiple source   Statement of proposition (name of author and of other research group members, if given; last name(s) of primary literature source(s) keyed to reference list in appendix B of this book)

In an attempt at a partial ordering of the propositions presented below, I have arbitrarily grouped them into a category set, reflecting their major focus. Although this schema was not used during the course of the project itself, it seems a convenient way of presenting such a diverse set of propositions.

Within the substantive set of propositions listed in this chapter, there are some that were stimulated by the literature but that were never formulated as part of a field design. They are presented here to indicate the kind of "potentially testable propositions" which were generated during the study and which might yet be incorporated in a field study and tested against empirical data or which might be used by a technology manager as a guide or check list in looking at idea flow in his organization.

The subsets of propositions that were actually incorporated into thesis, dissertation, and staff work and subjected to empirical testing are indicated by reference to the author's thesis, dissertation, or staff study (e.g., Pound Dissertation, pp. 24–76). The results of testing these propositions in the field studies are presented in those documents.

The choice of which particular propositions to incorporate into a particular design was influenced by many factors. Primary among them was the set of interests of the graduate students and/or staff members who were doing the theses, dissertations, and staff studies. A major constraint was the practical one of access to empirical data needed to test particular propositions. That is, some of the variables were very attractive in their *nominal* definition phase but posed tremendous problems in terms of *operationalization* and *data gathering* in the field. In other cases, the amount of field access was relatively limited (e.g., the amount of time we could spend with individual informants or subjects). Finally, in some instances, there appeared to be too little convergent support for the proposition in the literature to make it a good bet on which to spend limited field resources.

The propositions are as follows:

I. Factors that Influence the Idea-Generating Process and the Characteristics of Ideas

*1.1.   The perception of time pressures associated with the current research work encourages idea generation and development associated with the current work but stifles idea generation and development not associated with the current work.

   NU Sources: N. Baker, 65/4; Baker, Siegman and Rubenstein, 64/25.[†]

   Literature Sources: March and Simon, 1958, pp. 54, 185; Kaplan, 1960; Jones and Arnold, 1962,[‡]

*1.2.   In order that "free time" be used for idea generation, the associated effort must be perceived as resulting in rewards that are (a) the same as or more than those provided for effort on current work, but with which the researchers are not saturated, and/or (b) valued by the research staff but not provided from effort expended on current work.

   NU Sources: N. Baker, 65/4; Baker, Siegman and Rubenstein, 64/25.

   Literature Sources: Marcson, 1960, p. 113; Kaplan, 1960, p. 25; Bennis et al., 1950; Storer, 1962.

*1.3.   In order for idea generation to continue over time, previously submitted ideas must be perceived as having been positively or enthusiastically received and evaluated by the organization's reviewers.

   NU Sources: N. Baker, 65/4; Baker, Siegman and Rubenstein, 64/25.

   Literature Sources: Jones and Arnold, 1962, p. 55; Kaplan, 1960, p. 25; MacLaurin, 1955; Williamson, 1960.

*1.4.   Prior to submission, research personnel tend to screen their ideas according to their relevance.

   NU Sources: N. Baker, 65/4; Baker, Siegman and Rubenstein, 64/25.

   Literature Sources: Cohen, 1958; Blau, 1963, pp. 127–130; Morris, 1962; Kornhauser, 1962, p. 64.

*1.5.   Perceptions of organizational goals and needs stimulate idea generation and development congruent with these perceptions but stifle idea generation and development not congruent with these perceptions.

   NU Sources: N. Baker, 65/4; Baker, Siegman and Rubenstein, 64/25.

   Literature Sources: Bralley, 1960; Hillier, 1960; Gershinowitz, 1960.

*1.6.   The scope of ideas submitted depends on accuracy of perception and degree of enculturation of the originator (submitter).

   NU Source: Utterback, 65/10.

1.7.   Idea production is related to the actual consensual role of the researcher.

   NU Source: F. Baker, 62/13.

---

[*]Incorporated into the empirical field studies.

[†] POMRAD Document Number. See Appendix B (NU = Northwestern University).

[‡]The reader will note the age of the citations. Most of the literature searching and proposition generating for the idea flow study was done in the early 1960s. Unfortunately, the state of the art in this field has not advanced much since then.

1.8. The type of research ideas generated and the evaluations of his own ideas by the researcher are functions of the researcher's role-ideal and his conception of the consensual role.

NU Source: F. Baker, 62/13.

*1.9. As an individual's accuracy of perception of the importance attached to certain evaluation criteria used by those influential over the objectives and activities of the laboratory increases, there will also be an increase in the number and the perceived relevance of his ideas.

NU Source: Pound, 65/12.

Literature Sources: Gage and Exline, 1953; Raven and Rietsema, 1957; Pepinski et al., 1959; Steiner and Dodge, 1957; Secrist, 1960; Hillier, 1958; Livingston and Milberg, 1957; MacLauren, 1955; Bralley, 1960; Jones and Arnold, 1962; Kaplan, 1960; Levering, 1958; Kornhauser, 1962.

1.10. The more enculturated the researcher, the more accurately will he perceive management criteria and the more will his criteria be in agreement with them.

NU Source: Siegman, 65/22.

1.11. The more enculturated the researcher, the more will his ideas fit into present organizational goals; the less enculturated, the more will his ideas tend to be divergent from present needs and goals.

NU Source: Siegman, 65/22.

II. Factors that Affect Communication of and Consultation About Ideas

*2.1. The number of ideas communicated by a researcher is directly related to his state of enculturation and his accuracy of perception.

NU Source: Utterback, 65/10.

*2.2. The more relevant an idea originator perceives his idea to be, the more likely he is to communicate it to other members of the organization.

NU Source: Pound, 65/12.

Literature Sources: Davis, 1953; Childs, 1962; Dearborn and Simon, 1958; March and Simon, 1958; Thibault and Kelley, 1959; Festinger, 1954; Newcomb, 1960; Asch, 1955; Avery, 1960; Kornhauser, 1962; Marcson, 1960a, 1960b; Pelz, 1956; Precker, 1952; Banta and Nelson, 1964; Homans, 1961; Bush and Hattery, 1956; Zaleznick, 1958; Hillier, 1958; Kaplan, 1960; Shepard, 1954; Collins and Guetzkow, 1964; Jackson, 1959; Newcomb, 1956.

*2.3. The greater the agreement on criteria between one individual and another in a laboratory, the greater will be the frequency of idea communication between them.

NU Source: Pound, 65/12.

---

*Incorporated into the empirical field studies.

Literature Sources: Homans, 1961; Jenkins et al., 1964; Barnes, 1960; Collins and Guetzkow, 1964; Rubenstein, 1953; Hertz and Rubenstein, 1953; Zaleznick, 1958; Bush and Hattery, 1956; Festinger, 1960; Blau and Scott, 1962; Katz and Lazarsfeld, 1955; Precker, 1952; Gage and Exline, 1953; Schellenberg, 1957; Allport et al., 1951; Hoffman, 1958; Burns, 1954; Triandis, 1959, 1960; Runkel, 1956; March and Simon, 1958; Newcomb, 1956; Homans, 1961; Thibault and Kelly, 1959; Rubenstein, 1957b; Lerner and Becker, 1962; Mellinger, 1955; Festinger, 1954; Avery, 1960; Jackson, 1959; Gullahorn, 1952; Backman and Secord, 1961; Collins and Guetzkow, 1964; Simon, 1954; Berkowitz and Bennis, 1961; Simon et al., 1950; Pigors, 1949; Barnes, 1960; Homans, 1961; Katz and Lazarsfeld, 1955; Davis, 1953; Martin, 1956; Shepard, 1954; Torrance, 1957; Columbia University, 1958.

*2.4. The greater the accuracy of perception of one individual (A) in a laboratory by another (B), the greater will be the frequency of idea communication from B to A.

Literature Sources: Same as 2.3 above.

2.5. Researchers are more likely to consult initially with their peers about their own ideas.

NU Source: Rubenstein, 63/16.

Literature Sources: Blau, 1954; Jacobson and Seashore, 1951.

2.6. Intragroup communication about ideas will be greater than cross-group communication in a highly structured organization.

NU Source: Rubenstein, 63/16.

Literature Sources: Blau, 1954; Jacobson and Seashore, 1951.

2.7. At least one liaison person will be found in each successful laboratory group who can transmit ideas outside the group.

NU Source: Rubenstein, 63/16.

Literature Sources: Blau, 1954; Jacobson, and Seashore, 1951.

III. Factors that Influence Submission Behavior

3.1. The less the researcher perceives his ideas will be accepted, the less will he attempt to "sell" them to management. This may occur even if, in fact, his criteria are the same as those of the organization and the organization would accept his ideas (if submitted).

NU Source: Siegman, 65/22; Utterback, 65/10.

3.2. A researcher with a high organizational and low professional orientation would tend to propose ideas that were more closely related to the current, obvious needs of the organization than would a researcher with high professional and either high or low organizational orientations.

NU Sources: Siegman, 65/22; Goldberg, 63/26; Rubenstein, 63/16.

---

*Incorporated into the empirical field studies.

3.3. The less enculturated the researcher, the less will he communicate his ideas to management.

NU Source: Siegman, 65/22.

3.4. Lack of understanding of the goals and criteria of the organization may result in submission of ideas not acceptable to the organization, as well as retention of potentially excellent ideas.

NU Sources: Utterback, 65/10; Pound, 63/24.

3.5. As a researcher becomes enculturated, the number of radical or new ideas he will submit will decrease.

NU Source: Utterback, 65/10.

3.6. A lower motivation to become accepted by a group may lead to the submission of a greater number of radical or new ideas.

NU Source: Utterback, 65/10.

3.7. The new recruit hired from a similar job may have the same set of criteria when he is hired as does the organization; he may submit the kind of ideas desired by the organization with little training or interaction having occurred.

NU Source: Utterback, 65/10.

3.8. An inventive person may submit ideas that are very useful but are not the result of any effort to meet the demands of the organization.

3.9. If an idea originator's (O) superior (S) shares O's orientation and has influence with his (S's) superiors at the next level, O will more likely propose his idea to S than if these conditions did not hold.

NU Source: Rubenstein, 63/16.

Literature Source: Pelz, 1951.

3.10. If S does share O's orientation but is not perceived by O as having influence with S's superiors, O will seek other channels for proposing his idea.

NU Source: Rubenstein 63/16.

Literature Source: Pelz, 1951.

3.11. R&D groups in a "low-science" environment, which are established with a "high-science" charter, will tend to drift into proposing ideas of a lower scientific level over time, unless their charter is continuously reinforced. Reinforcement can occur through combinations of the following kinds of mechanism: (a) an early, widely recognized success, (b) an independent source of funds (independent from "clients" in the organization with immediate, low-science problems), (c) adequate insulation from outside pressure, and (d) an internal *real* (as contrasted with more honorific) reward system.

NU Source: Rubenstein, 63/16.

Literature Sources: Weiner, 1960; Selznick, 1949; Alpert and Weitz, 1961; Avery, 1960.

3.12. If an idea originator can successfully rationalize to himself the rejection of his past ideas by the organization, he will continue to propose additional ideas.

NU Source: Rubenstein, 63/16.

Literature Source: Festinger, 1954.

*3.13. The flow of perceived ideas for changes in the firm's production process (many, but not all of them, coming from R&D) is influenced by the following: (a) the number of employees in the firm, (b) the risk-taking propensity of idea evaluators, (c) the ages and formal education levels of evaluators, (d) the evaluators' dissatisfaction with sales and cost levels, (e) exposure of evaluators to outside sources of information and contacts, and (f) the firm's freedom to alter its products.

NU Source: Martin, 67/29.

Literature Sources: Carter and Williams, 1958a, 1958b, 1958c; Mansfield, 1961, 1963; Bright, 1964; Myers, 1965; Chandler, 1966; Healry, 1954; Rogers, 1962; Strassman, 1959; Enos, 1958; Burns and Stalker, 1961; Grilliches, 1957; Danhof, 1949; March and Simon, 1958; Dill, 1962; Campbell, 1965; Sayles, 1964; Edwards, 1954; Hess and Miller, 1954; Hoffer and Stangland, 1958; Wallach et al., 1962; Wallach and Kogan, 1961; Baumol, 1961; Lionberger, 1960; Bell, 1962; Marsh and Coleman, 1955; Hildebrand and Partenheimer, 1958; Lapp, 1966; Coleman et al., 1964; Katz, 1963; Bourne, 1957; Gross and Taves, 1952.

IV. Factors that Influence Disposition of Submitted Ideas

*4.1. If the organizational reviewers perceive an idea to be relevant they are more likely to receive the idea positively and enthusiastically than if they perceive it to be nonrelevant.

NU Sources: N. Baker, 65/4; Baker, Siegman and Rubenstein, 64/25; Avery, 60/10.

Literature Sources: Kaplan, 1960, p. 26; Levering, 1958.

*4.2. The number of ideas selected from those submitted will increase as the accuracy of perception and enculturation of the originator increase.

NU Source: Utterback, 65/10.

4.3. Probability of acceptance of a submitted idea is a function of the congruence of the role-ideals of the submitter and the evaluator.

NU Source: F. Baker, 62/13.

*4.4. The greater the perceived relevance of a particular idea, the more likely it is to be accepted as an R&D project.

NU Source: Pound, 65/12.

Literature Sources: Barnard, 1938; Childs, 1962; Lionberger, 1960; Baker and Pound, 1964; Festinger, 1954; Menzel, 1960; Martin, 1956;

---

*Incorporated into the empirical field studies.

Mees via Jewkes, 1958; Churchman and Schainblatt, 1965; Runkel, 1956; Banta and Nelson, 1964; Allen, 1965.

*4.5. The greater the agreement on criteria among the persons involved in the evaluation of a given idea, the more likely the idea is to be accepted as an R&D project.

NU Source: Pound, 65/12.

Literature Sources: Same as 4.4.

4.6. The more enculturated the researcher, the higher the ratio of his submitted ideas that will be accepted by management.

NU Source: Siegman, 65/22.

4.7. The higher the organizational position of the idea originator, the more likely his ideas are to be accepted.

NU Sources: Utterback, 65/10; Pound, 63/24.

4.8. Organizations with a tradition of heavy dependence on science are likely to have a more diversified R&D portfolio of accepted projects in terms of time horizon and scope than organizations lacking such a tradition.

NU Source: Rubenstein, 63/16.

Literature Sources: Mansfield, 1962.

4.9. Organizations with a history of successful results from R&D are likely to include riskier projects in their portfolios than ones without such a history.

NU Source: Rubenstein, 63/16.

*4.10. The acceptance of an idea for changes in the firm's production process is influenced by the six factors listed under proposition 3.13, plus: (g) the fruitfulness of the idea source, as perceived by the idea evaluator, (h) the riskiness of the idea, (i) the perceived cost of implementing the idea, (j) the perceived availability of human resources to implement the idea, (k) the projected profitability of the idea, (l) the urgency of the problem to which an idea is a potential solution, and (m) the degree to which the idea is a potential solution to a problem for which the evaluator has actively been seeking a solution as compared to the idea being a potential solution to a general problem, where the evaluator is always looking for ideas, or an unsought idea.

NU Source: Martin, 67/29.

Literature Sources: Same as those cited under proposition 3.13.

V. Other Propositions—Relations Between Dependent or Independent Variables and Parameters

*5.1. Perceptions of organizational needs and goals are influenced by (a) the perceived reception and evaluation behavior of reviewers and (b) interaction with other laboratory personnel.

---

*Incorporated into the empirical studies.

NU Source: N. Baker, 65/4; Baker, Siegman and Rubenstein, 64/25; Avery, 60/9.

Literature Source: Houton, 1963.

*5.2.   Degree of enculturation and accuracy of perception (of organizational goals and needs) are related.

NU Source: Utterback, 65/10.

*5.3.   The amount of discrepancy between the role-ideal of a new researcher and the actual consensual role will decrease with increasing organizational age.

NU Source: F. Baker, 62/13.

Inspection of the many models developed from the above propositions suggests that a very large number of propositions can be generated which contain two variables, three elements (two variables and one parameter), or a larger number of variables and parameters. At one time or another, several hundred such combinations were written on blackboards, in notebooks, or in rough draft memos during the life of the idea flow project. Only a few survived confrontation with the literature, colleagues' criticism, and self-criticism. The propositions given above constitute an intermediate stage of the sorting out, combining and refining, and rejecting which took place during the project. They have been presented for illustrative purposes to indicate the kinds of potentially testable propositions that can be generated out of the models and literature related to the phenomenon of idea flow. Those few indicated by asterisks (*) are the ones that went through the entire process of operationalization, instrument development, data collection, analysis, and proposition testing. The reader interested in pursuing research on idea flow or attempting to perform some design work in an operating organization (hopefully his own R&D laboratory) can readily generate additional ones.

The economics and practical constraints of field studies in operating organizations severely limited the number of propositions that we could actually carry all the way to empirical testing.

6.2.10   Some Models of Idea Flow

As in the case of propositions, a large number of models and conceptual schemes were generated during the course of the study. Many of them disappeared in the dust of erased blackboards during our "paradigm" sessions. Others survived these sessions and found their way into memos, working papers, and in a few cases finished theses and publications. In this section a small sample of these models is presented. Some of them went all the way to the stage where propositions generated from them were subjected to empirical testing. Others merely served stimulational and conceptual purposes as propositions were being formulated and research designs were being developed. The models developed were of two types. The first type is a process flow diagram that postulates a sequence of events or a set of subprocesses that represent significant portions of the idea flow and project selection processes. The second type is the shorthand presentation of a set of

---

*Incorporated into the empirical studies.

relations between variables involved in the idea flow process, but which do not necessarily represent a sequence or flow of events.

Several attempts were made by members of the group to elaborate on and operationalize the early flow model of Hannenberg (Figure 6-3). This schematic left many questions about the idea flow process undeveloped and raised questions of operationalization and measurement of the elements it included.

Bolen identified a number of stages or states through which an idea "passes" on its way from the mind of the originator to its disposition by the organization's evaluators. It was necessary to identify these stages in order to get an initial fix on an individual idea—that is, to determine its "existence"—as well as to trace it through time and space in the organization. Figure 6-6 indicates the stages developed by Bolen (63/25).

We successfully demonstrated that lists of ideas could be obtained and categorized by means of these stages and that idea flow could then be studied by observing the "movement" of ideas from stage to stage over time.

One weakness of such a "stage" model is that it tacitly assumes that an idea undergoes little change or development between the time of its inception and final disposition, that is, that an idea keeps its original identity during its life history. Also, the stage descriptions are restrictive in that they emphasize the role of the idea originator at the expense of others who may be very influential in the history of a particular idea.

Pound, following up the Hannenberg flow diagram, developed a communication network model (Figure 6-7) in which he portrayed the flow of a hypothetical idea. The nodes numbered 1, 2, . . ., 12 represent people in the R&D laboratory and are positioned in the pattern of the organization hierarchy. Thus, 12 is the head of the lab, 10 and 11 are senior researchers or project leaders, and so on. Nodes 13 and 14 are included to indicate that ideas may be originated by or communicated to persons outside the laboratory.

Stage 1	The idea is known only to one person, that is, the person who "thought up" the idea. The only ideas of this form that any given person will know about are, of course, those that he originated himself and has *not* mentioned to anyone else.
Stage 2	The idea has been informally mentioned to other people by its originator, but the idea has not been formally submitted to the organization for approval.
Stage 3	The idea has been formally mentioned to other people by its originator (i.e., it has been submitted to the organization).
Stage 4	The idea has been formally submitted to the organization by the originator, and it has been evaluated and/or acted on by means of committee consideration, feasibility studies, or other procedures, but no acceptance or rejection decision has been made.
Stage 5	The idea has been formally submitted to the organization by the originator,[*] and it has been evaluated and/or acted on and been rejected or put on hold.
Stage 6	The idea has been formally submitted to the organization by the originator, and it has been evaluated and/or acted on and has been accepted as a project.

Figure 6-6. Stages or states of an idea. (* In some cases, the idea is mentioned (discussed) formally or informally and/or submitted by a surrogate—someone acting for the originator(s)— for example, a supervisor.)

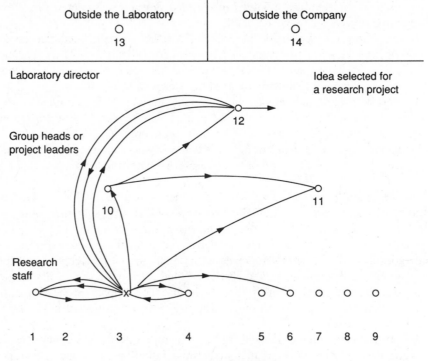

Figure 6-7. A communication network model of idea flow. X denotes the idea originator.

The illustrative idea is shown to originate at node 3. From here it is communicated (as indicated by the directed branches) to nodes (persons) 1, 4, 6, and 10. Following this, there ensue a series of communications dealing with the idea, which eventually result in a decision at node 12 to select the idea as a new project.

Many persons are expected to participate in the idea flow process in one or more capacities over time. Thus, the model used to describe the process must allow for considerable flexibility. A full description of the history of a particular idea by means of this model would include at least the following information:

1. Identification of the originator(s) of the idea.
2. Description of the communications dealing with the idea (direction, form, persons involved, etc.).
3. A description of any changes in the idea over time.
4. Identification of the "final" decisionmaker(s).
5. Specification of the nature of the "final" decision.

6.2.11 Probabilistic Models

At the same time that Pound was developing his communication network model, N. Baker was attempting to represent idea flow as a probabilistic process. It is interesting to note that if probabilities were included in Figure 6-7, to indicate the probability that a communication will occur between nodes i and j, then the two models would be very similar. The first model developed by Baker was a model structured around the theory of finite Markov chains.

After his pilot studies and design of his full-scale field study, Baker revised a number of his models and generated new ones (in nonprobability terms), which then served as the basic models for his dissertation. Figure 6-8 represents one such revised model of the overall idea flow process.

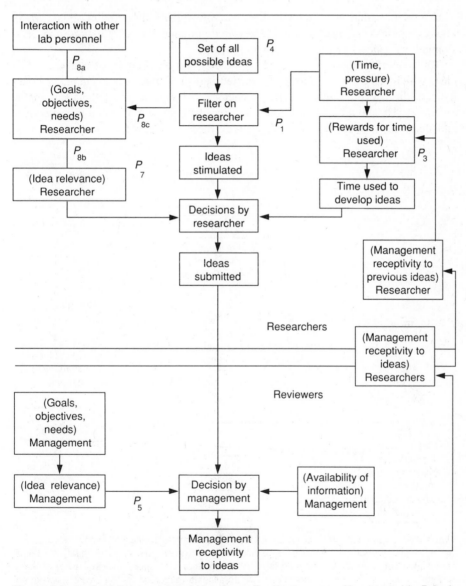

Figure 6-8. Revised Baker model of idea flow. P_i denotes proposition i tested in Baker's dissertation. ( ) is read "as perceived by."

Baker's empirical data came from a somewhat nontypical idea flow process, namely, from special idea generation groups. The function assigned to these groups involved an accelerated process of idea generation and flow. However, the models and propositions that Baker used were not developed specifically for these groups; rather they were developed for the idea process in general.

6.2.12   A Dependent Variable Model of Idea Flow

This model, which was developed as part of Pound's dissertation, is shown in Figure 6-9. It is intended to emphasize the communication, evaluation, and decisionmaking aspects of the idea flow concept.

The oval-shaped boxes are used to indicate the distribution of a given set of ideas which has occurred during a specified period of time. The rectangular boxes indicate the decisions that operate to change the distribution of the ideas in the given set. (It should be noted that the existence of communication, selection, and disposition behaviors may only *imply* the existence of a decision, that is,

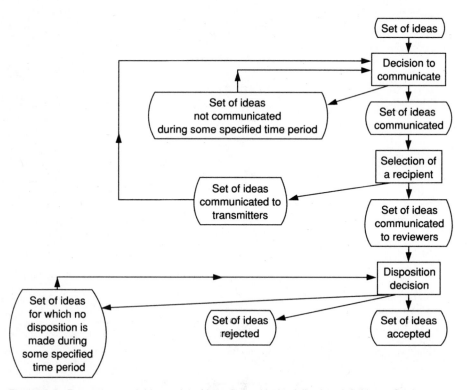

Figure 6-9. Dependent variable model of idea flow with the following definitions: Originator— the individual(s) who first introduces a particular idea. Transmitter—Any individual who takes part in idea communication, but who is not the originator of a specific idea. Reviewer—Any individual who has the organizational authority to allocate resources for the investigation of an idea. (Source: Pound 65/12.)

decisions as used here may be implicit as well as explicit.) At these decision points, other variables are expected to influence idea flow. Thus, the model is termed the "dependent variable" model of idea flow. Some "independent" or influencing variables are given earlier in this chapter in Section 6.2.9 on propositions.

Finally, Martin developed, as the basis for his dissertation, a flow model and a mathematical model for the evaluation of ideas for changes to production processes. Figure 6-10 illustrates his summary flow model. Following that, he developed a mathematical model for testing his two multivariate propositions.

6.2.13 How Might the Technology Manager and/or His Manager Use the Ideas in This Discussion of the Idea Flow Process?

Even though the several dissertations, theses, and staff studies described above were able to sort out some of the "static" in the literature related to idea generation and flow, they fell far short of settling the kinds of issues reflected in the questions at the beginning of the sections on idea flow. Much was learned, but much more insight is required before we can say that we fully "understand" the idea flow process, let alone feel competent to predict its course with accuracy or "make it happen" in ways we prefer.

Despite this, the very statement of propositions—reflecting the research-based literature—and development of conceptual models should be of value to practitioners who are concerned about "what may be wrong" with their firm's idea generating and/or communication process or, worse yet, who think that they are fine, when they are far from it.

Like the R&D manager who inspired our studies initially, many managers may find, on reflection, that their organization is not presented with enough ideas (in

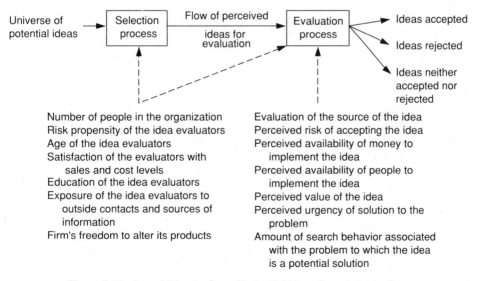

Figure 6-10. A model for the flow of industrial ideas through evaluation.

This questionnaire is for the study of innovation and new product development we are now conducting for the company. Our purpose is to obtain, with your assistance, a picture of the production and the utilization of technical ideas originated in the company over the past 3–5 years. We would like to obtain a better understanding of the factors that might be significant to the "flow" of ideas from the individual to the organization, and to the selection processes as to which ideas should be worked on by R&D and ultimately utilized by the company. The information supplied to us will be protected and treated as company confidential (we all have signed confidentiality agreements with the company).

INSTRUCTIONS

There are two parts to this questionnaire: the first part asks you to list the ideas for projects which have been originated in the past 3–5 years in the company—by you or someone else; the second part is a set of statements to be checked against each listed idea.

An *idea* for purposes of this questionnaire is one that:

(a) *is of a technical nature.*

Ideas for such things as department reorganization or hiring new personnel should be excluded.

(b) *implicitly or explicitly suggests research and/or development work to be done.*

An idea should be included if it entails or proposes some research effort which you or another person thinks should be undertaken by the company.

(c) *requires the new allocation or reallocation of company resources (time, funds, personnel, and/or equipment).*

The research effort suggested by the idea should necessitate the commitment of resources to a new research project or the addition of or shifting of some resources on a current project in a direction that is in some way new.

Please *do not exclude* ideas that fit the above criteria but that on some basis you believe are not necessarily "excellent" ones.

PROCEDURE

1. On the first page list the ideas you are aware of—both those you essentially consider are your own and those originated by others. To list an idea, number it and give it an identifying title. The title should be brief. If you believe any of the brief titles you use will not be sufficient to identify the ideas separately at a later period, please add enough detail to identify clearly the ideas.

2. After listing all the ideas, turn to parts A–D and check off the statements for each idea.

3. Finally, please complete Part E.

List the ideas for projects on this sheet. To list the ideas, use the numbers and give each one a brief title or description that will help identify it. If additional space is required, please use additional sheets.

Idea No.	Date of Origination*	Name of Originator (if known)	Idea Title or Brief Description	Idea Type Code[†]
1				
2				

Figure 6-11. Idea flow questionnaire.

Idea No.	Date of Origination*	Name of Originator (if known)	Idea Title or Brief Description	Idea Type Code†
3				
4				
5				
6				
7				
8				
9				
10				
11				
12				
13				
14				
15				

*Date you originated idea (if yours) or when you first heard of it.

†Please mark one of the following codes:

A—Idea is for a new product
B—Idea is for improvement of an existing product
C—Idea is for a new or improved process
D—Idea not covered by above (describe it)

Place a check or "X" in the appropriate boxes for each question for each idea you listed previously.

	1	2	3	4	5	6	7	8	9	10	11	12	13	14	15
A. How did the idea start?															
1. I thought of it first															
2. Another co-worker suggested the idea															
3. Suggested by a person outside the company															

Figure 6-11. *(Continued)*

	1	2	3	4	5	6	7	8	9	10	11	12	13	14	15
B. Who knows about the idea?															
1. Myself only															
2. A group or project leader (or any other supervisor or manager)															
3. People in the R&D organization but none in other parts of the company															
4. People in R&D and other parts of the company															
5. People in the company and also people outside the company															
C. What has happened to the idea?															
1. I did or am doing some work on the idea															
2. I expect to do some work on the idea															
3. I expect to be responsible for a major portion of the work															
4. Another person did or is doing some work on the idea															
5. One or more other persons will do all the work															
6. Concept design underway															
7. Concept design completed															
8. Prototype mold made															
9. Fabrication, test, or debugging completed															
10. Presented to customer															
11. Approved by customer															
12. Tooling completed															
13. Manufacturing started															
14. Some work was done on the idea but it has been dropped, put on the back burner, or fallen between the cracks															

Figure 6-11. *(Continued)*

	1	2	3	4	5	6	7	8	9	10	11	12	13	14	15
15. No work has been done to date on the idea															
D. What approval is, was, or will be needed before work is done on the idea? 1. Some form of approval must be obtained to work on the idea															
2. No formal approval is needed to begin work on this type of idea															
3. Permission was requested but not given for working on the idea															
4. Approval to initiate work has been given															
5. The idea has been submitted for formal approval as a project															

E. If the idea was dropped, put on the back burner, rejected, has fallen between the cracks, or is not likely to be approved or used by the company, please explain why for each such idea (please number the ideas).

_____

_____

_____

_____

_____

_____

_____

_____

_____

_____

_____

_____

_____

_____

_____

_____

_____

_____

_____

Figure 6-11. *(Continued)*

total) or not enough *good* ideas, after screening out the nongood ones, to enable it to develop or maintain a project portfolio that can improve the chances of obtaining more cost-effective results from their technology and R&D expenditures. On the other hand, some may find themselves drowning in large numbers of ideas which must be sorted out.

With the help of some of the ideas in this chapter and the ones about project evaluation in Chapter 7, the technology manager may be able to cut through the chaff and put resources behind the most promising ones. He may also be able to send signals to potential idea generators and communicators that good ideas are welcome and less-than-good ones are not. One tool that might help in this attempt at improving the idea generation and flow process in the firm is a version of the *idea inventory* which we used in our field sites to obtain a picture of what ideas were "around" and what had or was happening to them. An illustration of this form is given in Figure 6-11, which was adapted for a client by my consulting group, IASTA, Inc., aimed at identifying promising ideas for development.

6.3 RELATIONS WITH UNIVERSITIES

There have been periodic waves of interest on the part of industrial firms in exploring the possibilities and mechanisms of cooperating with universities in R&D. Some of the interest has been very specific—sponsorship of a particular project or line of research that is beyond the capabilities (time, talent, or both) of a company; retention of particular faculty members as consultants; or recruiting of graduates. For some very large technology-intensive firms, the search has been for continuing and broad relations with a group of leading universities or individual departments that have the chance of providing a range of benefits to the company and to the university.

In the late 1980s, another crest of this wave of interest is occurring, spurred by a number of factors, including rapid advances in the states of the art in biotechnology, materials, and computer software; increased competition for oustanding graduates; downsizing, cutbacks, and/or elimination of company research staffs, especially those in the CRLs (see Chapter 2); increased foreign competition in high-technology areas; reduction in government support for university research in some areas; and other economic and technical factors.

Making a connection with a university seems pretty straightforward in principle—just call them up, get together, and make music together to mutual benefit. In practice this courtship is seldom that smooth or straightforward. A large (or even smaller) company and a university are quite different kinds of institutions in terms of values, preoccupations, operating rules, legal constraints, operating styles, and organizational structure. This latter factor is characterized in Figure 6-12, which was inspired by a series of incidents that occurred in the mid-1950s when I was on the faculty of M.I.T. and very active in its Industrial Liaison Program—a highly organized mechanism for bringing industrial companies together with the M.I.T.

faculty. Figure 6-12 represents the difficulties and frustrations of members of each organization trying to locate, interact with, and "place" or calibrate a technical member of the other organization in terms of power and authority to finalize a decision or reach an agreement.

In this section we take a brief look at some major issues that must be dealt with in order to achieve effective industry–university collaboration and to provide a useful and continuing source of technology from the university for the industrial partners. I have been on both sides of this interface many times—representing the university, of course (Columbia, M.I.T., Oslo, Berkeley, and Northwestern), and representing many client companies attempting to establish effective university relations in a wide range of fields, such as electronics, computer science, chemicals, machinery, and materials. I have also served as a catalyst in attempting to bring university and industry closer together via governmental incentives and mechanisms (on behalf of the National Science Foundation, The National Aeronautics and Space Administration, The National Bureau of Standards, the Department of Agriculture, EPA, and several parts of the Department of Defense). Therefore, the comments below represent a view from several sides of the question of how industrial firms and universities can cooperate in the area of providing sources of technology for the firms and helping the universities to pursue their own agendas (including staying afloat financially).

Figure 6-13 lists several major issues—some of them composites of many subissues. The following subsections discuss each in turn. Some of these issues constitute or identify *barriers* to effective collaboration and others emphasize *facilitators* to such relations. They are not clean-cut: one firm's barrier may be another's facilitator and even the same firm can turn a facilitator into a barrier through inattention or specific blunders. An example might be the relative openness of the university with respect to exchange of information. This can be an advantage to a firm eager to receive ideas and specific know-how. It can also constitute a barrier to protection of proprietary information.

6.3.1   The Wide Variety of Possible Mechanisms

The wide range of mechanisms and forms of relations can provide a facilitator to cooperation in that high flexibility can help in selecting a potentially effective mechanism. It can also present a barrier in terms of delays in decisionmaking and extended "studies" of the "best" way of cooperating. My feeling is that a search for the optimal pattern may be self-defeating, since most mechanisms have their limitations and advantages. Furthermore, I believe that a healthy relation should involve multiple mechanisms and linkages between the firm and the university. One-shot deals for a specific piece of technical cooperation may be of short-run benefit to a particular industrial project or particular university financial need, but they may not build the kind of relations that can help both in the long run. Rather than search for the "optimal" means of cooperation, several apparently attractive linkages should be tried for a period of time to see which ones survive and which are synergistic. This means avoiding an initial set-in-concrete approach that leads

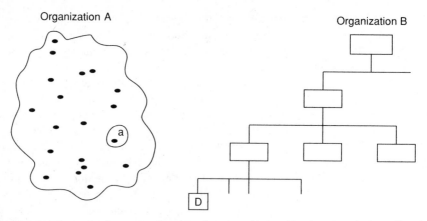

Figure 6-12. Contrast of structure between, for example, a university group and a military or industrial group.

to a "do or die" test of whether a firm and a university can work together. It takes time for relations to develop and mature and for mistakes to be made and overcome, sometimes on the order of years.

Figure 6-14 lists some of the many possible mechanisms for industry-university cooperation. Some of them are aimed specifically at research and others are more general—involving both the educational and research roles of universities.

After looking at this possible range of mechanisms for industry-university cooperation, let us take a step back and ask "why *should* they cooperate" beyond the traditional modes involving the university's role in providing students and knowledge and the industry roles of providing jobs and, in some cases, contributions to educational costs.

In most of the discussions on this subject, euphemisms abound for what appears to be a very central issue for both parties to potential cooperation: financial viability. This may appear to be a somewhat cynical view and, if attempting to get to basic motivations is cynical, then it is. The chief motivation driving universities toward seeking closer relations with industry is their need for financing to remain solvent and to continue their educational and research roles. Specifically, this situation is exacerbated by the double pressures of falling government support of university research and graduate training in many areas and the significant increases in costs of staying in business and continuing to innovate.

---

1. The wide variety of possible mechanisms
2. Differences in style and organization
3. Ownership of the fruits of R&D innovation
4. Differences in roles and interests along the R&D-to-application path
5. Individual consulting relations
6. Size and flows of money
7. Potentially fatal flaws

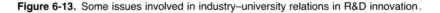

Figure 6-13. Some issues involved in industry–university relations in R&D innovation.

Corporate Contributions to University

- Undirected corporate gifts to university fund
- Capital contributions: gifts to specific departments, centers, or laboratories for construction, renovation, and equipment
- Industrial fellowships and contributions to specific departments, centers, or laboratories as fellowships for graduate students

Procurement of Services

- By university from industry: prototype development, fabrication, testing; on-the-job training and experience for students; thesis topics and advisors; specialized training
- By industry from university: education and training of employees (degree programs, specialized training, continuing education); contract research and testing; consulting services on specific, technical, and/or management problems
- Industrial associates: single university, usually multiple companies; industry pays fee to university to have access to total resources of university

Cooperative Research

- Cooperative research projects: direct cooperation between university and industry scientists on projects of mutual interest; usually basic, nonproprietary research. No money changes hands; each sector pays salaries of its own scientists. May involve temporary transfers of personnel for conduct of research
- Cooperative research programs: industry support of portion of university research project (balance paid by university, private foundation, or government); results of special interest to company; variable amount of actual interaction
- Research consortia: single university, multiple companies; basic and applied research on generic problem of special interest to entire industry; industry receives special reports, briefings, and access to facilities

Research Partnerships

- Joint planning, implementation, evaluation of significant, long-term research program of mutual interest and benefit; specific, detailed, contractual arrangement governing relation; both parties contribute substantively to research enterprise

Figure 6-14. Types of university–industry relations. (*Source*: Prager and Omenn, *Science* 1/80, quoting Smith and Karleshy.)

The motivation driving industrial firms toward considering (not necessarily engaging in) more cooperation with universities in the R&D/innovation process is the need to sustain and improve their capabilities to do R&D and to produce technological innovations in the face of increasing competition and rising costs of production and distribution.

The time perspectives of both potential parties to cooperation may coincide in particular instances—for example, a university department or faculty member may need money to support a student this year and a possible industrial provider of support may need some advice or other input (e.g., experimentation, model building, theory development, training, troubleshooting, or ideas) this year also. In that case, given that a host of potential barriers to cooperation can be effectively dealt with, they are in phase and a "deal" may be struck.

Frequently, however, the two parties are out of phase and may differ significantly on time perspective. Industry may need results "yesterday" and the university people may need "long-term support for a continuing research and educational effort." If a reasonable quid pro quo can be worked out, this mismatch can be overcome. Often, however, it cannot. Many research efforts are long-term, multiyear efforts that cannot be chopped up or shut off neatly within the typical industrial budget cycle. In addition, if we accept the idea that effective cooperation requires personal relations and trust between the parties, the turnover and role changing of industrial personnel may be too destabilizing to provide for long-term support in terms of multiyear commitments.

6.3.2 Differences in Style and Organization

More than 25 years ago, we first started to do serious research in an area of R&D management which we have since labeled LINCOTT—an acronym for the many labels that have been used in this field, including *L*iaison, *I*nterface, *C*oupling, *T*echnology *T*ransfer. Our first approach involved attempting to understand the forms and sources of difficulties which appeared to beset attempts at cooperation between university research groups and industrial sponsors or members of project teams.

One of the early results of that inquiry, which has been going on all the years since then in a wide variety of settings, was Figure 6-12, representing perhaps the frustration of each party in trying to find, make contact with, and work effectively with a member of the other organization. At the time this figure was drawn, we were expanding our interest to include cooperation between universities and military organizations—for example, through the sponsorship of R&D at universities by the military.

In addition to apparent differences in organizational structure reflected in these mutual perceptions, there are also many potential differences in style of operation between members of a university faculty, their students, and other associates and members of an industrial company that is trying to interface with it. It is not that university people are necessarily or even largely more "laid back" than industrial people. On the contrary, many of the university people who do engage with industry are hard to distinguish, in some meetings, from their industrial counterparts, pipes and tweeds notwithstanding. There are hyper individuals on both sides of the interface and also people who appear overrelaxed or unconscious of time and economic pressures.

By and large, however, there can be tremendous style differences that can bug people on both sides. Some of them can be characterized as follows:

- Insistence on worrying a problem to death versus getting a quick if not ultimate or "supportable" answer.
- Consultation patterns and decisionmaking procedures within research teams and other parts of each organization—for example, too much democracy or too much autocracy.

- Following of procedures and protocols for communication and dissemination of information.
- Concern with keeping information closely held versus disseminating it quickly and widely.
- Styles of personal and interpersonal address, communication, dress, degree of formality, and so on.

Exaggerated or not, some of these differences can cause difficulties in cooperation, despite general good will and a mutual interest in the joint effort. Neither group is likely to change fundamentally in the short run to accommodate the other. Therefore, adequate provision has to be made in terms of time and effort to allow the parties to become mutually acquainted and to develop mechanisms and patterns for accommodating styles and modes of operation.

6.3.3 Ownership of the Fruits of R&D/Innovation

Many of the industry–university situations in which I have been concerned — several dozen from both sides — involved the issue of who owns what. In the cases of truly uncommitted, no-strings support for basic research provided by industry, this has been less of a problem but still lurks in the background.

Aside from clearly charitable contributions to education and research, which is really not the focus of this discussion, there is generally the issue of what happens to the spoils "if we get lucky and something useful comes out of the research." The uproar and battle of the attorneys revolving around the Interferon case (*Science,* 1980) appears to concern a particular issue — the ownership of research materials. However, many other lawsuits and not-yet-legal conflicts have arisen in connection with other aspects of research output in industry–university relations. Who owns inventions, technical data, software, algorithms, research equipment, test results, manufacturing know-how, drawings, and unpublished reports that result from an industry-sponsored project? This becomes complicated primarily when there are significant economic stakes in the situation — such as royalties or profits. If the research fails to produce a potentially practical result, or if the sponsor has no interest in it or ability to exploit it, it may become a truly "academic" question.

However, when the smell of revenue and profits is strong in the wind, each side typically maneuvers for an advantageous position. This can occur despite an apparently clear and unambiguous prior agreement, and litigation over patent ownerships and commercialization rights may go on for years or decades.

Universities are pulled in several directions on this issue. Idealistically, many members of the academic community feel that the fruits of their research should be made available for the public good, no matter who sponsors it. Not so idealistically, many inventors in the university want a "piece of the action" of a potentially commercial property. A few universities have done very well financially as a result of their research outputs and the income they bring. Their industrial partners have also done well. Where there are no fruits to distribute there may be no conflict, but there may still be disappointment and resentment for one reason or another.

The industry–university relation on this issue of "who owns what" is another version of an even larger problem in the university–government arena, where access to the results of government-sponsored research may be too loose for some university researchers and too tight for others.

I see no neat solution to either problem, because both parties—sponsors and doers of research—have legitimate claims. Again, each situation must be negotiated on its own merits. This should be done within a general policy framework that does not allow such partnerships to stumble blindly into legal and personal conflict because adequate "front-end" effort was not devoted to attempting to anticipate the possible outcomes of a cooperative effort.

6.3.4 Differences in Roles and Interests Along the R&D-to-Application Path

Historically, and right up to the present time, the division of labor between university researchers and industrial researchers has been relatively clear. The former are supposed to contribute to general knowledge and the latter are supposed to use it. Of course, there have always been exceptions and fuzzy regions between the two. Many university people, especially in the "high-technology" areas of Boston, California, and New York, have moved relatively deep into application and moneymaking (or at least trying to make money); and some large industrial firms have managed to contribute significantly to scientific knowledge in a whole range of fields.

By and large, however, the division of labor has been pretty straightforward and has given rise to an environment in which discontinuities of interest, capabilities, and research behavior have tended to leave large gaps between universities and industry in terms of potential research collaboration.

That situation, however, may be changing enough to reduce or overcome the historical gaps and to make collaboration more likely in some areas. In almost all the companies where I am currently or have recently consulted on the management of technology or the R&D/innovation process, there is renewed or new interest in support of generic technology and a perceived need to "refill the technology barrel" or "strengthen the technology base." Of course, I may be observing a biased sample, since their retaining of me as a consultant rather than one of the large general management consulting organizations may reflect an affinity for university people and the more basic end of the R&D/innovation spectrum.

However, given the possibility of that bias, I see a modest trend, in several technical fields, toward making alliances with universities in order to help the industrial companies go beyond the solution of individual problems in the direction of enlarging and improving the technology base they operate from in support of their products and processes.

Where the university or group of universities considered for this technology-base-building role dovetails neatly into the individual firm's plans, they can make beautiful music together. Where there are misconceptions of how far the university researchers can and/or will reach "downstream" and how far the industrial people are willing and able to reach "upstream," both parties can end up disappointed. Although it is not possible to spell such matters out in complete detail, a serious attempt should be made to get through the clichés and ambiguous wording of agreements or even preliminary discussions. Even within my own specialty, the distinctions between "basic" research and "applied" research are not very clear. Some of my colleagues consider "applications" to be much more upstream than others do.

6.3.5 Individual Consulting Relations

Academic and other (freelance) consultants who are at the leading edge of a rapidly moving research field can play a number of roles in industrial R&D:

- Idea generators.
- Working team members, although less than full time.
- "Teachers" and information sources, conducting seminars and acting as "information gatekeepers" to the literature or unpublished information.
- Linkages with the scientific community in general.
- Linkages with leading universities in particular.
- Reviewers of scientific work in the lab or company.

Many industrial companies, however, have not learned to use academic and other consultants effectively. Some of this is because many of the consultants "pop in" only occasionally, throw in some radical new idea or bit of advice, and then "pop out" again with little or no follow-through. In addition, there are often psychological problems for the professional staff involving the use of consultants who behave in a distant and superior manner and leave the staff feeling inferior or resentful. Still worse is the case where, from the staff's viewpoint, an outsider is paid a high fee (higher than their equivalent salary) even though they feel that he either does not have superior knowledge scientifically or, if he does, combines this with a very impractical view of what is feasible within an industrial R&D or business context.

Some consultants work very well on two general levels: (1) as a general backup for the regular R&D staff in a broad field or set of fields; they can perform a number of the roles listed above on a variety of topics and can be a "sounding board" for staff ideas, problems, and uncertainties; and (2) as highly specialized sources or "idea leaders" in a field which is new to the staff and in which they are trying to gain competence. Both of these modes of interaction require careful preparation and a receptive attitude on the part of both the staff and the consultant. If he acts too much like a professor lecturing and judging his "students" in the company, their self-confidence and independence of thought may not develop adequately.

There is a third mode of interaction that is very rare. This is a radical method of using consultants as working members of the company's R&D team, although not as full-time employees. It has worked for a few high-technology companies (e.g.,

Bell Labs) who are on the cutting edge of their fields and who want to integrate the most advanced academic thinking into their ongoing work, beyond an occasional visit, seminar, or consulting study.

The arrangements will vary with the individual and the rules of his university for outside activity. Almost all leading universities have a policy that allows for one day per week of outside consulting that is professionally related to the faculty member's work. For some "idea source" people, this may be about all that an industrial R&D group can absorb of his time without suffering from "information overload." Where this is not enough time for the consultant to become integrated into an ongoing project or disciplinary group's general activities, other arrangements can be made. One possibility is the kind of agreement that Monsanto and Exxon have made with Harvard and M.I.T. for a proportion of faculty members' research time. This is more suited to basic research in which the faculty member stays in his own lab or university and meets occasionally with the sponsoring organization.

This may not be adequate, however, to create the type of "team membership" discussed above for selected outside consultants. Instead, there are possibilities, if the faculty member and the university are willing, to go beyond the one-day-per-week consulting arrangement and contract with the faculty member for several days of his time, having him take a proportional reduction in academic salary for the period during which he is on a reduced academic load. This may not be agreeable to the university, because it may not see the benefit to them. Another approach would be to contract directly with the university for a percentage of the faculty member's time. This is not likely to be agreeable to most leading faculty members, since the funds in that case would go directly to the university, without giving them additional income and with an obligation that reduces their own freedom in using their time.

Sabbaticals or "summers in industry" have been proposed and used by a number of companies—not always successfully. The key factors in success appear to be (1) that the faculty member have a definite assignment in the lab that is of high interest to the lab and the company and that they are not merely given space to "do their own thing," unrelated to lab interests and activities; (2) that there be a prior relation between members of the professional staff and the candidate faculty member so that months do not have to be spent in "getting acquainted" and achieving a working relation. Given these two conditions, even short-term arrangements during the summer can be fruitful, if there is a clear short-term effort that can culminate in a usable idea or report. A common mistake is for the faculty member to undertake a vague piece of work that is left unfinished at the end of the summer and that is not usable in any way by the lab staff.

Twelve-month or academic-year sabbatical appointments are very promising under the conditions mentioned above and with the understanding that the most outstanding and busiest people in the field will not likely be interested, but that some of their close associates may be—for example, a postdoctoral researcher.

This approach of having outstanding faculty members become part-time employees or independent contractors for a limited time can provide a number of advantages in helping to build the scientific image of the company—via joint presentations, publications, and other mechanisms. It can also significantly augment the regular staff in important areas where it is not possible or feasible to hire the best people full time. If image building is an objective, care must be taken not to take on less than outstanding people in this consultant role. If their reputations in the scientific community are not on a high level, they can be ineffective or even damaging to the firm's scientific image. For the most outstanding people, there is also the possibility that they will not want to work exclusively for any specific company. In many fields, the leading companies have to be content with sharing the leading consultants, based on the faith that such people will protect any proprietary information they pick up in their labs. In some cases, the level of consultation does not involve getting into company proprietary information; but in the "team member" mode, this would be a factor that has to be weighed carefully.

Of course, with the intense competition for the top talent in, for example, computer science, biotech, materials, and related fields, the firm may not be able to get the very top individual or more than one such person to consult regularly. In this case, some of the associates (junior faculty members or research staff) of the top people may be available and may provide an indirect link to the top person, as long as there are no conflicts of interests involving direct competitors.

6.3.6 Size and Flows of Money

One general problem that may not be obvious to people outside the university community—industry sponsors, for example—is that the value system of the university community is far from monolithic.

One issue that brings this to a point of potential conflict is the competing objectives of obtaining funds for "general university purposes" versus obtaining funding for very specific research projects of individual faculty members. In one form, this potential conflict may be exacerbated by block grants to universities for subsequent dispersal internally. University research centers and interdisciplinary programs have been used as "dispersal" mechanisms in this connection. Some faculty members, especially those who may not have had much experience or success in raising outside project money themselves, may prefer this two-step procedure. Others—including perhaps the more entrepreneurial practitioners of "grantsmanship"—may be negatively inclined toward this system. They may see it as reducing the amount of money they can get for their own projects and programs from a given source and increasing the amount of bureaucratic "hassle" in which they must engage.

In a more general sense, the financing of higher education is split among several approaches. The view of the university administrator and the trustees is likely to be more focused on general operating revenue with as few strings attached as possible. Individual, research-oriented faculty members may strongly prefer earmarked funds for research, especially if they are earmarked for their own projects and programs. These differences also apply at the departmental level, where conflicts are not uncommon over division of the available resources between general purposes and specific research projects. The industrial partner to a relation with a university would do well to become familiar with these potential differences and be prepared to negotiate within that context when he has views of his own as to how any funds he supplies are to be applied.

Another subissue in this area of funding concerns the question of size of a research grant or contract. Rising costs and rising expectations among faculty members in science, engineering, and other primarily government-funded fields have set a lower size limit below which many "high roller" grantsmen may not be interested in going after funding. One reason may be the "big buck" habit we (the author included) have gotten into as the result of generous federal funding and, in some cases, foundation funding. Another contributing factor is the rising costs of supporting even one graduate student. In some private universities such a student, along with his or her tuition and university overhead and benefits, can cost almost $25,000 per year. A grant of $20,000–25,000 would therefore leave virtually nothing over for the principal investigator's summer or released time salary, equipment, travel, or other research-related expenses. Still other factors are the potential prestige and leverage with deans and department heads which larger grants and contracts may bring.

When we contrast this with the level at which some very large companies are currently providing or offering to provide support—for example, $25,000–75,000 for a "bundle" of projects—we see that industrial funds may be an order of magnitude off in providing the kind of support that the federal government has been supplying to leading researchers for many years through agencies such as NSF, NASA, NIH, and the Department of Defense.

One of the recent mechanisms which attempts to obtain broad research support from industrial firms aims low and sometimes even misses that target. For example, a typical engineering school may have a half dozen research centers focused on particular areas of science and technology. Each tries to get industrial sponsors to contribute on the order of $25,000 per year to the center. After center, school, and university overheads and benefits are deducted (in some cases such contributions are treated as "gifts" and the full university overheads of 50–75% on direct costs are reduced or waived), there is not much left from an individual grant to support the actual research. This means that the center has to go out after a large number of sponsors, spend a lot of time "servicing" them, and spend a lot of time and funds in going after renewals and replacements for defecting sponsors. Few faculty members have the motivation and the stamina to conduct such a continuing campaign for funding and the life cycle of such centers can be very short.

Figure 6-15, which comes out of one of our consulting engagements with the Department of Transportation in connection with the ill-fated "car of the future" project, deals with an issue that can help industrial firms to understand better the motivations and constraints involved in attempting to do business with universities in the area of technology. It is based on interviews with engineering school faculty members and identifies very specific types of mechanisms for getting faculty "attention."

1. *Freedom to choose research problems* (typical in universities, but also true of many researchers in other kinds of organizations)
2. *Subsidies* for research projects and programs already under way
3. *Venture capital* (for university-based, small high-technology firms and individual inventors)
4. *Fellowships* (and research assistantships)
5. *Internships* in industry for undergraduates and graduates
6. *Sabbaticals* in both directions—from and to university and industry
7. *Continuity of funding*—almost a universal desire for most research organizations in the face of fluctuating funding levels
8. *Access* to technical information, equipment, and facilities
9. *Proprietary rights* to research results and royalties or fees from their technological consequences
10. *Easier, less time- and effort-consuming and less frustrating procedures* for applying for and getting funding
11. *Thesis* and other academic *research* topics and *ideas*
12. *Closer relations with industry* in general
13. *Consulting* opportunities
14. Research *grants* as opposed to *contracts*
15. *Recognition and publication rights*
16. *Involvement in commercialization*
17. *Risk sharing* in the research and early development phases
18. *Faster procurement and granting process* to allow for better planning
19. A *piece of the action* in terms of sharing in the equity of a new research-based venture

Figure 6-15. Kinds of needs or wants of potential university performers of automotive research which might be addressed by particular incentives.

6.3.7 Potentially Fatal Flaws

In summary, and adding some ideas not mentioned before, there are a number of ways in which industry–university relations can fail, based on single factors, single events, or a combination of factors and events. Success, on the other hand, may require a complex set of favorable factors and events if the relation is to go beyond a hand shake, a series of lunches or committee meetings, or a one-shot grant that is not followed up. Some possible fatal flaws are the following:

- Mismatch of personalities and styles of the participating organizations' "point" or liaison people.
- Too little funding in individual packages or in aggregate to engage or sustain the attention of the most attractive university people.
- A "flabby" trial of one mechanism with a drawing-back if it does not click right away.
- Failure to clarify, as far as possible, the roles, contributions, and obligations of each party.

- Lack of ability of either or both parties to communicate technically and substantively about the joint efforts.
- Intrusion of the industrial company into a hornet's nest of intrauniversity differences and conflict without being prepared to deal with them.

I welcome the apparent upsurge of interest in industry–university research collaboration, which is being fostered by the federal government, some trade associations, and some individual firms and executives. However, at the same time I hope that these relations will be approached thoughtfully and systematically, so that they are not destroyed or seriously damaged, possibly for whole generations of the management of the firm, by some fatal flaw or flaws that could readily have been avoided or "fixed."

6.4 MAKE OR BUY TECHNOLOGY?

Figure 6-16 lists a number of factors that influence corporate and divisional management to search for technology outside the firm, that is, to "buy" rather than "make their own." For the division manager, many of these factors reflect the time and other pressures on him and the need to reduce uncertainty on such matters as cost, timing, and performance for new and improved products and manufacturing technology. Frequently, in his search for more certainty and a "guarantee" of delivery and performance, he ends up out on a limb with no backup for a failed or ineffective technology acquisition.

My bias shows through in this section. I am certainly not against all forms of outside technology acquisition. To the contrary, some external sources may help

---

1. Uncertainty of costs of internal development—preference for a fixed price
2. Uncertainty of waiting times for internal developments—preference for avoiding the front-end time commitments in the overall R&D/innovation process
3. Uncertainty of acceptability and performance of the technology—preference for guarantees, warranties, prior use, or demonstration of feasibility ("look and buy")
4. Preference for sharing costs of development
5. Lack of specific capability to carry out a given project or program
6. Lack of general capability to support a given program or multiple programs
7. Turning away from corporate research laboratories (CRLs) as sources of specific technologies and heavier reliance on operating divisions or business units
8. Preoccupation of operating divisions or business units with maintenance R&D and incremental improvements at the expense of aiming for radical developments or new departures
9. Disappointment with the track records of the firm's existing R&D units in generating new products or technologies
10. Pressures to acquire software quickly and/or at fixed costs (a growing special case)

Figure 6-16. Factors influencing the search for technology outside the firm (buy versus make).

overcome many of the limitations of the DM's and the firm's own technology that have been discussed up until now (and will continue to be discussed in later chapters). The bias is toward a *combination* and *blending*, where feasible, of the division's and the firm's *inside* technology capabilities with the *best attainable outside* technology. This approach can help assure that technology acquisition from outside will not be a series of on-off, episodic transactions with little learning and skill development inside the company. With this preamble, we launch into brief discussions of some key factors influencing the search for external technology.

6.4.1 Factors Influencing the Search for Outside Technology

Uncertainty of Costs of Internal Development—Preference for a Fixed Price. In many cases, "pieces" of technology can be purchased for an agreed on price in advance and delivered for that price. The best example, of course, is the "imbedding" of technology in a piece of equipment, a device, a new material, or an entire production line. This is a well-trod path and many very large firms and very large divisions have traditionally bought even their very specialized production equipment from vendors who can "quote a price, deliver on time, and warrant that the equipment will perform as expected."

Where reliable and technologically advanced vendors of this sort are available— whether on an exclusive basis or not—this can be an excellent way to reduce uncertainty of costs. Exceptions are for "very special" designs, which may include provisions for add-ons, cost overruns, bonuses, and/or price renegotiations. This is a situation the military and other government agencies find themselves in when they are asking a vendor or contractor to develop a radically new or very specialized piece of equipment that has never been done before—by that vendor or by anyone. Fixed price purchases of this sort can be very comforting to division managers who have capital budgets with very little slack (or combinations of capital and expense budgets which may be used together in procuring a combination of hardware and software). A potential problem with this approach, when relied on completely or too heavily for new technology, is that the vendor must "cut the suit to fit the cloth" and is unlikely to go beyond the agreed upon specs to add features that might be useful but that would cost him more, without assurance of reimbursement. "Buying in" by vendors, in order to get a foot in the door, can also result in a less than adequate or less than feasible design and/or the necessary or desirable supporting technology—software, manuals, maintenance and repair features, full service warranty, and other features.

So this is a good strategy for buying "technology packages," which can be neatly packaged, which rely on current technology, and which do not have too many "loose ends" sticking out that can generate excessive costs *after delivery*. One common loose end is the endless delay encountered by the division's plant(s) in trying to start up the new equipment and integrate it smoothly into its existing equipment. They may never get the (vaguely promised) yields or a trouble-free steady-state performance. That is, a fixed price at the "front end" may generate many costs and much *uncertainty* of those costs downstream in the implementation and start-up phases. When this certainty of cost is sought in purchasing a new *product*, the situation can lead to even more uncertainty, both up front and downstream.

The uncertainties of cost in commercial development of new products are found at all stages of the overall R&D/innovation process. Buying a product design or even a prototype can reduce the uncertainties of that early stage, but there is no guarantee that the costs of engineering, design for production, start-up, and "climbing the manufacturing cost and/or quality learning curve" will not skyrocket and make the initial cost look pale by comparison.

We continually encounter cases of the need for complete redesign of products purchased from outside vendors or inventors who have no experience with or feeling for the manufacturing process in the purchaser's plants. On the contrary, many purchasers of new product designs keep the vendors out of their plants and away from their engineers and production people in the interests of secrecy or proprietary methods.

If indeed the DM's internal technology people have become obsolete, lazy, out-of-touch, noninnovative, or overcommitted to firefighting, he may have no alternative but to go outside if he needs new products. However, a pure or dominant *cost* motivation may backfire in many ways during and after the procurement.

Uncertainty of Waiting Times for Internal Developments—Preference for Avoiding the Front-End Time Commitments in the Overall R&D/Innovation Process. Where the purchased technology is "off the shelf" or "ready and waiting," this objective can certainly be accomplished. Contracting for a piece of equipment, a production line, or even a new product can save time if it is a catalog item in the vendor's shop. However, this can mean compromise with the motive of getting unique, tailor-made, or proprietary technology.

Even where the design is special, unique, and proprietary, the features are custom, and the vendor has not developed this exact thing before, he can often deliver faster than the DM's own technology people. This can be achieved in a number of ways. He can "flog" his people, provide them with incentives to work faster, and cut out time-wasting and nonproductive bureaucratic procedures like coordination meetings, approvals, and design reviews, which may be required in the buyer's own organization and which are often part of the formal development process. There may be a price to be paid for such speed that will not be collected until the purchased technology is actually received and the downstream time lags in actually getting it to plant or market take hold.

The frequent redesign mentioned above in connection with cost can also have tremendous impacts on the *total time* or the "life cycle" of a new product or process development. A more extended discussion of this technology development life cycle is presented in Chapter 8 on commercialization.

The main point to be made here is that reducing the front-end (R&D, design) time for a development can, but often does not, reduce the overall life cycle time. This occurs especially when the purchased new technology is not phased in gradually *during* its development (again, see the discussion in Chapter 8) but is "sprung" on the technology people in the division. They must often start from scratch to get acquainted with it, to begin to feel comfortable with it, and to adapt it to their division's specific needs and circumstances—for example, production methods.

It is easy to understand the DM's frustration with waiting what seems like endlessly for his own technology people to come up with an improved or new piece of technology. Despite his tight control over the divisional lab and engineering groups, he may not know how to or be able to "speed them up" and get them to create, invent, or innovate on a schedule. Such frustration is leading increasingly to the search for more certainty of development times. The search typically leads outside the company. New product design in electronics is being enhanced by a newly prominent group of "product design shops" in places like Silicon Valley, which can come up with product designs on demand and on schedule. It is not yet clear how such deals have been impacting internal operations in the buying organizations. An alternative to going outside for the main purpose of reducing lead time and the uncertainty of that lead time may be found in the revived interest in internal technical entrepreneurship or "intrapreneurship," which is currently in vogue. It can be too rich for the blood and style of most division managers (as we have been describing them in this book) but this is an internal path worth exploring. It also is discussed in detail in Chapter 8 in the general consideration of commercialization.

Uncertainty of Acceptability and Performance of the Technology— Preference for Guarantees, Warranties, Prior Use, or Demonstration of Feasibility ("Look and Buy"). Along with cost and time guarantees or promises, assurance or belief that the purchased technology will perform as needed or expected is a big attraction for the DM. While his internal people promise (explicitly or implicitly) to "do their best," he knows from past experience that they often fall short of the mark and leave him unhappy and disadvantaged in the marketplace or on the P&L statement.

He is even willing to trade off a proprietary position or a unique product or process for more certainty that it will meet minimum requirements. Where the technology has indeed been around for a while, he may be on safe ground. Where the vendor is sticking his neck out with warranties and performance promises, the DM may find himself "holding the baby." The vendor may be trying to get rid of the piece of technology after finding that it really is not what other customers want or that it is far from state of the art. That is, part of the trade-off may be almost a guarantee of near-term obsolescence against assurance that the technology will work.

Preference for Sharing Costs of Development. For "off the shelf" technology or technology being developed for a consortium of buyers, development costs savings can be high. If the DM is not concerned about proprietary or unique or custom technology, he can benefit greatly in terms of cost by this approach. Piggybacking onto an ongoing development project in a vendor's lab can not only save time but also significant costs. In some industries and technology fields such joint development or "client-sponsored development" is common. The vendor or an inventor or product developer gathers enough up-front money to begin the development project and may, subject to the agreement of his clients, gather more travelers along the way. There may also be spin-off advantages from sharing the application and implementation experience of other consortia members. This can help avoid costly mistakes when the technology is brought in-house.

Training costs can also be shared and, subject to the terms of the agreement, improvements made by one or more customers may also be shared with the rest of the consortium. One disadvantage to this approach, in addition to loss of exclusivity, is that disagreements, delays, and cost overruns may occur because a member of the buying consortium has "special requirements" that he wants included in the design or gets into disputes with the vendor about design or performance specs or other matters not "exactly covered" in the contract. The DM who enters such a joint development program with outside firms would do well to check out his partners for a history or tendency to haggle and delay projects and to avoid sharing improvements and experience.

Lack of Specific Capability to Carry Out a Given Project or Program.
Of course, if the division cannot handle it,،on its own, or in an "internal consortium" with the CRL or other divisions, the DM has little choice regardless of cost, time, and other factors. If his division urgently needs the technology and he lacks the internal capability (in the division or the firm), he is constrained to go outside and "buy." Some DMs will not give in to this constraint and will try to get their divisional technology people to "stretch themselves" in terms of time and technical capability. This can and does frequently backfire, since it gets the divisional technology people in over their heads and can lead to further costs, delays, and technical glitches down the line. Many DMs have reported that, after trying to do it themselves, they have been forced to go outside for technical help and "packaged technology." By that time, a market opportunity has often been missed, or costs have been running down the drain in the plants.

In my advocacy of "making" technology in-house *where feasible*, I must caution that a rigid policy *against* going outside can be self-defeating. Again, where feasible, a combination strategy can have both short-term and longer-term payoffs through getting the technology needed and improving the internal capability for the next time around.

Few vendors are delighted about the prospect of training customer *technology* personnel as part of the deal. (They often gladly train *production* people to *use* the purchased items.) This can delay the project, run up costs through wasting vendor personnel time, give away vendor's secrets, and cut him out as a source next time around. However, such provisions can be built into the contract and paid for appropriately enough to overcome the objections.

The "black days" that many divisional technology people face are when a purchased "black box" arrives and they are expected to unwrap it and "get it working in their plants forthwith."

Lack of General Capability to Support a Given Program or Multiple Programs.
Aside from the *specific* capability for handling a particular project, the divisional lab and other technical groups may just not have enough slack to give their attention to "another whole program" or to pursue several major ones

at the same time. Where this is the case and no help is available from inside the company (this should, in my opinion, always be the place to look first), then again there is no alternative to going outside. Beyond the specific disadvantages of going outside mentioned earlier, however, some longer-term and potentially more significant problems can arise.

Delegating or contracting out significant whole programs to outside vendors involves both long-term risks and short-term risks that can damage the division and the career of the DM. We have observed a large and increasing number of cases of contracting out for software and hardware–software combinations which "never get done" or are delivered too late at too high an add-on cost and too "off-spec" for the changed conditions in the firm since the order was let or the contract terms formulated.

The industrial scene is littered with system design projects—for manufacturing systems, information systems, and others—which failed because the vendor went broke, division managers changed and tried to renege on or drastically modify the system design contract, or for other reasons. The apparent control that a buyer can exert over his vendors fades away when the vendor gets into trouble, is acquired or merged, loses its key people, or runs out of working capital.

Promises! Promises! Without a strong and direct line to the people actually developing the technology (not the salespeople or managers) the buyer may not even know that his project is in trouble until it is too late to salvage it. At least in his own division or even in the company at large, he can yell and scream and kick around the circle to make things happen or at least find out what *is* happening. With an outside vendor he typically does not have that leverage. The most he can do is cancel the contract and/or sue. But this does not deliver the technology he needs and he may now be back to square one with a lot of time and money down the drain. Wait a minute—is this a complete dead end: You cannot do it yourself and you must go outside because you need it, which is a risky proposition? No, the alternative is to go outside *very carefully* with proven vendors or ones with good reputations and a solid track record for delivering what they promise and when they promise it and the ability to *recover* if they get into trouble. Unfortunately, this "play it safe" approach to contracting out may eliminate some very innovative and exciting vendors who could, given the right circumstances, put the buyer way ahead in the market or in manufacturing technology. The DM will have to weigh the risks and take his choice of strategies. One way of hedging the bet sounds inefficient and expensive—running a "parallel" project in-house at a modest level. This *can* be done in an economically and politically efficient way if the DM is willing to look broadly inside the company for such a monitoring group or individual to keep the vendor honest and to provide early warning of problems and potential disasters.

Contrary to the beliefs of many DMs there *are* many people in the CRL who can do a great job of this parallel tracking. They can form a temporary team to do it and fade away when the project seems assured of success and on time delivery, thus cutting off further billing to the division. We have seen this work very well in many cases of acquiring systems for manufacturing, information, transportation/distribution, and others. It is up to the DM to decide, however, whether a modest "double" investment is worth making to protect his major investment in the procurement. This is a procedure very familiar in the aerospace and defense industries, or was until cost cutting reduced in-house staffs and capabilities severely. It can help the DM have his cake and eat it too—go outside for technology and also use internal capability to monitor and enhance the technology for the longer term.

Turning Away From Corporate Research Laboratories (CRLs) as Sources of Specific Technologies and Heavier Reliance on Operating Divisions or Business Units.   This general tendency, which follows the reduction in size, scope, and mission of existing CRLs and the elimination of some entirely, reduces the firm's options for "making" their own technology. Aside from the almost routine product and process improvement that the typical divisional lab undertakes, there is little interest in or slack for developing major new items. There is good logic in attempting to differentiate clearly the missions of a CRL and the divisional labs, respectively. The general differentiation of thrusts and responsibilities described in Chapter 2 makes sense *in general*. However, if it is used as a barrier to joint development, it can be self-defeating.

Lack of access to corporate-level technology people for specific divisional requirements—products, manufacturing processes, materials, testing procedures, specialized equipment—may occur for a number of reasons. If the CRL and other corporate-level groups are designed and controlled so as to assure that they do not have the capability to do such "practical" and specific things, then divisions have little alternative but to try to do it themselves, go out and buy it, or do without it. Top management and divisional management may be making a serious mistake if they let this situation develop due to lack of attention and understanding of its long-term implications. Such a situation has a *positive* feedback effect, reenforcing attitudes and practices toward make or buy that can severely erode the firm's technology capabilities in the long run and not even provide the best available technology in the short run.

Many CTOs try to fight this tendency and, if they have the personal influence or top-level backing to do so, can keep it under control. Where they are not able to control it, the company may be spending a lot of capital on duplicative procurements by independent division managers and may open itself to competitive inroads because the divisions are not technically advanced enough or have lost their capability to strike back quickly and effectively to meet competitive technical threats and opportunities.

Preoccupation of Operating Divisions or Business Units With Maintenance R&D and Incremental Improvements at The Expense of Aiming for Radical Developments or New Departures.   Muscles need to be used to keep them in tone and to develop them further. Where divisional and SBU technical staff are completely occupied with short-term, firefighting activities, their "innovation muscles" go slack. When a radically improved or new piece of technology is needed, such staffs are often "not at home." They have lost, if they ever had, the ability to move out into new areas technically. This self-fulfilling prophecy may force the DM to go outside for technology and generally does. Prevention of this spiral requires that the divisional lab and other technology groups in the division be allowed to spend or be *required* to spend a certain amount of their time and energy in probing and pushing the technology limits of the division's products and processes and pushing their own personal technical capabilities.

Disappointment With the Track Records of the Firm's Existing R&D Units in Generating New Products or Technologies. If the existing labs have not been successful in coming up with needed or expected or surprise technical advances, a rigid policy of going outside frequently is most likely to perpetuate that state of affairs. If there is general agreement that the in-house labs are not up to the task, that should be taken as a signal for *renewal*, not *abandonment*, of those units. The situation can only deteriorate further as the divisional and even CRL researchers perceive that they are not trusted with such important projects. Their morale, which may already be low due to lack of success, may deteriorate further and reduce any motivation to fight the trend. Renewal can be traumatic, expensive, and time consuming. However, it may be the most cost-effective strategy for the longer term. A thorough technology/innovation audit (see Chapter 7) can reveal the needs and opportunities for such renewal and whether it is likely to be cost effective in a given situation.

Pressure to Acquire Software Quickly and/or at Fixed Costs (a Growing Special Case). Until recently, software projects were indeed a special case. In the 1980s, however, hardware and software costs for many manufacturing systems and even individual pieces of equipment have "crossed over." This is also true for many products with built-in microprocessors or other "smart" electronic components.

Software now costs as much or more than the hardware it serves. This topic is discussed in some detail in Chapter 9, since it has been the subject of a 2-year research study in our group at Northwestern in the mid-1980s. Make or buy of software was the focus of a Ph.D. dissertation by Bruce Buchowicz associated with that project and has stimulated some of the ideas in Chapter 9.

For this discussion of general motives for make or buy, we can note that buying software packages has become dominant in most firms we have observed and worked with. Many of the factors discussed above contribute to this trend, and there are special ones peculiar to the nature of computer software.

The capability to write innovative and large-scale software has not yet been developed in most manufacturing firms, with certain exceptions. Encouragement to R&D and other technical people to become programmers can be offset by impatience at how long it seems to take, how much it seems to cost, and how off-target much of it seems to be when finally delivered.

I say "seems" here, because DMs and other company executives do not yet have good measures of what software *really* costs and how good it really is. They also find the *total life cycle time* for development, implementation, and integration of the software into routine operations hidden from view and not readily quantifiable. "The system is 'almost' ready; we just have to do a little more debugging, write a little more code, get the manual written, try a few more runs and...." Lack of competent and enough programmers can leave the DM and the firm as a whole with no alternative but to go outside for software. However, a parallel strategy can again help provide for the future when the firm and division can become almost self-sufficient in developing software or at least in evaluating, monitoring, and selecting vendors.

Some additional, specific reasons for going outside for technology, from a POMRAD dissertation by Falguni Sen, include the following:

- In-house work may have run into a patent (or exclusive license) held by someone outside.
- Desire to provide a window on technology and a chance to pick up a new skill for the inside technology staff.
- A license offered that will save much time, money, and effort to fight or try to work around.
- A division manager's circle of friends and business colleagues suggest an attractive source of technology; or the recommendation of a colleague in another company that "these guys are good and they'll run rings around your own R&D people."

6.4.2 Modes of Buying Technology

Figure 6-17 lists some of the more common modes of acquiring technology from outside the firm. Many of them are familiar to the manager who has had some

---

1. Licensing
2. Joint ventures
3. Limited R&D partnerships
4. Minority interests in firms with R&D programs
5. Contracts for R&D to other companies and research institutes
6. University contracts, grants, and consortia
7. Bilateral cooperative technology arrangements
8. Hiring individual specialists or teams of specialists
9. Stepping up technical intelligence activities
10. Buying technology imbedded in products, material, equipment, or processes (at a premium, if necessary)
11. Increasing pressure on suppliers to innovate
12. Persuading customers to share innovation
13. Acquiring small high-technology companies
14. Other modes (limited only by imagination and legal constraints)

Figure 6-17. Modes of buying technology.

experience in the technology area, even if it has been primarily with "inside" or "make" rather than buy. Many representatives of other companies, brokers, agents, licensing specialists, inventors, and others will have been around or in touch to push their wares and see if a deal can be made. Trade journals and the mail abound with stories and announcements of new product and process opportunities.

I will not go into detail on the characteristics of all these external modes of acquiring technology. Many of them are straightforward, at least in terms of the underlying concepts, and are well covered in specialized books, journals, and reports on specific modes—for example, licensing, joint ventures, and limited R&D partnerships. I discuss some of them in more detail where they are not in as common use as others and where there are special aspects that pertain to the main focus of this book and this chapter on sources of ideas and technology acquisition for the decentralized firm.

Limited R&D partnerships—some, modeled after older real estate ventures—appeared on the scene over the past decade or more. They are viewed primarily as a vehicle for investors to participate in high-technology start-ups without being directly involved in their management and with a large element of risk sharing and tax saving. They still exist and some large firms continue to participate, but changes in the tax laws and disappointing results have made them less attractive for many "arm's-length" investors.

The underlying risk-sharing feature and the ability to participate without direct involvement still hold attractions for the serious investor—individual or corporate. Unless the firm takes on the role of general partner and does in fact get directly involved in management decisions and policy about the venture, it has little appeal to the "hands-on" division manager. He is generally looking for a specific piece of technology in response to a specific, short-term need or opportunity. He seldom invests in a portfolio of such ventures. In fact most companies do not permit DMs to maintain their own investment portfolios.

On the other hand, the corporation itself may be very interested in such investments as part of its general *investment* portfolio or its specific *technology* portfolio. It is like the practice of buying minority interests in high-tech firms or firms with an interesting product line and technology base. In some of them, the size of the investment requires less than "full involvement," that is, a seat on the board, active participation in management, or assuming the primary role of "deep pockets" for future trips to the financing well.

In this case, the firm can view the investment as a "window on technology" or a toe in the water of a new field where they have not yet made up their minds to take the plunge. This approach—either through the limited R&D partnership or the minority interest—can be a comforting part of the firm's overall technology portfolio without it creating internal management problems, for the moment (see the section below on acquiring small high-tech firms).

A number of such deals, and other variations on them, do contain "hooks and levers" whereby the investing firm can increase its percentage of ownership, get first refusal rights on any results from the research, or otherwise share beyond the formal investment itself. A number of large firms are doing deals of this sort independent of their CTO and their corporate or divisional technology people. In those cases, the investments are made through a separate corporate "new ventures" or "mergers and acquisitions" group, frequently reporting up a quite different line from the firm's in-house technology activities. And herein lies a tale of conflict, duplicative efforts, and power struggles just below the top of the corporation.

We have been involved in half a dozen firms where this investment strategy is separate from and, in some cases, more favored than the firm's in-house technology. The level of consultation or even courtesy is low between the two groups and the CTO and his staff are not privy to the investment until after it has been consummated. The pretense for such lack of coordination and integration with the firm's own technology portfolio is often based on "legal" considerations, sometimes only thinly veiled as "mind your own business."

Where such acquisitions or involvements are indeed part of a "pure investment" strategy and where the conditions of involvement are clearly "hands off the venture" by the firm's management, such a separate path may be justified. This is especially true where the "only" motive for the investment is to broaden the firm's general investment portfolio and to hedge its bets. However, where there are mixed motives, such as preparing for possible future direct involvement in the new field, then the firm's technology staff and especially the CTO should be involved from the start.

In several companies we have observed such investments being sprung on the CTO and his long-range strategic technology planners, much to their dismay. Such moves can seriously impinge on their longer-term plans to directly enter a given field through in-house R&D, to participate in other ways, and to start on the learning curve internally to be ready for active participation in it. This has been the case for several of the new technologies in the fields of biotechnology and materials, for example. The complaints we hear from the *technology* people revolve around the issue that "the *acquisition* people are really not R&D types and don't know what they are doing" in terms of evaluating a possible investment and anticipating the longer-term impacts of it on the firm. The acquisition people complain that if they "did allow the R&D types into negotiations too early, they would start poking around and asking so many questions that they might queer the deal." In other firms, investment in or acquisition of high-tech or technology-based firms is considered "different" from the usual acquisition for general objectives, and the technology people are very much involved from the early stages. The Ph.D. dissertation by POMRAD member F. Sen shows these stages and gives some data on involvement by R&D people in a sample of U.S. and Indian firms.

Such outside investments as these limited R&D partnerships and minority interests pose some dilemmas for the firm's CTO, his staff, and the technology people in the divisions most likely to be interested in or impacted by the technology represented. Such outside, indirect involvements as well as some others on the list in Figure 6-17 (5, 6, 7, 12, and 13) can have important implications for the firm's or a division's technology program. The dilemma is whether the CTO of the firm or the division can safely count on such an involvement as part of his own or the firm's technology efforts and avoid duplication. Should he assume it is "in the family" and set up strong liaison or coordinating mechanisms?

The CTO and his CRL people may have many technology irons in the fire, such as with universities, government agencies, or "finders and brokers" of technology, and arrangements with other domestic and foreign sources. They may view these investments as just part of the general technical intelligence and technological forecasting efforts in which they are continually engaged. When a specific opportunity for further and deeper involvement occurs, their liaison efforts can be stepped up to a higher level of participation.

For the division manager, on the other hand, the question is more black and white. Does he or does he not (will he or will he not) have access to this new technology within a mode and time frame that are compatible with his plans and operations? If not, then he can readily turn away from it and seek technology from other sources even if the duplication is apparent from the viewpoint of the overall firm.

In other words, consistent with the perceptions and constraints on him, he is not generally interested in such indirect sources of "potential future" technology. When he needs it or wants it, *now* is the time and he wants *direct access* and as much *control* as he can get. These considerations—the impacts on both corporate and divisional technology—should be seriously considered before such a program of indirect technology ventures or acquisitions is undertaken or even before the first, sometimes opportunistic, one is made.

Relations with universities and indirectly or completely acquiring small high-tech firms are discussed in this chapter. Each mode of acquiring technology has some special features that make it both very promising and somewhat hazardous for the firm and, especially, for a division manager acting alone. The discussion above about minority interests in technology ventures concludes with a brief case study from a very large (top 50) U.S. firm.

Chairman: We want to set up a separate venture fund, independent of our internal technology efforts and our other, regular acquisition programs.

Venture Capital Specialist: What size investment are you talking about?

Chairman: Around a hundred million for openers—then we'll see.

Venture Capitalist: How will you select and manage the portfolio?

Chairman: That's just it; we want to approach this as a pure investment strategy in high technology and we want an independent portfolio manager who will do the selection and management without interference from us.

Venture Capitalist: Will you give him any guidelines, other than financial ones, about the kinds of companies he looks at and invests in?

Chairman: Not really, only we would like them to be somehow related to our likely future technology areas.

Venture Capitalist: Pardon me, but that sounds like you *are* planning to manage this portfolio from the start.

Chairman: Well, you don't expect us to let someone just go after anything he feels like, even if it has no possible relation to the firm's technology interests.

Venture Capitalist: How will you evaluate his performance?

Chairman: Why, just like any other fund or investment portfolio.

Venture Capitalist: But you may be stacking the cards on him before you even get started.

Chairman: You don't understand, he'll have a free hand to monitor and dispose of the investments as he sees fit; this is just an investment with us.

Venture Capitalist: But you said you wanted companies in fields of . . . never mind. Will your corporate technology people and CTO be involved in setting up the guidelines for the portfolio or monitoring the companies in it?

Chairman: No, I told you, this is a separate venture and has nothing to do with internal technology activities.

Venture Capitalist: But how will the firm benefit if one or more of the portfolio companies comes up with something of direct potential use to one of the divisions or the corporate technology people?

Chairman: We'll worry about that when it happens.

Such a lack of contingency planning has resulted in lost opportunities because, frequently, the internal technology people are not clued in and ready to pick up on outcomes from these "marginal" investments. Sometimes, the firm has to go into the marketplace and try to buy such outputs at arm's length when they could have had an inside track and the internal capability to exploit results as they were produced. Again, I make a plea for integrating such investment strategies into the firm's (and some individual divisions') long-term strategic technology planning.

Hiring individual specialists or teams of specialists as a means of acquiring specific technology is somewhat different from the usual recruiting of personnel for the CRL and other technology groups in the firm. In the latter case, the expectation is that such hires are "permanent" and that the general skills and experience of the recruit have continuing relevance to the firm's businesses and technology interests. Going after people in a field new to the company, where it is not yet clear that it will be one of continuing interest, is a different matter. Here there is a question of good faith and long-term intentions. Where there is a free marketplace and no one makes any promises other than "we want to look into this area to see if it's one we want to get involved in," then the acquired people are on their own. Frequently, a premium is paid to them as an inducement to leave the university or government lab or to switch from another company. This should be a warning to the recruits that, although they appear to be highly valued now, they may be expendable later.

Recent involvement with one such major recruitment campaign in the area of materials demonstrated the uncertainties for the firm and the hazards for the recruits. Several major nonmaterials firms made major forays into exotic new materials in the late 1970s and early 1980s. Most of those efforts have been scaled back or closed down by the late 1980s. In this particular case, not unrepresentative of the others, a major recruiting campaign was launched to get the *best* material science people in the country, from whatever source. People were attracted from universities, government agencies, and other firms by handsome salaries and promises of exciting programs with a "free hand to explore and exploit new materials." For other reasons—independent of the quality and progress of the research going on in the new lab—the whole materials venture was canceled and phased out within 2 years of its initiation.

Several of the key technical people hired by the lab vowed "never again to be suckered into a start-up deal like that." A year after the shutdown, some key people still had not found equivalent jobs.

Although the monetary cost to the firm of this aborted effort was not negligible (there were some expensive employment contracts that had to be negotiated and the output of the lab was essentially nil after the shutdown was announced), the real damage was to its reputation. It will be a while before top-level technology people will consider the company as a good bet for employment. One, perhaps, minor spin-off of this case and that of a number of others is that university professors— the prime source of top new graduates—will no longer recommend that their best students consider these firms as places to further their careers.

Where both parties (recruiters and recruits) go into such a relation with their eyes open and understanding that it is an exploratory thrust that might not be sustained, then this can be a very effective way of acquiring specific technology and building an internal capability quickly. It is a device many universities use to get into a new field quickly and at a critical level: it is called "buying a group." Typically, the senior person in the group has tenure or a long-term contract. The junior people and technicians brought along may not, and they are vulnerable to cutbacks and reverses if the "soft money" on which many groups exist dries up. That's considered part of the academic game and high-quality people have to remain fairly mobile if they are going to play the "move and rise" game of going for the higher salaries and better conditions (equipment, facilities, low teaching load, prestige surroundings). Also, changes in university research thrusts, especially when tied to teaching programs, are much more difficult to make and take longer to achieve by university top managements. Some curricula and research programs that have been marked for extinction hang on for years or even decades while the academic decision process drags on.

For really good technology people, hired by the firm as a probe into new fields, backing out of the field may not be a disaster for the people brought in. In the months or years they are in the company, often in the CRL, they can establish reputations as generally good people—scientists or administrators—regardless of their exact specialization. Many such people remain in the firm and enrich it with their infusion of new ideas and enthusiasm across the science/technology spectrum. This "residual" mode has been followed in a number of companies that started and aborted new programs in computers, materials, physics, biotechnology, manufacturing technology, and other specialties. Indirectly, this mode of "technology acquisition" has been a source of general technical renewal, even where the specific technology did not pan out.

Modes 11 and 12 in Figure 6-17 are very common. Most division managers are well versed in using them. Generally, they are more successful in exerting leverage on suppliers to undertake technology projects on the firm's behalf than they are with customers. However, innovation projects of both kinds are common ways of extending the capabilities and scope of the division's own technology efforts.

Customer-and supplier-stimulated technology projects are widespread and major modes of technology transfer in some industries. For many years, the textile and construction industries, for example, were accused of not spending enough on R&D and neglecting to innovate. That accusation was literally true and the amounts spent by individual firms in these sectors were very low relative to that in other sectors and relative to the potential for innovation in their own sectors.

However, when such sectors are viewed on an overall basis, it is clear that large amounts of R&D/innovation are under way in the labs of their suppliers—the chemical, fiber, materials, equipment, and other firms—on behalf of those "low-innovation" sectors.

We have done a number of studies of innovation among suppliers to the automotive industry and find that they, along with other suppliers, have a number of barriers to contend with in trying to innovate "from afar" for their customers. Published papers on the results from that series of studies suggest some general patterns that may apply to many other sectors and kinds of companies. (See Appendix B.)

Research on the sources of technical ideas over many years at places like M.I.T., Northwestern, and other universities have looked closely at the question of *where* ideas originate. Many of these studies strongly suggest that customers are a major source of ideas for new and improved products and processes.

However, one must note that the typical "idea" received from a customer is not generally complete, in the sense of an "idea" as defined at the beginning of this chapter. Ideas from customers generally contain the *need* portion of ideas in some form or other—often vague and general. But they seldom contain the other part—the scientific or technical *means* of solving the problem or filling the need. That is the job of internal R&D by the supplier who may in fact be overblessed with too many needs and not have enough ideas for the means or the resources to pursue them.

A division manager who does not fully exploit the potential for acquiring technology from both his suppliers and customers is missing a good bet for leveraging his own, often modest, internal technology activities. This mode of technology accession is worth a careful strategy of its own.

6.5   WHY DO SO MANY HIGH-TECH ACQUISITIONS GET INTO TROUBLE?

A very attractive way of "leapfrogging" technology and avoiding the time lags, front-end expenses, and uncertainties of getting into new fields involves acquiring a small high-tech firm or set of new firms. These are special types of acquisitions and mergers and need special attention in the large decentralized firm. The motives for the large acquiring company may be very clear in general. They may want the people, the products, the raw technology, the market entry, the "panache," or some combination of these. Their motives may be complex combinations of the factors mentioned earlier, which persuade them to buy instead of make technology in-house.

The motives of the generally small, generally new high-tech firm may also be complex and combine several factors. Figure 6-18, from a study by the National Science Foundation, suggests a number of motives, reflected in the kinds of problems such firms face. Added to these factors, which they think a larger firm can help resolve, are the personal factors of "need and greed." Many of the founders of these companies have worked hard and sacrificed much to launch and sustain their ventures. They may now want to cash in and to achieve some security in exchange for the sacrifices and risks they undertook.

In any event, there is a clear and active market generated by the motives of these two kinds of firms. The number and rate of such deals accelerated rapidly during the 1970s and 1980s, supported by tremendous advances and promised payoffs from research in biotechnology, computers, materials, electronics, and other fields. This was not the first wave of interest by large firms in these small high-tech firms. In the 1950s there were several dozen "raids" by "old line" companies into the New England nurseries for such firms. Most of them failed, due to poor fit. The more recent acquisitions have benefited from earlier experience and *some* acquiring firms have become more sophisticated and more cautious in making such acquisitions and trying to make them work within their own management systems. But the failures continue and many other acquiring firms keep making the same mistakes

Problem Areas	Percentage of Firms Considering It a Major Problem
1. The ability to provide competitive salaries and benefits to key personnel	69%
2. The ability to maintain an adequate level of R&D activity	68%
3. Obtaining venture and working capital	66%
4. Attracting and keeping personnel	63%
5. Ability to obtain capital equipment as needed	62%
6. Marketing a product once it has been successfully developed	60%
7. Undertaking high-risk R&D projects	60%

Figure 6-18. Problem areas of major concern to high-technology firms. (*Source*: National Science Foundation. *Problems of Small, High Technology Firms,* December 1981.)

and go into such deals with the same naiveté that an earlier generation of acquirers displayed.

At the outset of this discussion, let us make it clear that there is no sure path to success of such acquisitions. By nature, they involve many problems and issues that are unlikely to be resolved between the typical large firm and the typical small high-tech firm. But we *have* learned something over the past three decades of close involvement with this kind of technology venture about what causes problems and how to address or alleviate them.

6.5.1 Some Problems/Issues

Figure 6-19 presents a list of the problems/issues most frequently and severely encountered in the scores of cases we have observed or been directly involved in. Many of these factors are self-evident and, with certain variations, appear in most such acquisitions. The damage they can do and their influence on the degree of success of the acquisition vary widely from case to case.

The "elephant and mouse" syndrome is almost a definition. The acquiring firm is typically very large and the acquired firm is typically very small. All the jokes and wisdom about "how does a mouse sleep with an elephant" are relevant here. A slight twitch by the elephant may obliterate the mouse. Also, elephants, except

---

1. The elephant and mouse syndrome—negotiating power
2. Mismatches of culture, style, and perceptions—how much control
3. Contact points and "fitting in" to the larger organization
4. Career paths of leaders (usually founders) of the acquired company
5. Lack of clear understanding of who does what to whom, when, where, how, and why
6. The "post-courtship brush-off" where the original acquirer (sometimes a CEO) is no longer accessible
7. How to drop the other shoe, when something is ready for commercialization
8. From whose pocket does continuation funding come
9. How can the acquired entrepreneurial people be rewarded within the large company policies and "fairness" environment
10. What to do for an encore
11. How many can a large firm handle
12. What does "independence" mean within a large company context
13. What substitutes for the "piece of the action" the acquired entrepreneurial people are accustomed to
14. Can and how can the acquired people be integrated into the large firm—old boy/new boy conflicts
15. What happens when the large company changes its mind

Figure 6-19. Factors affecting the success of small high-technology firms acquired by large companies.

when enraged (which can happen due to the frustration of such pairings) move slowly. Small high-tech firms are used to moving quickly, and indeed that is one secret of their success and why they were attractive to the large firm in the first place. However, differences in pace, style of operating, and culture (that mysterious complex of factors) can doom the venture from the start, if not understood and addressed.

How much control will the firm and its staffs exert over the new group's plans, behavior, and freedom of action? How does it fit into the hierarchy, the pecking order, and the line for allocation of resources? What about the career paths of the new people? What obligations and privileges do they have or can they try to get? Who are their contact and pressure points? If they are subjected to the "postcourtship brush-off" by the top manager(s) who were in on the original deal-making, where do they turn for their "champion" or advisors or sources of power?

Few of these acquisitions are thought all the way through or "gamed" into the commercialization and end-of-life scenarios of their potential outputs. There is frequently no clear path for the downstream phases of the project or projects they bring into the firm or launch soon after they arrive.

If they are placed within an existing division or product group, it may turn out to be the wrong placement on technical or personal, even if not on marketing grounds.

Their expectations that the large firm would be able to move quickly and enthusiastically to place their product(s) into production and commercialization may be unfounded. The placement may have been a matter of administrative convenience rather than technical compatibility and "dovetailing."

One of the reasons for being acquired is frequently that the small firm does not have the experience or self-confidence in its ability to produce and market effectively. Finding themselves stranded at critical junction points in the product life cycle can be bitterly disappointing to them and their parent company. One major issue is the question of "from whose pockets does continuation funding come." A major motive in seeking to be acquired may have been shortage of capital. The owners may have the perception that the large company has very deep pockets and is willing and able to let the new group dip into them often and quickly. This might be the case in principle, but it may operate quite differently in practice. After the initial outlay for the acquisition itself, the CEO or other top management people may assume that now that "they have done their job" it is up to the divisional or group managers to take it from there. Herein lies the tale of failure of many such acquisitions. The division and group managers (many of the latter having no funds "of their own") have other fish to fry—namely, their ongoing operations. They may place the new venture very low on their priority list for working capital, personnel, facilities, equipment, or even their own precious time and energy. "Welcome to the club! But you're on your own now and will have to get in line for resources and favors."

Some firms are very aware of this "placement" problem and try to keep such an acquisition in an "incubator-like" environment for some time after it enters the firm. However, this poses some problems also.[*] The CEO has only a limited time to spend nurturing such ventures (whether "made" or "bought") and his moral support of them may not be enough to provide them the material support they need for going down the R&D/innovation path all the way to market. Besides, how many of these can a firm handle without throwing its regular businesses off track by the examples of special treatment for acquisitions? How many of these can even a very large firm handle? There is ample evidence in the wake of several massive failures of this approach that the number is small and that isolating them from the mainstream has its severe disadvantages as compared to "mainstreaming" them.

The career paths of the new people are seldom thought through beyond the "a good man can find lots of opportunities here" stage. If they stick to the high-tech path, what do they do for an encore to keep in the limelight and to show that they were not just one-shot wonders?

For a group used to a "piece of the action"—equity ownership and options of various types in the small firm—what can substitute as incentive and reward? What does "independence" promised them during the negotiations really mean in the large firm? How can they be integrated into the firm and used for more than their very specific technical specialties or "start-up" talents? Finally, what happens if and when the large firm changes its mind and decides it wants to get out of the relation. How can the losses be constrained and the potential gains be realized?

6.5.2   How Can Such Issues/Problems Be Resolved?

Figure 6-20 addresses, but does not solve, these problems and issues. Some of the ways of avoiding or dealing with them may seem obvious. If so, why don't most firms pursue them? Some of them seem very expensive in terms of time and resources. When compared with the potential benefits and the increased chances of success, these expenses may shrink significantly.

The first and last recommendations are the same: "if you don't mean it, don't do it." This should probably be repeated several more times in the list of how to deal with this kind of acquisition. Most very large companies are probably not good bets to make a success of such acquisitions. This pessimistic view is based on the fundamental incompatibility of the very large firm, with all it implies, and the stereotype of the very small entrepreneurial firm which may be exaggerated but frequently reflects the essence of its actual characteristics and behavior.

There are many exceptions, of course, or there would not be a track record of scores of such acquisitions that have lasted and paid off for both parties. The many more nonsuccessful ones, however, suggest that special conditions may have prevailed in the successful cases. This is what the list of policies and practices in Figure 6-20 is about—how to make this unlikely event succeed, or at least keep from failing early on.

I must hasten to add certain general exceptions to the "low likelihood of success" prediction. Large firms with a strong and advanced technology base and run by

[*]See the Miller Chemical Case in Appendix C.

1. If you don't mean it, don't do it.
2. Carefully work through the "life cycle" scenario for the acquisition, with disposition options if things do not work out as planned.
3. Fully discuss and disclose the reporting, status, resource, organizational placement, and other relevant factors before the deal is concluded.
4. Provide an incubator or "greenhouse" environment initially and make provision for mainstreaming later.
5. Let it be known that this is a serious move and not just a flyer that can be readily written off.
6. Think through and follow through the application and commercialization aspects.
7. Reach a clear understanding of career path options with the key acquired people.
8. Avoid shutting off access to top management if the original entry was at that level.
9. Provide technical and other support as needed.
10. Study the experience of other large companies with such acquisitions and extract lessons that may be relevant in your case.
11. Agree on milestones and decision points based on the realities of such a venture rather than on standard corporate practice.
12. If you don't mean it, don't do it.

Figure 6-20. Some ways of avoiding or dealing with the problems of acquired small high-tech companies.

technical people, or at least counseled by people with experience in and feeling for technical entrepreneurship, have a much better track record than those who do not. Several major electronic firms have essentially been built by a series of such acquisitions, although there is also a record of failure even among those firms.

A second kind of exception is one that we are very familiar with—"the buy-out, sell-out" ventures that at one time were indeed parts of larger firms. (Chapter 8 discusses such internal technical ventures.) Managers in many such firms profess strong antipathy for the large company environment, which they abandoned in a "buy-out" or "run out" when the venture did not work in the large firm. Despite this attitude, the founders of the venture may be anxious to cash in their chips and go back into the womb of the large company for many of the reasons mentioned earlier. But more to the point, even though they may have appeared to be "rebels" prior to their spinning off, many such entrepreneurs are used to and know how to accommodate to the large company environment again. In fact, they may make a triumphant reentry as "successful independent entrepreneurs" (since they sold their company for a handsome profit) and gain new status in the large company pecking order.

With such general exceptions and the special ones where the CEO of the large firm takes them under his wing for a variety of reasons (nostalgia?, ego gratification?, vicarious adventuring?) the track record is not encouraging. The recommendations in Figure 6-20 are intended to help improve that track record, allowing for a certain level of "sure failures" that should not have been undertaken in the first place: "if you don't mean it (or can't handle it), don't do it."

A number of the recommendations in Figure 6-20 fit in with the main theme of Chapter 5 on planning (items 2, 3, 6, 7, 10, 11). If there is no real planning culture in the firm, or an effort to think things through and consider contingencies and alternative scenarios "gives top management headaches," then the odds of failure rise steeply. Such planning of resource allocation, end-of-life scenarios, action options, commercialization strategies, career paths, milestones, and decision points need not be developed completely. They also need not and should not be rigid and inflexible. Conditions change, unforeseen events (good and bad) occur, and opportunities or problems may arise beyond the foresight of anyone. So the plans have to be "living plans" that are subject to frequent and serious review and modification as circumstances warrant. But they have to exist as guidelines, targets, and boundaries for action. Without them, the follow-your-nose atmosphere takes over and threats, needs, and opportunities can creep or gallop up undetected.

Another set of these recommendations relate to the main themes in Chapter 2—on the organization of technology—and in Chapter 8—on commercialization. Item 4—the "greenhouse"—applies to both internally generated technology ventures and acquired ones. It is discussed in more detail in Section 8.4 on technical entrepreneurship inside the firm. At this point, the major issue is that the usual, normal organizational structure and mode of operation will probably not accommodate entrepreneurial high-tech ventures, whether internal or external.

Too many constraints on policy, decisionmaking, and action have built up to allow the pace and flexibility needed for such ventures when plunked down into an existing structure. Just one aspect of this is the need for technical and other support in a time pattern, at a pace, and an intensity that strains the usual service load and leads to conflict and loggerheads between the venture managers and the service groups.

The final set of factors improving the odds of success relate to the attitudes and behavior of the buyers, champions, or protectors of the acquired venture. If the acquisition was made by or with the direct involvement of the top management, then the extra clout provided by that level of interest must somehow be sustained if it is to contribute to the odds of success. Shaking hands with the CEO and then never having access to him again can severely dampen the enthusiasm of the new people.

It is common to assign responsibility for "following" the acquisition to an assistant, a staff group, or a line executive (corporate, group, or division) after the closing ceremonies. For many not-really-significant high-tech acquisitions, this may be necessary and it may indeed be preferable to have the CEO insulated from such small and potentially time-consuming projects. For potentially significant acquisitions, however, which were originally viewed as crucial to the firm's future or even "saving the store," such brush-offs may be fatal.

We are now back to the recurring motif of this section—if you don't mean it, don't do it. If top management has no intention of providing sustained momentum to the venture after its acquisition, and if such support is necessary to its success or even survival, then it was a bad deal from the start.

The reasons for this argument are relatively straightforward. The acquisition may represent a radical departure from the firm's current products, markets, technologies, or mode of doing business. It is therefore likely to be treated as the "foreign body" that it is and rejected by the organization or surrounded by anti-bodies that essentially immobilize it (please pardon the biological metaphor). The organization may close ranks around it and keep it from moving forward and challenging conventional wisdom, existing territorial rights, and traditional ways of doing business. No one likes a wise guy and many of the founders and managers of these small high-tech firms have had to be or become "wise" to survive. Like their counterparts in internal technical ventures, they are not generally "rule followers" and "team players" unless they are captains of their teams. On the contrary, they rock the boat and perform all the other metaphorical acts that make managers and their staffs in the well-ordered, conservative company uncomfortable and even angry. Going after small high-tech firms and trying to bring them into the large firm is a risky, annoying, and frustrating business. It needs to be carefully thought out and monitored closely as a mode of acquiring technology from outside.

A word is probably needed here on the concept of "loyalty to the company." Newly acquired management and professional people (scientists and engineers) have no trouble with exhibiting loyalty to their new company and putting all their energies to work on its behalf. However, this may be a very temporary and situational loyalty, which has very brittle and close-in limits. Some years ago we did some research on the mobility and characteristics of engineers and scientists in industry. We found, in the literature, a dichotomy between professionals. One category was called "locals"—representing, among other attributes, a long-term dedication and often unquestioning adherence to the firm's policies, goals, and procedures. They often had contributed personally to developing and maintaining them. The other category was called "cosmopolitans"—professionals who owed their primary allegiance to their professions—doctor, lawyer, scientist, engineer—and looked to it for values, approval, procedures, standards, and rewards.

In our empirical research on several hundred engineers and scientists, we had difficulty sorting them into only these two groups. They seemed too rigid. But we did find an interesting group of technical people who seemed to exhibit both high local *and* cosmopolitan tendencies. They looked to their professional peers, often outside the firm, for their values, codes of conduct, and career paths while at the same time they worked very hard for their company and its goals. The trick was that these "local" loyalties were very mobile! We found cases of professionals, including the new breed of planners, information systems and management science people, who could and did switch loyalties over the weekend from one company to another. This is nothing for the loyal company management person to become upset about. The good "local/cosmopolitans" are rare birds and should be appreciated and effectively used *while they are on board*. Many of them are "agents of change" and during their sometimes brief sojourn in the company can breathe in a lot of fresh air and have a very beneficial impact on the firm's technology capabilities and outlook.

7

EVALUATION OF PROJECTS AND PROGRAMS: BEFORE, DURING, AND AFTER

7.1 INTRODUCTION AND CHAPTER OVERVIEW

Evaluation is a touchy subject. Everyone is in favor of it in principle, yet most people resist being evaluated themselves. Scientists, engineers, and their managers are no exceptions. Too frequent and too intrusive an evaluation procedure can turn people off and cause them to devote more energy and time to dealing with the evaluation than with the technical work itself—the technology projects they are engaged in or proposing. Too *little* evaluation shortchanges everyone. It suggests an attitude of indifference or neglect and can lead to sloppy planning, doing, and implementation of technology projects. In this chapter the state of the art on using sophisticated techniques for evaluating technology projects and programs is reviewed briefly and found wanting. Evaluation practices in industry are also examined and found falling far short of what is feasible and even necessary in terms of improved evaluation.

The evaluation process discussed here is a continuous one, starting from the initial screening of ideas and the acceptance of projects into the portfolio, through the "doing" phase when projects are underway and require monitoring, to the after-the-fact phase when it is important to know to what degree a project was successful and what factors influenced the degree of success.

These phases of evaluation are not all economic and not all quantitative. They include judgmental factors about the technology involved in the project, the way it was conducted, and the longer-term impact on the firm's operations. For example, some projects are "skill enhancing" and add to the firm's future technology base and general "smarts." Others are "skill depleting" and "use up" the knowledge, time, and energy of talented people in a less than cost-effective manner.

289

Finally, a new evaluation technique is presented, which we have used successfully in a number of industrial firms and some R&D organizations in the not-for-profit sector. It is a technology/innovation audit (T/I audit) of an organization's technology operation—as a whole or in its component parts. It is designed to assess how the technology operation(s) helps the firm in the marketplace and on the balance sheet.

7.1.1 Some Issues/Questions

- Why evaluate?
- Who should/can evaluate?
- What techniques are available for evaluation?
- How much is too much evaluation?
- What are the differences between monitoring, control, and evaluation?
- How much can formulas and mathematical models help?

7.1.2 Main Body of Chapter

This contains a discussion of material on project selection/resource allocation (PS/RA), the technology innovation (T/I) audit, and *control* versus *evaluation*.

7.1.3 Implications for Management

Evaluation of technology projects, especially for new products and processes, is a continuous process. The criteria and techniques used have to be adapted to each stage of the project/program—before it is undertaken (project selection and resource allocation), while it is in progress (monitoring), and after completion (evaluation).

7.1.4 A Watch List

- How do proposed projects/programs fit into company plans and the overall portfolio of R&D/technology projects?
- How do proposed projects relate to the resources available: funds, people, facilities, skills, and time horizon for completion and implementation or commercialization?
- Does anyone know authoritatively "how a project is coming along"?
- Do projects consistently miss targets: time, cost, technical progress?
- Are projects adjusted to meet changing needs and resource constraints?
- How do you know that a completed project/program was cost effective?
- Have careful records of costs and revenues been kept so that a reasonable job of evaluation can be done?

7.2 WHY DO WE NEED IMPROVED METHODS OF SELECTING, MONITORING, AND EVALUATING TECHNOLOGY PROJECTS/PROGRAMS?

Figure 7-1 lists some of the reasons why improved methods are needed for decisionmaking and action-taking during the various stages of technology projects—from the latter phases of the idea flow process (see Chapter 6) through the project's life cycle and postproject evaluation. The current state of practice in many large firms, as well as in most small- and medium-size ones, is not much more advanced than it was two or three decades ago, even though the theoretical and academic state of the art has progressed rapidly, with increasing mathematical sophistication and increased potential for computer-based systems. Later on this chapter presents some surveys illustrating the low level of use of quantitative methods in this vital aspect of technology management.

In this section we focus on why improved methods are necessary and desirable.

1. *To Sharpen Up the Project Selection and Resource Allocation (PS/RA) Procedures in Technology Management and Make Them More Relevant to the Firm's Long- and Short-Term Interests.*   Given that a "sufficient" number and quality of technical ideas are generated internally or obtained from outside, it is important that the most relevant and most promising ones are actually converted into projects. Frequently, the actual project portfolio of a given division, a corporate lab, or the company as a whole is not a good reflection of the technical needs and interests of the firm. Many projects continue on from year to year or even

---

1. To sharpen up the project selection and resource allocation (PS/RA) procedures in technology management and make them more relevant to the firm's long- and short-term interests

2. To improve communication between all concerned parties about project objectives, focus, likely outcome, and relation to other company activities

3. To identify barriers and facilitators to individual innovation projects and to the overall R&D/Innovation process

4. To continually weed out and improve the project portfolio by terminating dead ends and adding promising new opportunities (subject to the time response capabilities of the R&D/Innovation process)

5. To better relate innovation projects to long-range strategic technology plans

6. To identify off-track projects and signal corrective action

7. To prepare downstream activities, such as tooling, training, procurement, equipment and facilities design and installation, production, marketing, planning, and selling

8. To add a "technical accomplishments" dimension to the usual time and cost control variables

9. To demonstrate the value of the outputs of technology projects

---

Figure 7-1. Why do we need improved methods of monitoring and evaluating technology projects/programs?

from decade to decade without a clear link to the changing needs and objectives of the firm or the new opportunities and threats posed by outside forces in the marketplace and the economy.

Individual projects and whole programs are often pursued independent of shifts in the threats, needs, and opportunities (TNOs) with which the firm and individual divisions must deal. Sometimes the momentum for continuing irrelevant or unpromising projects and programs comes from the interests and skills of the R&D staff. Often, however, it comes from the poor linkage, discussed in Chapter 5, between business planning and strategic technology planning (STP).

I am not advocating a "zero-based budgeting" (ZBB) approach to the project portfolio, in which every project or program is placed in a "prove yourself or die" position every year or more frequently. That would contradict some of my earlier arguments about the need for continuity and stability in technology projects and an assurance to the R&D people that they can pursue ideas and radical technical approaches for a "reasonable" period without fear of being cut off in midstream. I *am* advocating a project selection and resource allocation (PS/RA) procedure that reviews the portfolio at intervals (annually in connection with budget making or at other times when major changes in objectives or focus are occurring in the firm). This review should consider the whole portfolio—ongoing as well as new projects—in terms of how they fit with divisional or company objectives and how much they are likely to contribute to those objectives—both short term and long term. For example, it may be that too much emphasis is being placed on technical service and minor improvements to existing product lines and not enough on addressing "future market missions" (see Figure 7-30). On the other hand, some CRLs may be spending too much time on pursuing exotic new fields when the company has in fact turned away from those kinds of diversifications to concentrate more on "core" business areas in its traditional fields of competence. A vital and sophisticated PS/RA procedure can signal such disparities and identify gaps in the portfolio as well as mismatches with company or divisional objectives and directions.

2. *To Improve Communication Between All Concerned Parties About Project Objectives, Related Company Activities.* Given that the match-up between the R&D portfolio and company objectives—both short and long term—is reasonable, this must be adequately and continually communicated. Too frequently, marketing, production, customer service, financial planning, and other groups are left in the dark about what is in the technology pipeline and how it might affect their activities and plans. We have observed many situations where new technology is being developed in R&D or engineering and other groups in the firm or division, which must implement and integrate it with existing technology, only have a vague idea of what is being done and what its impact is likely to be on them.

Another potentially important aspect of such communication is the opportunity provided to other groups to question and challenge the makeup and focus of the technology portfolio and to influence changes in it. For example, it is quite possible that a long-term company objective might well be served by a new technology initiative—a project or program. Top management of the division or firm might agree on this and R&D work might be started. However, the way in which technology planning occurs in many companies does not have enough "feedback loops" or participation by all interested and/or affected groups in the firm. This can then lead to the common stonewalling or redoing of projects when they hit these groups in the transition stages from laboratories to implementation in plant or market. While potentially distracting and seemingly time-wasting to management and technical people, early and continuous communication can be cost effective. It should cover what is in the portfolio, why it is there, and how it is likely to impact other functions.

3. *To Identify Barriers and Facilitators to Individual Innovation Projects and to the Overall R&D/Innovation Process.* Part of the methodology for selecting, monitoring, and evaluating technology projects and programs should be a procedure for identification and assessment of factors that might inhibit, block, or enhance projects. Some of these factors might simply slow down the progress of a project or reduce the level of accomplishment it might achieve. Removing or alleviating these could total up to a worthwhile improvement in the chances of success and actual outcomes. In addition, however, there might be some critical ones that could constitute "fatal flaws" and cause the project to fail or fall *far* short of its goals. This existence of fatal flaws has been continually brought home through my involvement in the venture capital field over the past three decades. Many ventures that looked good from a business and even a general technical viewpoint turned out, on closer examination, to have some built-in assumptions or "overlooked glitches" that made them infeasible or highly undesirable. A not untypical case involved a promising new computer printing process, employing several advanced technologies, which upon closer examination appeared to be violating several laws of physics and running up against several barriers in chemistry which were "immovable" at the then state of the art. The infeasibility was compounded by the optimistic lead times, costs, and performance goals set forth by the entrepreneurs. It was not funded when these flaws were identified and understood.

4. *To Continually Weed Out and Improve the Project Portfolio by Terminating Dead Ends and Adding Promising New Opportunities (Subject to the Time Response Capabilities of the R&D/Innovation Process).* A systematic, defensible, and "transparent" procedure for adding or deleting projects can help improve the portfolio value. I do not want to draw too direct an analogy with an investment portfolio, for reasons mentioned throughout the book—time lags, need for stability, and so on. But it is quite feasible to continually improve the portfolio through judicious adding and deleting of projects, as new threats, needs, and opportunities are recognized and analyzed. Again, management must be careful not to be too quick on the trigger.

5. *To Better Relate Innovation Projects to Long-Range Strategic Technology Plans.* This has been discussed at length in Chapter 5 in a general planning context. The emphasis here is on the need for the decision process and its mechanisms to convert technology plans and ideas into actual ongoing projects. The response to our deliberately naive question "why are you doing this R&D?" is

often "don't be silly" or "because it's what we know how to do" (if the respondent is honest) or "why not?" If the links between technology plans and a given technology program (not necessarily every individual project) cannot clearly be demonstrated or at least argued for, then deeper probing is indicated and "fixing" may be needed.

6. *To Identify Off-Track Projects and Signal Corrective Action.*   Section 7.4 on the monitoring and review process presents some methods of identifying and correcting the trajectory of projects and programs. The annual reviews are not sufficient in most cases to perform this vital function. So-called "reviews" are often after the fact and do not occur frequently enough or in the right time pattern needed for actually making course corrections of a major type. Certainly, top management or even top technology management cannot afford to get involved in micro-management of projects. That is what the technology management hierarchy, including project and program managers, are for. But there is also need for a constant oversight of ongoing projects to make sure, in a gross way, that they are moving in the general directions agreed upon or that have evolved from the work itself and *subsequently* agreed upon as proper directions. "Killing" projects is one of the problems technology managers mention most frequently and few of them have a completely satisfactory way of performing that frequently onerous chore.

7. *To Prepare Downstream Activities Such as Tooling, Training, Procurement, Equipment and Facilities Design and Installation, Production, Marketing, Planning and Selling.*   Chapter 8, on commercialization, discusses the general transition phases of projects from R&D to production and other downstream activities. However, an important part of the early-stage PS/RA procedure is to give early notice to these other functions in order to help them plan their future activities. It is important to alert such groups to be ready to receive the results of technology projects, especially where radical innovations are involved. Many of the downstream functions will appear to ignore long lead times reflected in early notice of projects in the portfolio, because they are too busy or not confident that the projects will actually be completed when scheduled or "ever." However, periodic reminders of what is in the pipeline can help gain their critical cooperation in implementation and help them plan their own activities to accommodate insertion of new technology and products into their ongoing programs.

8. *To Add a "Technical Accomplishments" Dimension to the Usual Time and Cost Control Variables.*   Unfortunately, high levels of uncertainty characterize many technology projects, especially attempts at radical improvements or new products/processes. This is often used as an excuse for falling back on cost and time as the major, if not the only, criteria of progress. If the schedule of the project appears acceptable and it is "within budget," then many managers are willing to assume that the project is "on track." Improved monitoring and assessment procedures should include level and rate of *technical* achievements as critical parts of the procedure. Otherwise, without that critical dimension, many projects go merrily along on schedule and on budget, but way off target in terms of what they were intended to accomplish technically. It is very difficult, some will say

impossible, to assign degrees of technical *achievement* to a project that can be used for measuring progress. Most frequently, technical *activities* are substituted for technical *results* or *progress*. For example, number of tests run, samples made, prototypes developed, and other *activities* are counted as technical *progress*. Certainly, they are necessary for technical progress. That is, they must be done if the project is to succeed. But they are not in themselves degrees of achievement of the technical *goals*—the new product features or operating characteristics of a system or piece of equipment.

An effective monitoring and review system must have this dimension of technical accomplishment as an integral component.

9. *To Demonstrate the Value of the Outputs of Technology Projects.* Finally, a complete system must have the ability to evaluate the actual outputs and impacts of the project or program that has been completed. More on this will be said later in Section 7.5. However, criteria and measures of value must be incorporated at every stage of managing technology projects, from the initial idea generation through the project's end-of-life stages.

The above reasons for improving the procedures for managing technology projects and programs relate to three important phases of the overall R&D/innovation process. These three phases are closely linked and are hard to separate cleanly. The next three sections of this chapter propose methods of dealing with them, with some inevitable overlap.

1. *Before* projects are initiated, when ideas are being evaluated and selected for inclusion in the portfolio.
2. *During* the life of the project, when management has to assure that they are proceeding as originally intended or later agreed to.
3. *After* the project has been completed and its results and impacts have been realized.

Many companies have improved parts of this process, with major emphasis on the before, during, *or* after. Few have complete systems for coping with all three phases and making sure that they are consistent. That is, criteria for evaluating projects should not necessarily remain rigidly the same throughout the project's life. But the criteria used at the three stages should be compatible or comparable so that one can trace the objectives of the project throughout its life cycle and continually relate it to the business and technical objectives that gave rise to it. Figure 7-2, which I published 30 years ago,[*] sums up the "control" and "evaluation" aspects of the overall R&D/innovation process presented in this chapter, differentiates between them, and relates them to the objectives management has in establishing and maintaining a technology program in the first place.

---

[*]Hence some of the "frustrations" discussed in the Preface.

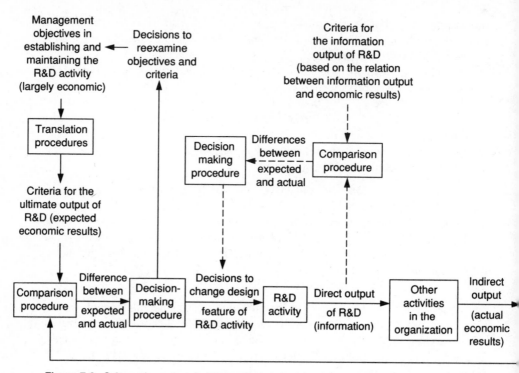

Figure 7-2. Schematic portrayal of the control and evaluation procedures, including the occasional procedure of reexamining objectives and criteria.

7.3 BEFORE: PROJECT SELECTION AND RESOURCE ALLOCATION (PS/RA)

This section includes three sets of recommendations for designing or improving the PS/RA system in the firm or in a particular division. Before that, however, some basic aspects of the PS/RA process are introduced briefly. Figure 7-3 illustrates the cash flow or cumulative cost/return curve that is generally implicit in evaluating potential projects, even if not an explicit part of the supporting documentation. It has several interesting features that can be useful in the "during" and "after" phases of monitoring and evaluation. The first is that it gives a picture of the shape of the cost/return pattern (see Figure 7-4 for a histogram version of such curves). It therefore enables people to plan for expenditures and returns in terms of their capital budgets and operating funds. The second is the possibility of superimposing a management decision criterion such as time horizon or time to payoff (the shaded vertical line), which signals the length of time management is willing to wait for the cash flow to turn positive. For example, if an overriding criterion for selecting and continuing to support projects is that "they must pay off in X months"—the shaded line—then clearly project A has met or is predicted to meet that criterion while project B has not or will not. The irony in the example chosen for Figure 7-3 is that project A will be preferred over B according to this time-horizon criterion even

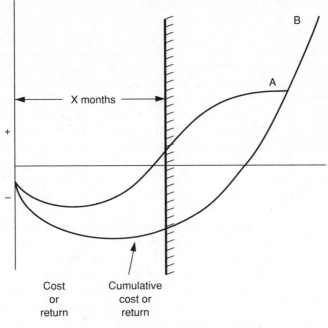

Cost
or
return

Cumulative
cost or
return

Figure 7-3. Cash flow pattern for projects.

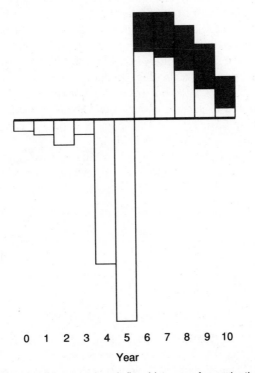

0 1 2 3 4 5 6 7 8 9 10

Year

Figure 7-4. Discounted cash flow histogram for evaluation.

though the long-term cumulative payoff from B far exceeds that from A. A third feature of such diagrams, not shown in Figure 7-3, is the ability to superimpose the *cost* of capital or rate of return. This would have the effect of shifting both curves to the right—indicating a longer period to return the original dollars invested plus the cost of the money invested (spent) to date. A final feature of such curves is the ability to "move along the curve" in time, as the project proceeds, and correct both the estimated cash flow figures with actual costs/returns to date and to reestimate the path of the curve for future periods. Unfortunately, many people who use such curves do not consider them "living documents" to be continually updated, as they must be, in order to serve as successful PS/RA and monitoring/evaluation tools.

Figure 7-5 displays basic formulas for calculating the value of a project, at any time in its life cycle, but especially during the PS/RA stage. W_i is the worth of an individual project (this can be cumulated for the entire portfolio). R_i is the cumulative return and C_i is the cumulative cost. These three entities are used in different combinations—sums, ratios, or other formulations, but they are the building blocks of any evaluation formula or attempt to quantify the economics of the project. The upper versions do not contain terms for probabilities; the lower ones have associated probability estimates to reflect the degree of uncertainty in cost components and return.

Figure 7-6 is a bridge between the gross budgeting for technology and the way the budget is allocated. Many technology programs and R&D labs suffer from having more ideas they want to work on than there are funds to support them. Others find it difficult to justify even the guaranteed or "easy" access level of budget with enough credible good ideas that are both feasible and desirable.

An important part of the technology budgeting and PS/RA process is this "matching up." Historically, many technology budgets have just continued at the same level, perhaps adjusted for inflation. Insufficient analysis has been done to assure that the funds required to support essential technology projects and programs are available and that, in turn, sufficient technically good and commercially relevant projects are available to justify the expenditures.

Figure 7-7 illustrates point 4 (weeding out and improving the portfolio) in the previous section. Subject to the ability of the technology staff to respond to such

(1) $W_i = f(R_i, C_i)$

(2) $W_i = R_i - C_i$

(3) $W_i = \dfrac{R_i}{C_i}$

(4) $P_{ci}C_i = P_aC_a + P_bC_b + P_cC_c{}^*$

(5) $P(W_i) = f(P_{R_i}R_i, P_{C_i}C_i)$

Figure 7-5. Basic formulas for project selection. (*These are components of the cost.)

List of Projects		Available Budget	
Must do	A_L	Guaranteed	1_B
Should do	B_L	Could get	2_B
Nice to do	C_L	Might get	3_B
Only if excess $	D_L	Not available	4_B
Don't do	E_L		

Figure 7-6. Project ranking versus budget.

changes, due to inertia and the "frequency–response characteristics" of the whole portfolio, projects can be upgraded or downgraded as conditions change so that the funded projects constitute an "optimal" or at least "pretty good" portfolio. Figure 7-7 shows a project originally rated well below the cutoff line being upgraded to replace one that was originally well up in the ratings. Changes in external conditions (market, economic, technical) or internal events (focus, success/failure, technical problems) can signal the need for such changes, again subject to the ability of the technology function to respond without major disruption or side effects.

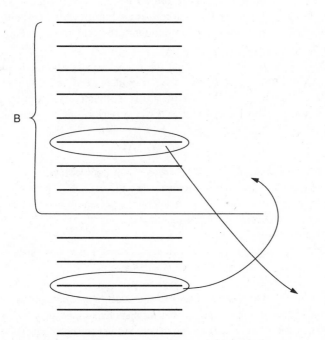

Figure 7-7. Substituting projects in the portfolio.

7.3.1   What Are Some of the Characteristics of an Effective PS/RA System? (See Figure 7-8)

1. The PS/RA system should be *simple* to conceptualize by the users (direct and indirect), simple to operate, simple to explain, and simple to update and change. This does not mean that it need lack sophistication, but that it should not be needlessly complex in terminology, instructions-for-use, interpretation, and evaluation. It is very important to know, before introducing any new PS/RA systems, the history of ones previously tried or used, including what happened to them and whether or not they have been used or are being used by any of the current management who are involved in the overall R&D or planning aspects of the company. Even more important is calibrating the reaction of those key individuals who have been subjected to such systems and who are expected to be involved in new ones. If management's prior experiences have "turned them off" to formal methods of PS/RA, then we have a significant problem of implementation and gaining "real" acceptance as compared to honorific or superficial acceptance and later lack of user cooperation with the new system. Simplicity/complexity is not the only factor that leads to a negative reaction by potential users and contributors, but it is certainly a major one. Another way of looking at this dimension is from the viewpoint of "face validity." If decisionmakers cannot satisfy themselves that the system is "making sense" of the information fed to it, then they are less likely (in the extreme, quite unlikely) to participate in its use and assure its success. Certainly, experienced managers are not unprepared for "counterintuitive" results from an analytical approach to decisionmaking, but they also certainly do not want to be victimized by a system that is yielding both counterintuitive and incorrect or unacceptably risky results.

2. Point 1 leads to the notion that the PS/RA system be developed in an *evolutionary* mode, building up from fairly straightforward and intuitively acceptable methods like "ranking" to eventually, fairly sophisticated "optimization" or, at least, "satisficing" models/procedures. Over a period of 2–3 years it may be fea-

---

1. Simple to conceptualize, operate, explain, update, and change
2. Evolutionary, building from simple methods
3. Emphasis on information for decisionmaking rather than specific plans as such
4. As close to "real-time" operations as possible
5. An interactive system
6. Flexibility in adapting to individual variations in needs and constraints
7. Use of existing information systems where feasible
8. Emphasis on acceptability to decisionmakers
9. Within a portfolio framework rather than individual "egg-sorting" decisions
10. Based on the concept of R&D as an investment rather than as an expense
11. Important role in cutoff and deletion rather than just adding of projects and programs

---

Figure 7-8. Characteristics of an effective project selection and resource allocation (PS/RA) system to support R&D/innovation planning.

sible to help the users of the system evolve from ranking (or Q-sort* or rating) methods that are not drastically different from what they do now, perhaps in a less formal and systematic manner, to more and more sophisticated multiple-criteria decision methods. These latter can help account for some important subtleties of the PS/RA process, such as trade-offs, combining multiple probability distributions, and compensating systematically for biases in estimation.

3. In all discussions and selling of PS/PA systems, emphasis should be placed on *information for decisionmaking* rather than "optimizing" or "solving" or "automating" the PS/RA process. This is related to the universal issue that *managers* make decisions and not computer programs or formal procedures. Many managers react negatively to the notion that they might be replaced by a formal (especially computerized) procedure, or that anyone of importance thinks that they might be.

4. The PS/RA system should be as close to *real-time* operation as possible. That is, the decisionmaking, monitoring, control, evaluation, information acquisition, and transfer aspects of the system should not be driven by the calendar only or even primarily. It should instead be driven by important *events* and changing situations, internal or external to the company.

5. The PS/RA should be an *interactive* system, which permits decisionmakers and their supporting staff people to ask questions, pose alternative scenarios, change criterion functions, simulate interaction between the important variables, do sensitivity analyses, add information, delete information, and so on. One of the most attractive features of some of the newer real-time or interactive information/decision systems is that there are many bonuses to a conscientious user—for example, information storage, reminders, communication channels, and analytic and computational aids. All these features may not be warranted initially, but there may be a computer component in the system (ideally a time-sharing network with individual input facilities—typewriter or voice or hand written). An ultimate aim might be to build toward a real-time interactive system with many of the above features (which are now technically feasible and are temporarily cost constrained). None of the above remarks should be interpreted in terms of fulfilling the wildest dreams of a computer enthusiast. If you can do it better, faster, cheaper, more flexibly, and/or more accurately by hand, that is the preferred method. However, the intrinsic nature of the information and decisions involved in PS/RA in a large organization provides a prime opportunity for the careful introduction of computer-based information/decision systems.

6. One of the major causes of failure of most of the PS/RA systems that have been introduced or proposed in large industrial and government R&D organizations has been their lack of *flexibility* and adaptability to variations in individual's needs, preferences, and styles of operation. This idea does not lead to the necessity of a different system for every decisionmaker. Quite the contrary, all systems used must, at some point, share common information, criteria for decision, decision premises, time perspectives, and other major aspects. The notion of a flexible system would allow for several options in a number of the subsystems of the PS/RA system. For example, we have found that some people are quite comfortable with

*A technique of ranking items by successive sorting of subsets of them.

and prefer Q-sorts for ranking or rating. Others prefer explicit scales. Some want to work alone on their ratings and analysis; others prefer or are willing to attend meetings. Some want the full range of uncertainty laid out in as much detail as reasonable (giving the "states of nature" and information about them); others prefer to lump uncertainties into "black box" transformations, such as a simple Monte Carlo or other simulation routine (not necessarily computer based). Finally, some people like or can develop a taste for direct interaction with computers; others prefer to be insulated by human interfaces. We have had some modest success in offering a small number of options related to these aspects of the system, leading to an appearance of "tailor-making" of the individual systems, without leading to incompatibility.

7. Whenever possible, *existing information and decision systems* already in routine use for technology should be incorporated into the new system. This can save much time, much money, and much conflict with people and organizational units which have a favorite system already in place that can fit in as a module to the new system. Assuming such modular components are reasonably effective and based on valid assumptions and design, such a strategy should increase the chances of acceptance of the new system.

8. Throughout the whole development and shakedown phases of the PS/RA system introduction, close attention should be paid to the *acceptability* of the system and its various demands on the time and attention of users. Even after the system is considered operating at a steady state, systematic monitoring should be done of the reactions to its use so that modifications can be made before they threaten its continued and effective use.

9. Although some of the components of the PS/RA system, as described in the next section, deal with individual projects, the emerging view in the field is that project selection and research allocation in R&D/innovation are most effectively done in terms of *portfolio decisions*, into which feed the analyses and preliminary decisions on individual projects and subprograms. PS/RA is treated, in much of the literature and in many organizations, as a one-shot, "egg-sorting" procedure without regard to the place of groupings of projects in a balanced portfolio. This leads to loss of synergy between projects and the side effects of cancellation or failure of a given project.

10. Related to point 9 is the emerging notion, in some organizations (private and government—e.g., the view apparently reflected in some tax regulations and increasing federal support of basic research) of *R&D as an investment,* rather than as an expense. Unfortunately, the tax laws have historically encouraged the current expense view and many non-R&D managers let that influence their overall view of expenditures on technology.*

11. The procedures developed as components of the PS/RA system have to be useful in *cut off* and *reallocation* decisions in R&D, not just in putting projects

---

*Some recent legislation has been aimed at encouraging an "investment" point of view, but its impact on funding of R&D by large firms does not appear significant. That is, the general level of R&D funding is not as sensitive to such incentives as it is to the business considerations that have historically influenced it.

into the portfolio. This issue of "how do you cut off a project" is one of the most popular ones (and one of the most plaintive in discussions among R&D directors). If properly used, along with a "sunk-cost" philosophy, an effective PS/RA system should provide straightforward answers to that perennial question.

7.3.2 What Are the Components of an Effective PS/RA System? (See Figure 7-9)

In view of point 1 in the previous section, which emphasizes the desirability of simplicity in the PS/RA system, this section identifies components of the PS/RA process which might be separately developed, packaged, and used as part of an overall strategic technology plan or STP system. Ideally, all components should eventually be incorporated into the PS/RA and STP systems developed for the company. However, development, conceptualization, introduction, and adoption may be considerably simplified if the several components are handled separately, with continual conceptual and operational integration in the design, as the components are clarified, developed, pilot tested, introduced, and debugged. The following components do not necessarily represent sequential phases of the system. Some of them can be developed at any time and introduced at any time. There is high interdependence among some subsets of components and one of the important tasks of the development phase of the PS/RA system is to be sure that these interdependencies are clearly recognized and accounted for in designing and installing the system.

1. Identification of *future market missions* (FMMs) and the capabilities and skills needed to attain them. Such a procedure is needed for each major business unit that participates in the STP and PS/RA systems in order to systematically explore, identify, calibrate, and establish targets for entering radically new fields or fields that are more than mere extrapolations of what is currently being done. This component of the PS/RA and overall STP process requires frank, open participation

---

1. Procedure for identification of "future market missions"
2. Monitoring component for all phases of idea-to-evaluation process
3. Idea generation and flow component
4. Product life cycle component (including end-of-life strategy)
5. Procedure for continuously relating portfolio to budget
6. Component for evaluating new product opportunities
7. Flow model and control component for decisions and actions
8. Decision and information checklist
9. Technical alternatives component
10. External intelligence system
11. An action and follow-up plan for these components

Figure 7-9. Components of an effective project selection and resource allocation system.

of the highest levels of each business area that wants to participate and should include a small number of top functional specialists such as marketing, engineering, R&D, production, and finance. This is a very unanalytical area and techniques for assessing judgments and perceptions will be the main methods used—for example, Delphi-type techniques. As major inputs for this component, the participants would require good technological forecasts and other assessments of what the environment is likely to be in the time frame contemplated—whatever target period is chosen for the FMM exercises (see Figure 7-30).

2. A *monitoring* component for the overall "idea-to-review" phases of the individual project selection procedure. Project selection is a continuous process, involving multiple decisions (screening, evaluation, placing in portfolio, funding, assigning people) and continuous reviews. Some of the key decisions encountered in managing technology projects are listed in Figure 7-10.

3. An *idea generation and flow component*, which directly addresses the front end of the project selection process—getting good ideas into the system for

---

1. Organization
    (a) Decision on whether to set up a special organization—for example, a project group
    (b) Full entrepreneurial responsibility to project leader or less
    (c) Set up separate government product division or group
    (d) Separate/integrated organizational form
    (e) Project/functional setup of R&D and related innovation activities
    (f) Level in the organization (how important is the project)
    (g) Decision to set up program in organization on major footing
2. Personnel
    (a) Whom to assign
    (b) How much resources to allocate
    (c) Key person assignment or less capable person
    (d) Hire specialists
    (e) Assignment of personnel to project phases
3. Marketing
    (a) Market research—degree of effort and commitment
    (b) Set up new distribution system or change existing one
    (c) Representatives (Reps) versus direct selling; other forms of distribution set up on a project basis
    (d) Entry into a new field or just moving slightly to one side
    (e) Search or devoting selective attention to opportunities
    (f) Bid high or low—buy into for sake of follow-on or building credibility or reputation in the field
4. R&D
    (a) Decision to bid on a development contract

Figure 7-10. Illustrative list of general decision or action points in the R&D/innovation process (not in order of importance or sequence).

evaluation and decision. This component will help assure that good ideas get into the system and that they are not filtered or blocked out for extraneous reasons (many of the psychological, organizational, or political reasons that kill off or discourage potentially valuable ideas before they even become projects). See Chapter 6 for an extended discussion of this component.

4. A *product life cycle* component (sometimes called end of life). At this stage we are dealing nose-to-nose with the product managers and other business people in the firm who may or may not be reluctant to spend R&D funds in anticipation of and to help prevent obsolescence (technical or economic) of a product or business line. (See Figures 5-6 through 5-8.) Here the use of various kinds of economic analysis (e.g., sunk cost, marginal analysis, return on investment) may help clarify the proper level of investment of R&D funds in *mature* products and lines. In addition to the analytics, techniques for communication of the results of the analysis and methods of decisionmaking and persuasion may be needed. (See Chapter 5.)

   (b) Initiate or accelerate R&D

   (c) Decision to innovate beyond the specific order or contract

   (d) Investment level and allocation to different phases of the R&D/innovation process

   (e) Pursuit of a request for proposal (RFP) or solicitation a bit afield from regular lines of business

   (f) Firm's awareness of RFP, extent and level in organization

   (g) Decision to engage in R&D beyond RFP delivery needs

   (h) Perceived opportunities and costs of specific procurement and commercial follow-up

5. Production process

   (a) One shot versus follow-on

   (b) Go into it on a full scale or not

   (c) Make or buy components, materials, services, facilities, products, equipment

   (d) New facilities or equipment

   (e) Merger or acquisition to obtain technical, production, or marketing capability

   (f) Tooling—new, extent, quality

   (g) Critical path behaviors or events: tooling, letting subcontracts

   (h) Decision to tool and so on for longer-run production

   (i) Perceived opportunities and costs of specific procurement and commercial follow-up

6. Finance

   (a) Optimize profit on a particular order

   (b) Investment level and allocation to different phases of the R&D/innovation process

   (c) Source of funds—cash flow, reserves, go to bank, long-term debt, equity

7. All

   Investment of time, personnel, money, executive attention in search/bid activities

---

Figure 7-10. *(Continued)*

5. A component is needed for continuously *relating* the *project list* to the technology *budgets*. Figure 7-6 in section 7-3 suggests that there is more than one budget and that its size depends on the list of projects or portfolio presented. There are a number of portfolio models and programs that are applicable to this component of the PS/RA system including models for handling "budget increments and decrements" which we developed some years ago.[*]

6. A component on *evaluating new product opportunities* in terms of market size, market share, competition, price, and profit. In this "total project evaluation" component, deep participation by marketing and business people is essential. They are the prime source of estimates of levels and probabilities of the various factors. This will require small group estimating techniques, simulation methods (hand or computer or combinations), and continuous refining, review, and revision of estimates. The reason that this must be kept up to date is that some of these factors (e.g., market share) may turn out to be "flip factors" that can influence "go or no-go" decisions on a project, product, product line, or business. All too often the estimates on which such decisions are made (e.g., going ahead with a new product or line or entering a new market) are based on one-shot, single-value, old estimates. This component is included to protect against the deficiencies common to such estimating. It should be used periodically in the life of a project.

7. A *flow model and control* component for the detailed decision and information activities required in the overall process involving individual ideas/projects and the entire portfolio. Figure 7-2 illustrates this component. We have used components like it in attempts to elaborate the actual nitty-gritties of the process. This component, of course, is intimately related to components 2 and 3 above, and to others mentioned in this section. It needs special attention as part of the steady-state operation of the PS/RA system.

8. A *decision and information checklist* component. Figure 7-10 illustrates the kinds of decision events and other events that should trigger decisions plus the supporting information for these decisions. In earlier reference to time-driven versus event- and decision-driven PS/RA systems, the point was made that such a system should be primarily of the latter kind, with the need for a decision, unexpected events, new information, and other factors triggering the various components of the PS/RA system—for example, all the components mentioned in this section. These decisions, events, and information items are the operational subcomponents of the monitoring component and the *flow model and control* components and are related to the other components in significant ways.

9. A *technical alternatives* component. One of several systems in current use is needed for detailed examination of technical needs, opportunities, threats, alternatives, options, and so on.

10. An *external intelligence* component. This is a procedure and organization that provides inputs on the wide range of factors external to the company which can

[*] See list in Appendix B for citations to the work of Souder, Maher, Rubenstein, and other members of POMRAD on this subject.

influence the desirability and viability of products, product lines, and businesses—both new and existing. Here the technological forecasting, technology assessment, futurology, social indicators, and other specialties can contribute to many of the components of the PS/RA and the overall STP systems.

11. An *action and follow-up* component. Without this, most of the other components will not be activated or continued.

7.3.3 How Much Would All These Components Cost?

A reader of this chapter might understandably say: "Your first recommendation about PS/RA systems is that they should be *simple*! And yet you present a mind-numbing array of techniques and procedures that are sure to take too much time and cost too much money." My response is that the front-end cost of developing and installing some of these components and indeed the whole PS/RA system is not low, that development is not quick, and that an effective system requires a lot of thought and persistence. These resources may be in short supply in a busy and fully committed organization—operating division or firm. The decision to deploy some of these precious resources to investment in a PS/RA system in support of the technology program may seem like a longer-term investment than many managers are willing to make.

If that is the situation in a particular firm or division, then the elaborate procedures and system designs recommended here are not appropriate. In those cases, however, many of the fundamental ideas behind the recommended PS/RA system may be useful as stimulants to see that *some* of the essential functions are performed in some minimally acceptable manner. These include looking at the environment, speculating about the future, making sure good ideas are generated and communicated, relating the technology portfolio to company or divisional plans, and seeing that projects are conducted in the manner intended.

Once the shock of a potentially high front-end cost is weathered, the CEO, division manager, CTO, or other responsible executive might recognize that some of these costs could very well be offset by avoiding some of the mistakes and project failures that his organization has historically experienced. One "saved" project can easily repay the whole cost of developing and implementing a PS/RA system with all the bells and whistles advocated above.

Short of that justification, another calculation might be that many of the components are already in place in the firm in some form or other. They may merely need tightening up, sharpening up, or placing-in-context of the PS/RA procedure. Certainly, in large R&D performing firms, some versions of all these components are most likely in place. The primary task would be streamlining them, eliminating redundancies and overlaps, integrating them, simplifying them, and communicating them to the concerned individuals and groups in the spirit of an *integrated* PS/RA system.

Although widespread support for this integration is needed to give the PS/RA system a chance to succeed, the actual personnel requirements may be quite modest. In many companies, one part-time staff member has been able to develop, refine, integrate, and put in place a PS/RA system in less than a calendar year. The critical resources are management support and the cooperation of the various members of the technology management community. As to upkeep—the hidden cost of many new management systems, especially computerized ones—it can be surprisingly inexpensive. At the corporate level, one member of the CTO's staff has been able to support such a system and also have time for special assignments. At the divisional level, one person-day per week might be enough to carry out most of the operations of the PS/RA system. We have observed one staff person supporting several hundred scientists and engineers in large laboratories in terms of the methodology, reports, and administration of such a PS/RA system. Of course, many of the components require the intellectual participation of many people in making estimates, preparing forecasts, gathering information, and other supporting activities. But they must do these things anyway in order to perform their own assignments. An integrated PS/RA system provides the framework and synergy for helping them to do their assigned jobs more effectively.

On the other side of the cost–benefit equation, Figure 7-11 lists some of the direct and longer-term benefits that can be derived from an improved and effective PS/RA system.

Expected Direct Benefits from Improved PS/RA Systems

1. More sharing of information relative to opportunities, threats, and plans for long- and short-range technology projects
2. Improved definition and disclosure of risks and assumptions underlying proposals and decisions
3. More rapid and definite early warning of impending factors that can affect the risk, benefits, costs, and timeliness of projects and programs
4. A more systematic basis for taking into account and sharing external intelligence about markets, economic factors, government actions, and marketplace changes
5. Providing a framework in which decisions and actions related to technology programs are related in a logical fashion so that important decisions do not fall between the chairs
6. Providing improved coupling between business strategy and plans and R&D strategy and plans so that they can all support each other

These Improvements in Information Sharing, Analyses, and Decision-Making Should Help To:

1. Avoid costly project or program failures
2. Reduce nondeliberate redundancy of R&D programs between parts of the corporation
3. Decrease the likelihood of missed opportunities due to inadequate communication and analysis
4. Improve the reaction time to competitive pressures and business opportunities

---

Figure 7-11. Benefits from improved PS/RA systems.

7.4 DURING: MONITORING AND REVIEW

7.4.1 A Monitoring Module

The naive set of questions in Figure 7-12 summarizes what various levels of management *should* be asking about the progress of the projects and programs in their area of responsibility. The "loss of control" over projects/programs that many managers complain about is often due to the lack of this kind of questioning. Periodic "dog and pony shows," which are commonly mistaken for monitoring progress, tend to stress overpreparation, anxiety-generation, best-foot-forward, and other superficial aspects of "what is really happening." Upper managements, several levels removed from the actual technical work, seem to prefer these periodic show-and-tell sessions to actually getting involved in the details of the work. For many nontechnical managers and even for a lot of busy technology managers who have espoused an extreme decentralization ideology, this arms-length contact with technology projects may be perceived as enough or may be preferred.

For managers at any level, however, who *really* want to know what is going on so that they can predict the course of a project or program and make arrangements for its next phases (tooling, production, market research, recruiting, training, facilities planning, marketing, etc.) such presentations are not enough. Depending on the level of technical sophistication and interest on the part of the questioner, this set of naive questions can yield a great deal of information for decisionmaking and action-taking. For example, although most formal presentations and written reports start off with an *objectives* section, such statements are often in terms of technical objectives and do not contain any or any adequate linkage with business and economic objectives of the firm. The "hooker" in the second half of the first question can disclose fuzzy objectives and nonconnection with business objectives or the stated long-term goals of the strategic technology plan (STP) itself, if there is one.

It is important to raise this question periodically and upon the occasion of potentially significant events (e.g., a change in the market, the entry of a new player, a competitor's introduction of a new product, loss of a key technical person

---

1. What are you trying to do in the project, task, or other unit of work and why are you trying to do it?
2. What has been accomplished so far?
3. How do the actual outcomes compare with the intended ones?
4. What are the reasons for the differences?
5. What can be done about it?
6. Who can and will do it?
7. How will you know effective corrective actions have been taken?

Figure 7-12. Summary of monitoring module.

involved in the project, or rumors of a setback in the project itself, which might require rethinking its potential payoffs). This is part of the "openness and sharing" discussed as desirable features of long-range STP (see Chapter 5). It also applies to monitoring the shorter-term technology projects. Many of the so-called failures in R&D and technology projects in general are not technical failures in that the researchers did not accomplish their *own* technical objectives or come close—for example, an improved material, a better yield for a process, an additional feature for a product. They are failures to meet the expectations of the people representing various business functions of the company and the financial criteria that are used to evaluate the final results. If these judgments have to wait until the "after" phases of the project (see Section 7.5), then there is no corrective mechanism available to convert potential failures into potential successes. The monitoring procedure recommended in this section addresses that "conversion" opportunity.

The manner in which the questions in Figure 7-12 are posed and the degree of elaboration of the responses can vary widely with the particular needs and circumstances of the questioner and respondent. If the CEO or division manager merely wants assurance that a project/program is indeed proceeding according to plan, this set of questions can be condensed into an idiosyncratic dialogue, such as: "How's the project going, Jack? Any problems with it? Any changes in direction? Any roadblocks ahead? How can I help? Are we still on target and on schedule?" (The dialogue, of course, involves letting the respondent answer fully and candidly and *listening* carefully.)

Project managers, project personnel, and even the managers of the project managers are not eager to disclose glitches they have encountered, mistakes made in planning or formulating the project and its targets originally, or even the details of what is happening when things are going well. They may resent it as snooping or lack of trust if managers ask too many questions too often, in too pointed a fashion. Indeed, "micromanagement" has become an epithet in some companies, whose managers do indeed lean over the project team's shoulders. So, short of a disaster that needs top-level fixing, and sometimes not even then, "real" indicators of progress or lack thereof are not likely to reach upper-management levels until it is too late to act other than in a crisis mode. Under these circumstances, the initiative must be taken to find out what is happening in projects and programs beyond the superficial "no problem" and the formal presentations and reports. Some form of dialogue based on the seven questions in Figure 7-12 is recommended for that purpose.

If the organization is large, if the number of projects is large, and/or if the "moseying-around" approach does not fit the style of the manager, he may prefer a more formal procedure for monitoring progress. Figure 7-13 is a form we developed for a more formal usage. It ties into the original Project Summary, which in most cases states the objectives, costs, and milestones for the project. The first line is an "alert" for the manager that the project he originally approved or agreed to has undergone some changes—perhaps in targets, schedule, costs, relevance to business plans, or other important aspects.

Project personnel may and do find the job of filling out such a form onerous. They may say they do not have the time to "fool around" with such paperwork;

1. Have there been any changes in the answers to any of the questions in the original Project Summary form?

   Yes _____   No _____

   If yes, make the changes on a fresh form (as indicated in the manual) and write the date the changes were made. Also indicate the Schedule where additional information on the changes may be found.

2. Have there been any changes to the original technical and commercial milestones?

   Original Milestone   New Milestone   Reason

3. Have there been any slippages on accomplishment of either the original or changed milestones?

   Milestone   Slippage   Reasons   Action

4. Have any new barriers emerged or are barriers that were originally identified still unresolved?

   Barrier   Reason   Action

5. Date of next review meeting: _____

6. Have any reports been written? If yes, indicate kind of report, how many, and in general to whom they have been distributed.

7. Budget *this year*   Budget for *total project*

   Actually spent to date: _____   Actually spent to date: _____

   Percentage spent: _____   Percentage spent: _____

   Why? (if significant deviation) _____   Why? (if significant deviation) _____

Figure 7-13. A brief monitoring form.

they have to spend all their time on the project. Sometimes, however, it is not the time but the "sweats" they object to—disclosing problems and changes they did not anticipate. If the monitoring procedure and its paperwork or dialogues are used in a threatening or punishing atmosphere, they are certainly justified in their reluctance and apprehension.

Effective monitoring is not a "we–them" system in which "we will watch those rascals closely and catch them out if we can." It has to be a cooperative system in which all parties are attempting to assure that projects and programs are proceeding as planned or that changes or difficulties are acknowleged and are used to guide future activities and expectations. A more formal and periodic (in this case, quarterly) monitoring form is presented in Figure 7-14. It digs a little deeper, emphasizing costs and forecasts more than the "touch base" form in Figure 7-13 does.

Figure 7-15 gives an overview of a reporting/monitoring system that can be adapted to different kinds of projects and specific circumstances. This system emphasizes, in Sections 3 and 4, substantive reports on what is happening in terms of actual *technical* results, not just budget and schedule information. It is a "down–up" system, in which events in the project trigger reports. However, the initiative for establishing and reenforcing the importance of the system is very much an "up-down" responsibility due to the reluctance of project people to "admit to problems," "wave them around," or publicize results until they are "absolutely" sure of them.

1. Plot, for each quarter, personnel utilization and total money spent—actual and projected.
2. Indicate Project Status
   (a) *Milestone accomplishment*

Milestone	Scheduled Date	% Complete	Remarks
(as identified in Schedule B)	(as indicated in Schedule B)		(reasons for slippage, etc.)

   (b) *Barrier status*

Barriers	Status	Action	Responsibility
(as indicated on cover sheet and Schedule G)	(solved/unsolved/ more complicated/ partially solved/ newly emerged/etc.)	(what needs to be done and by when)	(who has to do it)

   (c) *Cost performance*
   - (i) Forecasted costs to date
   - (ii) Actual costs to date
   - (iii) Total amount budgeted
   - (iv) Total budget balance
   - (v) Estimated total costs
   - (vi) Estimated cost to complete the project

   (d) *Changes*

Milestones	Milestones Now	Remarks
(as originally indicated)	(as they now stand)	

   (e) Next review point: (indicate date of review, members participating, etc.)
3. Indicate if any of the estimates (costs, dates, etc.) provided by you need serious revision. If yes, include the revised estimates and justification for the same (e.g., indicate additional funds needed, additional personnel).
4. Indicate any reports, publications, or memos that have been written or presentations made and those in the pipeline.
5. What is your best estimate of the chances or probability for the project to move on to the next stage (e.g., from exploratory to feasibility to development)?
6. Would you consider the following actions on the proposal, based on progress (or lack of progress) to date and your estimates of its likely ultimate level of success:
   (a) Major design review now
   (b) Major design review during next ____ months
   (c) Putting the project on hold, waiting for these developments and/or clarifications

   _____

   _____

   (d) Scrapping the project

Figure 7-14. Quarterly monitoring procedure.

Finally, for organizations with very large technology programs and very many projects, it is possible to design an even more formal monitoring system than the ones described above.[*] The specifics of such a system must be tailored to the firm's specific needs and circumstances, but a general "mind-numbing" outline of a monitoring/review procedure we developed for several government R&D

[*]Based on work by the author and Elie Geisler.

What Is to Be Done?	Who Does It? (or Is Responsible for Seeing That It Is Done)?	When Does He Do It?	Where Does It Go?
1. Update brief monitoring form	Project leader(s)	Quarterly and whenever new information comes in that warrants such an update	All top management, technical information services (TIS), program managers, leaders of interdependent projects, operating groups, and others who in the near future may provide inputs or take outputs
2. Update quarterly monitoring form	Project leader(s)	Quarterly and whenever new information comes in that warrants such an update	All managers on request and the discretion of the program manager and project leader
3. One-page note on results	Responsible project team member or project leader	Accomplishment of a technical milestone; overcoming of a technical barrier; slippages in completing technical milestone; delay in overcoming technical barriers; new barrier	Project leader(s), program manager, leader(s) of interdependent projects, people in a position to help out, interested individuals at the discretion of the program manager and/or project leader
4. Interim note on results	Project leader(s)	Completion of a set of related milestones; completion of a subproject in a program; whenever a finding of any significance has been established	Whomever the project leader(s) feels may be interested in using the results or providing valuable comments; others at the project leader's and program manager's discretion

Figure 7-15. Overview of reporting/monitoring system.

313

A. Identify objectives.

B. Identify clients.

C. Identify performers.

G. Select indicators.

I. Rank or rate indicators.

J. Obtain output data.

K. Compare outcomes with objectives.

L. Explain deviations.

M. Signal corrective action.

Figure 7-16. Necessary steps for a quick-response, first-cut assessment, for use in those cases where only a few days and very limited personnel are available for conducting a quick assessment of a project/program or other relevant unit of work.

programs might have some elements of use in this tailoring. Figure 7-16 outlines the necessary steps for a quick-response, first-cut review or assessment. The need for this may arise as part of a periodic review of programs; as the result of a change in management where the new manager(s) wants to know "what those guys are doing with my money;" or following a significant failure, success, or significant request for additional time and money. Presumably, most of the information is already available in the original project documentation or the organizational memory, so that it is primarily a matter of capturing it in a form suitable for quick review. Backing up this quick review procedure is a much more elaborate one, illustrated in Figures 7-17 through 7-19. This procedure involves three major stages:

One	*Identify the key organizational components and individuals involved in the project/program*
Two	*Perform the assessment/review*
Three	*Implement and follow-up*

Of course, the procedure described in Figures 7-17 through 7-19 is too formal and elaborate for most industrial organizations and even for many government agencies. However, it can serve as a checklist in any organization that is serious about knowing what is going on in its technology programs and doing something to improve or rectify them.

---

*A. Identify the technical, organizational, and commercial objectives of the program or project.

*B. Identify the client or sponsor of the program.

*C. Identify the performers and the managerial responsibilities for each program.

 D. Identify the linkage and communication mechanisms between the program and the clients/sponsors.

Figure 7-17. Stage One: Steps in establishing an assessment system. (*These items are considered the minimal necessary steps in a quick-response, first-cut assessment. The term "program" as used here includes whole technology/R&D programs, significant program components, or individual projects.)

E. Assign a team (can be an individual, full time or part time) for overall coordination.

F. Assign a person for the assessment effort for each major subprogram or related cluster of projects.

*G. Select indicators of the potential outputs of each program/project (P/P).

H. Operationalize and scale the indicators.

*I. Rank or rate indicators for importance.

*J. Obtain outcome and "progress, problems, plans" data for each P/P and its significant components (e.g., critical tasks,) including data on barriers and facilitators to progress and outcomes.

*K. Compare outcomes with technical and organizational objectives and identify deviations (positive and negative).

*L. Attempt to explain deviations via the barriers and facilitators identified.

*M. Signal corrective action.

N. Design monitoring (how we are doing) and evaluation (how did we do) forms.

O. Pilot test forms and procedures for at least one budget cycle.

P. Integrate use of forms into regular reporting responsibilities of technology managers.

Q. Design control charts for use in assessment and monitoring.

R. Establish routine signaling and follow-up procedure.

S. Hold orientation and training meetings on the monitoring/evaluation methodology and the entire assessment system.

T. Carry out periodic audits of the monitoring/evaluation system and evaluate the performance of the system.

U. Feed back results of the assessment effort and provide continual technical assistance to technology managers on self-monitoring and assessment methodology.

Figure 7-18. Stage Two: Implementation of assessment methodology. (*These items are considered the minimal necessary steps in a quick-response, first-cut assessment—see Figure 7-16.)

For example, one of the themes running through this procedure is that of identifying and confirming the *client* for the project/program. Many technology projects and even whole programs turn out to be "orphans" in the organization. This can happen in a number of ways. The original champion, client, or sponsor may have been promoted, terminated, or become involved in other things. There may never have been a specific client, just a *hope* that someone would be interested in

V. Feed outcome of assessment effort to the assessment team.

W. Analyze outcomes of assessment for each major program.

X. Feed back results of analysis to the management of each program in the form of a "watch list," which includes those aspects of performance that might require corrective action. Indicate potential corrective action that might be needed.

Y. Request, of technology management, that corrective measures be undertaken within, for example, 6–12 months. Provide a reporting procedure and format for the corrective measures.

Z. Apply budgetary and organizational rewards (positive and negative) for each program in the system. Publicize these actions within the technology management community.

Figure 7-19. Stage Three: Follow-up.

commercializing the results if the project/program were to succeed. Or the project itself may have taken a turn that is no longer of interest to the original or potential client at the time it was approved and initiated. Finally, external circumstances may have changed—for example, the price of materials, an available market niche, a customer problem, an energy crisis, an environmental problem, or another entry into the market—making the original objectives of the project obsolete or no longer cost effective.

So this procedure, which includes confirming that there is indeed a client for the results of the project, may pay off in just that way alone—by disclosing that the project is no longer relevant for the company or that even if it appears to be, in general, there is no one designated to pick it up and carry it through to commercialization. This can be a great payoff of the monitoring system itself.

Other kinds of payoffs, parallel to some claimed in Chapter 5 for strategic technology planning (STP) are listed in Figure 7-20. Many of these items tie the monitoring or "during" phase into the "before" (PS/RA) and "after" (evaluation) phases of technology management, as well as into the STP process that should give rise to all these phases. A dominant theme of these payoff items is that a carefully designed and maintained monitoring system for technology projects can help improve the overall quality and relevance of the R&D/Innovation process. It can provide a constant check on whether the technology programs are consonant with company objectives, needs, and opportunities.

---

1. Improved *institutional memory* and an increase in *capability for future assessments*
2. Identification of *opportunities and needs for R&D* as a result of relating project/program outcomes to needs of clients and other potential impactees and users
3. *Generation and communication of ideas, project formulations,* and *proposals* for new programs/projects/fields of activity
4. *Improvements in project selection* and *resource allocation* (PS/RA) of R&D/innovation activities
5. *Improving* R&D's *relations with relevant communities,* such as sponsors, clients, technical and scientific communities, and other downstream or supporting groups
6. *Decreased chances of false starts* on projects, programs, or fields that have intrinsically low relevance to the organization's and R&D's general mission, or whose promise is low in view of certain constraints that are not readily recognized by the proponents or champions
7. *Improved and clarified decision premises* and a better basis for many of the major decisions that must be made (e.g., termination or rebudgeting decisions)
8. *Improved information* to feed the decision systems mentioned above
9. *Early warning of off-target work* while corrective action is still feasible
10. *Reduction of ambiguity of goals and objectives,* internally and with respect to the organization's environment
11. Increased *ability to anticipate* and deal with (or recommend actions by others to deal with) *barriers* to further development, application, and implementation of R&D results
12. *Credible rationale for new initiatives* and for *discontinuing* or *reducing existing activities* that are not paying off

---

Figure 7-20. Specific potential contributions of the assessment system.

Many of the aspects of monitoring are carried out in various forms in most well-managed companies. However, lack of an integrated system that constantly checks *basics*—"why are we doing this and who needs it"—can make such efforts mechanistic and not adaptable to changing circumstances. Certainly, the direct management of technology should be left to the technologists and their technology managers. But it is the responsibility of top management and other functional managers to keep informed on how technology *is* being managed and how its outputs might impact their areas of responsibility.

7.4.2 Criteria for Evaluating the Monitoring System

Finally, Figure 7-21 suggests four criteria for evaluating the monitoring or assessment system *itself*. When we have proposed various levels of monitoring systems for technology projects, managers responsible for technology programs have asked a number of very practical questions about their characteristics and feasibility. Here are some of those questions and our responses.

> *Question:* What is the recommended frequency and intensity of monitoring/review?
>
> *Answer:* Ultimate control of quality, cost, and progress rests with the performers and managers of technology projects and their sponsors or clients, just as product quality can only be assured by the people who actually design and manufacture a product. However, in quality assurance it has been found that a group which is independent of the design/manufacturing process is needed to set quality standards, to monitor and review operations, to assess the outcomes of the process, and to signal for and take action as necessary.
>
> The frequency and intensity of review and monitoring and other aspects of *control* of technology projects depends on a number of factors, including:
>
> - Size of the project/program.
> - Newness of the project/program and its underlying technology—that is, radical departures from existing technology should be monitored more carefully (not necessarily more frequently) than simple extrapolations of known technology.
> - Criticality of the project/program to the organization and its urgency.
> - Intrinsic rate of progress or change in state of the art in the fields covered by the project/program; some fields move in an explosive fashion, while others hardly creep along.

---

1. Feasibility and ease of application
2. Replicability
3. Communicability
4. Face validity

Figure 7-21. General criteria for evaluation of an R&D assessment system itself.

- Turnover of personnel in both the project itself and the client/sponsor organization; new managers should ask for an early and thorough review to increase their own understanding, not necessarily to judge whether the project is progressing well.
- Newness (to the project) and general experience of the project leader.
- "Causes for concern," which arose initially or during previous progress reviews—items on the project *watch list*.
- Apparent stagnation or bogging down of a project/program for an "unreasonable" period (months, years) relative to expectations.

As a steady state is reached in operation of a new monitoring system, the need for calendar-driven reviews may diminish and they can be triggered by extraordinary events (full-dress ones) or the passage of a "reasonable" length of time since the last report (internal "how's it going" dialogues).

Question: What are some common reasons for reviews beyond the "periodic" and the "catastrophic"?

Answer: One is the "great man" syndrome. Project sponsors and top managers have often been stung by erratic or deteriorating performance on technology start-ups or by technology management people who suddenly or gradually go off-track without the notice of anyone in authority who can take corrective action in time to avert disaster. Careful, even if causal, monitoring can pick up subtle signals that "things are not right" and the project or its leader needs watching. Another is the "things are going fine" syndrome. People too close to a situation—the project team members, their leaders, and even the clients/sponsors—may be subject to overoptimism or even self-delusion. They may not be fully aware of or fully appreciate the significance of environmental changes that might threaten the viability of their project or the value of its outputs. Furthermore, they may be too close to the technical work itself to notice small but potentially significant drifting away from technical objectives or future milestones or lack of significant progress toward *commercial* objectives as compared to incremental *technical* progress.

Question: What are some "quick and dirty" techniques and mechanisms that can be used as part of the monitoring system?

Answer: Several such techniques are listed and discussed briefly below. Some of them may be designed as part of the overall monitoring system. Some can be used in addition to or in place of a formal system.

1. *The Watch List.* This has been very effective in helping us monitor and react quickly to aspects of a project or venture which might be crucial for its success and which gave earlier evidence of being questionable (e.g., leadership, choice of approach, key personnel, adequacy of funding, or realism of targets). We use it on a basis of "he's great, but has some weaknesses."

2. *The PPP Written Report.* If feasible, formal budgeting/financial reports should be augmented or supplemented by a very brief progress, plans, and problems (PPP) memo whose frequency should depend on some of the factors listed earlier. This can help flag chronic or acute problems and deviations from plans which a project has encountered or continues to encounter.

3. *Peer or Interested-Party Participation.* There is an increasing trend in industry to have some sort of outside participation by technical specialists in their program reviews. Some of these take the form of technical advisory boards and others of periodic overviews by task forces, including outside experts. This procedure is important in rapidly changing fields where an in-house group might be falling behind the leading edge of their technologies and related fields. It is also important for more static fields where outsiders can often suggest new approaches from other, sometimes indirectly related fields.

4. *Size and Composition of Program Review Groups.* Large audiences of marginally involved people may not be cost effective in program reviews. In addition to wasting valuable human resources, they can inhibit productive interchanges between the active participants and representatives of *interested* user organizations and individuals with specialized and relevant information and judgment. However, program reviews, if properly prepared and carefully followed up, can serve as an important communication channel between *doers* and *users.* This can help prevent the people involved from coasting or becoming too insular in their technical approaches.

5. *Random Scan or Audit.* A sampling procedure which in addition to "problem" cases does a quick scan of "nonproblem" projects, selected at random. The frequency and intensity of the scan depends on the factors mentioned earlier.

6. *Event-Driven Reviews.* Important internal or external events (e.g., news from the factory or market on failures or weaknesses in equipment or systems as well as technical breakthroughs or breakdowns relative to a project/program).

7. *Barrier Analyses.* Another scanning procedure, between formal reviews, in the spirit of "what can we do to help you overcome barriers to progress or success in your project/program?"

8. *Opportunity Analyses.* A method used in some large industrial companies to help technology programs to take advantage of opportunities to:

(a) Refill the technology barrel and maintain the skills of their personnel.

(b) Benefit from cooperating and exchange with other R&D and user groups.

(c) Relate their program objectives to long-range corporate interests.

(d) Move quickly into new opportunity areas by redeploying assets as opportunities arise without waiting for the annual budget procedures.

7.5   AFTER: ECONOMIC AND OTHER EVALUATION

All organizations and all managers do some sort of post hoc evaluation of their technology projects and programs. Few do it systematically and quantitatively. Mostly it is done informally or intuitively, with qualitative judgments or personal opinions dominating: "That project went pretty well and seems to be paying off." "That was a terrible project (program) and we should never have undertaken it in the first place." "What a waste of time and money." "That's been our best effort to date in that field." And so on.

The mysterious and uncertain nature of technology, especially toward the more research end of the scale, makes it difficult to apply the usual economic measures *before* a project is undertaken (PS/RA) and *during* its performance (monitoring). But those uncertainties and mysteries should have been resolved by the time the project is finished, and good management should require some sort of systematic evaluation to assure that the money was well spent and that the results justified the time and other resources devoted to the project.

Despite the management logic, formal post hoc evaluation is very rare, even among the best-managed firms. One argument in defense of this gap goes as follows: "Real successes and real failures are obvious, so why bother with formal evaluation—everyone can recognize them. The others are difficult to pick apart, since there are many fuzzy inputs and many nonspecific outputs that can't really be pinned down. It's not worth the effort." This may very well be the case: it may not be worthwhile to do detailed postmortems and quantitative analyses either for obvious successes and failures *or* for the residual. That is, unless there is something of value to be *learned* from such evaluation that can serve to strengthen the overall technology program in the company and lead to improved project selection and improved performance in the *future*.

Figure 7-22 lists some of the perennial issues involved with attempts to evaluate technology projects "after the fact."

7.5.1   Why Evaluate?

Generally, when evaluation of a whole R&D or technology program is asked for by top management, *that activity is in trouble*. Given the general reluctance to carry out formal evaluations, when the CEO asks for (demands?) one, he already has formed a negative opinion and may want the analysis to support his decision to make drastic changes, like closing down labs or replacing the CTO or other people responsible for the technology function. I have witnessed and participated in too many such "wakes" to be sanguine about the motives and values to the technology program itself of *hostile* evaluations. During many years of urging the R&D management community to do its own self-evaluation so as to forestall the confrontation that is involved in attempting to "prove its value" to people who have already made negative judgments, I have observed too few cases of successful defense.

1. *Why Evaluate?*
   - To prove its value
   - To improve its operation
   - To reexamine its objectives
2. *At What Levels Should Evaluation Be Done?*
   - Individual projects
   - Technical or thrust areas
   - The whole program
3. *Who Can or Should Evaluate?*
   - Technology management
   - General management
   - Auditors or analysts
4. *What Is Needed to Evaluate Effectively?*
   - Objectives for R&D/innovation
   - Criteria (agreed upon)
   - Data base (costs and returns)
   - Measurement method
   - Theory or model
   - Judgment mechanism
   - Ability to take long-term corrective action
   - Management and R&D cooperation
5. *From Where Do the Evaluation Criteria Come?*
   - Derived from broad objectives
   - Tied to strategic planning
   - Continually reexamined
   - Flexible enough to encourage risk taking
   - Quantifiable if feasible
   - General versus special (ad hoc) criteria
6. *Why Is There Confusion Between Long-term Evaluation and Short-term Monitoring or Control?*
7. *What Can Be Done With the Results of Evaluation?*

Figure 7-22. Perennial issues related to the evaluation of R&D/technology.

If, on the other hand, evaluation is used as a regular management tool by the CTO and other responsible managers in the technology program itself, then the credibility of their evaluation, when questions of cost effectiveness arise, is more likely to be accepted. That is, if an R&D lab or the technology program as a whole is continually engaged in self-evaluation, according to credible and transparent (i.e., understandable and replicable) criteria and procedures, then the likelihood of a hostile evaluation being requested lessens. Of course, no self-evaluation is likely to forestall the need for a technology evaluation driven by such circumstances as a new CEO, a merger or takeover, or the necessity of drastic cost cutting. Having the procedures and results available before such crises occur, however, can make things much easier for both technology management and top management, even if the end result—for example, reducing or eliminating a lab or program—is inevitable.

One CEO, famous for his negative attitude toward R&D in his company, says: "It's lousy and if I could do without it I would." When the CTO and his staff (greatly reduced when the CEO took over) belatedly tried to develop some data on the results and impacts of the firm's technology programs, both corporate and divisional, the CEO was not really interested in "those details." In such a circumstance, formal evaluation undertaken in defense of his operations by the CTO was too late. It is not clear whether he might have done better if he had had the data in place before the new CEO began making his opinion known; but at least he could not have done worse. He eventually took early retirement and all vestiges of corporate technology planning disappeared from the firm, despite its heavy dependence on technology for all its product lines and markets.[*]

There is a very beneficial offshoot of routine and continuous evaluation aimed at improving the quality of the technology programs in the firm. It is the opportunity, as part of the process, to occasionally or even periodically reexamine its objectives and to see how they are connected with the goals of the firm, as expressed in the business plans—explicit or implicit. Another advantage is to see how realistic the objectives for technology programs are with respect to the resources available and the constraints under which they must operate in the decentralized firm. Although other factors play an important, even critical role, the mismatch between what a particular corporate research lab (CRL) or other technology unit was trying to do and the realities of the constraints under which they had to operate has contributed to the decline and elimination of an increasing number of CRLs in large decentralized firms. Many of them were attempting to play the traditional CRL game in a highly divisionalized and highly diversified firm in which there was little support and few points of insertion for their results when some were achieved. (See Chapter 2 for a detailed discussion of this.) My point of view is clear here: evaluation should be done as a regular part of the the technology management process, primarily for improving its operation, but also to be ready to respond to questions about its past, present, and potential future value and its relations to company objectives.

7.5.2   At What Levels Should Evaluation Be Done?

The answer to this is simple: at all levels—projects, programs, special thrust areas. If evaluation is to make any sense, then all aspects of the technology function should be analyzed for their contributions and shortcomings. Techniques are available for "global" measures of the value of a lab or program. They generally involve gross comparisons of the money spent in a given period versus economic results achieved—sales, profits, return on investment. However, such global measures, unless they are overwhelmingly positive or negative, need the support of evaluations of the overall program's components.

---

[*] Some aspects of planning seem to have been reinstituted recently under a new VP of planning, according to my informal sources in the company.

7.5.3   Who Can or Should Evaluate?

As strongly urged above, the first and most important source of evaluation should be technology management itself. General management should also be continuously evaluating what they see as coming out of their investments in technology. But if top management does not do this except in a crisis mode or in a critical mood, then the evaluation context may be very unproductive. If, as they often do in such negative circumstances, they bring in outside auditors or analysts to support their negative opinions about the value of the technology program or the whole technology function in the firm, then the game may be lost before it is begun. Many of the outside auditors or analysts—often with a dominant accounting, financial, or production bias—have insufficient insight into the nature of technology problems and the R&D/innovation process to do an objective and appropriate evaluation. They often apply inappropriate criteria and procedures in such evaluations and overlook or deliberately discount many of the important outcomes of technology projects, such as skill-building, protection against technological surprise, increased "technical smarts," avoidance of mistakes and loss of customer confidence, and other noneconomic and nonquantifiable outcomes. In the next section, on the technology/innovation (T/I) audit, a broader set of criteria for a technology program is given than is generally used by outside auditors or people unfamiliar with the nature of technology programs. If top management insists on an outside audit, this broader framework may help avoid the mistakes of omission and inappropriate evaluation which are common.

Ideally, continuing or even episodic (e.g., crisis-driven) evaluation should be a *team* effort. It should include both technology managers and general management people, as well as representatives of other analytical groups such as finance and accounting, whether they are internal or external. Unfortunately, we continually observe crisis-driven evaluation as an *adversary* relation, with the outside auditors operating as inquisitors rather than members of a broad-based evaluation team, including the relevant and knowledgeable insiders.

7.5.4   What Is Needed to Evaluate Effectively?

Many of the factors are the same as those needed for the "before" and "during" phases of evaluation, which are discussed in the preceding sections. Most of the items on the list in Figure 7-22 are obvious and straightforward—objectives, criteria, and so on. Perhaps the need for a theory or model needs some explanation. The term *theory* often evokes a glassy-eyed response by many practical managers. Most of them do in fact have their own theories of how technology fits into the firm and their plans. These are not generally elegant or quantitative theories which would be recognized as such by their own scientists and engineers. However, they do imply sets of assumptions, decision premises, predictions, expected outcomes, "black box models," and other theory-like features.

They have general and even sometimes very specific ideas of how the technology functions in the firm do, might, or should convert the resources provided into benefits for the firm and its components—operating divisions, strategic business units, and other technical functions. For example, a common model or theory that many managers hold is expressed in the simple return on investment (ROI), present value, cash flow, or other formulation that relates inputs to outputs and shows how a component of the firm (factory, sales department, R&D lab, etc.) converts the resources provided to it into sales, profits, growth, and other measures of success.

There are also "partial" models or theories governing the *portfolio* aspects of technology projects—that is, the chances of success to be expected from investment in a given set of projects or thrusts. Some of these are analogous to or indeed identical with the kinds of models that venture capitalists use in judging the odds or that investment managers use in continually changing their portfolios. Whatever the origin or level of elegance of such models or theories, the evaluation process needs them as a framework within which data can be collected, judgments made, and corrective action taken, if needed.

For example, one dominant set of models that should be very prominent in attempts to evaluate technology are those dealing with the "imputation" of outputs to one of several functions or activities that have contributed to such outputs. The market and economic success of a new product can and should not be attributed solely to one group or function in the firm. It may have resulted from long and intense efforts by many groups and individuals in R&D, planning, marketing, tooling, production, sales, engineering, finance, accounting, and other functions. How much "credit" to attribute to each of these groups is of concern in such models involving imputation or allocating credit (and perhaps blame). "Joint cost" models are also needed to sort out the input side, so that technology is not charged with or credited for use of resources at an unfair level.

Many such models can only be employed in a fairly general way, since the data and the "money trail" do not permit precise attribution. But their underlying assumptions—for example, joint costs and returns—can be valuable in objective attempts to evaluate the cost effectiveness of a technology project or program.

7.5.5 From Where Do the Evaluation Criteria Come?

In general, they come from the planning and other organizational processes discussed throughout this book. In addition, they come from individual values, experiences, opinions, hopes, dreams, and biases. If the "cloud of criteria" is to be reduced to a common set that can be used to evaluate technology, then they should be firmly anchored in the objectives for technology, which in turn should derive from strategic planning of technology. They should be periodically reexamined to insure they are still relevant and valid. They should be flexible enough to encourage risk-taking, and quantifiable where feasible, but not to the exclusion of relevant nonquantitative criteria. Evaluation criteria can be developed and used at many levels, from really nitty-gritty measures of project progress in a particular dimension (e.g., cost reduction possibilities) to fairly grand measures of overall program contribution to the firm.

Figure 7-23 illustrates various levels of criteria and gives examples from a number of evaluations we have done in a wide range of industrial firms and technologies. The framework presented in these figures includes the general reasons for undertaking the evaluation; some general categories of outputs, benefits, impacts, and contributions; general targets for or potential beneficiaries of these impacts; and the association between outputs/impacts and targets (beneficiaries or impactees). The procedure also includes a matrix associating criteria or measures with beneficiaries or impactees. This matrix can be used for a quick scan of what has come out of a technology program and which functions have been impacted. Finally, Figures 7-24 and 7-25 provide data on the evaluation methods and criteria used by large industrial firms in two surveys we conducted—almost 25 years apart.

Some of the same firms participated in both surveys, although many of them had changed dramatically in size, focus, ownership, and management during the intervening years. It is clear from these two sets of data, even though they are not strictly comparable, that not much progress has been made in improving the evaluation of the R&D component of technology during that period. Many of the same issues and questions (Figure 7-24) are being raised currently as were raised then and even earlier when we did our first such survey in 1950, as part of Columbia University's First Annual Conference on Industrial Research.

7.5.6 Why Is There Confusion Between Long-Term Evaluation and Short-Term Monitoring or Control?

One of the reasons that there has not been more progress in developing and applying effective evaluation methods for technology is that there is commonly a confusion about what to expect at the different stages of the overall R&D/innovation process. As Figure 7-2 tries to show, there is a significant difference in the "outputs" at various stages of the process. The direct or "immediate" outputs of R&D, for example, are primarily in the form of ideas, information, insights, possible solutions to operating problems, and preliminary models or samples of products or manufacturing methods. Until much more work is done on these direct outputs by many other groups in the firm, nothing of real economic value is available to evaluate! It is not until these other efforts—such as tooling, engineering, marketing, selling, production, finance, accounting—have been applied, that economic benefits such as sales, market shares, profits, growth, and return on investment are realized.

However, the temptation is to attempt to evaluate *ongoing* projects on the basis of criteria that can only be properly measured *after* the project or program has been completed. Such measurement may only be reasonable when the total life cycle of the project has been completed, so that all the significant costs and returns associated with it (direct and imputed) can be marshalled into a cost–benefit formula or accounting model.

On the other hand, as indicated in Section 7.4, there are many "upstream" criteria that can be used for monitoring and reviewing projects and programs to

1. *Objectives for Performing Evaluations*
   - Identify outputs, benefits, impacts, and contributions of recent R&D/technology programs.
   - Relate output and impacts of R&D to performance of operating units where feasible.
   - Contribute to technology planning and business planning for operating units and the firm as a whole.
   - Provide a framework and method for future monitoring and evaluation of contributions to product lines, operating units, and other components.
2. *Some General Categories of Outputs, Benefits, Impacts, and Contributions (some in quantitative form, others in descriptive form, including specific cases and scenarios)*
   - Cost reduction in production.
   - Cost savings in product redesign and other areas (e.g., energy).
   - Prevention of loss of a customer or major sale.
   - Minor and major improvements in product and product line.
   - Increased productivity and resource utilization.
   - Improved product quality.
   - Reduced dependence on outside sources.
   - Improved performance of product/product line.
   - Maintenance/protection of lead or position in industry/market.
   - Facilitation of use by client.
   - Adequate response to environmental and other regulations.
   - Improved potential adaptability of manufacturing to new processes and methods (e.g., flexible manufacturing systems, robotics).
   - Facilitation of penetration into new market areas.
   - Improved competitive features of product/product lines.
   - Introductions of new products/product lines/processes.
   - Provision of technical information for management in areas such as licensing (in and out), joint ventures, and acquisitions.
3. *Some General Targets for the Impacts, Contributions, and Benefits of the Outputs of R&D/Technology*
   (a) Corporate
      - Corporate strategic goals and objectives.
      - Corporate image.
      - Competitive posture of the corporation.
      - Financial strength of the corporation.
      - Vertical integration.
      - Combined corporate market share.
      - Return on investment.
      - Cash flow.
      - Risk involved in business.
      - Cost-effective exit from market.
      - Entrance into new markets.
   (b) Operating Units
      - Product and product-line life cycle.
      - Manufacturing.
      - Procurement.
      - Product warranty.
      - Product liability.

Figure 7-23. Level of criteria and beneficiaries/impactees.

- Product/product-line market share.
- Profitability.
- Client satisfaction/customer loyalty.
- Regulatory requirements.
- Product replacement cycle.
- Service and maintenance.
- Interface/assistance to marketing function.
- Interface/assistance to manufacturing/production function.
- Manufacturing capacity utilization.
- Competitive posture of the operating unit (leader/follower).

4. *Description of Impacts: Some Specific Indicators of Outputs, Impacts, Benefits, and Contributions*
   - Cost savings in manufacturing, procurement, warranty, and liability.
   - Correction of quality problems.
   - Reduction of exposure to product liability.
   - Reduction in warranty costs.
   - Facilitation of use by client for improved performance of product.
   - Preservation or extension of market share.
   - Basis for increased prices and revenue.
   - Improved competitiveness.
   - Improved customer loyalty by improved product quality, introduction of new products/ lines, and effective service and maintenance.

   For each Target (Section 3) try for description of impacts based on the categories of outputs in Section 2: for example, impacts on and contributions to *product liability protection*
   - Respond adequately to regulatory requirements.
   - Provide technical information/basis for management and other staff.
   - Improved performance of product/line.
   - Improved quality of product.
   - Improved use of resources and facilities (e.g., testing).

5. *Impacts of R&D Outputs on the R&D Units Themselves*
   - Problems with "image" of the portfolio, between research and engineering/service/ maintenance. Obtain "credit" for the nonresearch activities.
   - Impact on morale of the R&D units. Apply techniques to improve morale and reduce sentiments of "being abandoned by management."
   - Relation of R&D outputs and contributions to the strategic and business aspects of the divisions *and* the corporation. Bring such relation to the attention of the R&D performers.
   - Of the various indicators of R&D contributions, emphasize such indicators as customer satisfaction and loyalty, by "caselets" of strong relations between R&D/engineering people and customers.
   - Emphasize the importance of the R&D units and their outputs to the business aspects, sales, manufacturing, profitability, survival, and so on.
   - Emphasize the relation of the R&D activities to the business plan of the divisions and the corporation, by identifying the *technology policies* and components of the plan, and by caselets accompanied by numerical (dollar) examples, where feasible.
   - Need for support from the business management of the operating divisions to clarify and *enhance* the role played by R&D, perhaps with the use of a mechanism such as joint research committees. If already in use, add the elements of enhancement of R&D image.

Figure 7-23. *(Continued)*

	Yes	No
1. Use of *formal* system for evaluating R&D output	8	22
2. Quantitative system*	15	15
3. Qualitative system*	18	12

	Less than 5	5–9	Over 10
4. Length of time system in use (in years)	3	5	16

	Annually	Semiannually	Quarterly	More Frequently
5. Frequency of review/evaluation	16	3	4	4

6. Kinds of methods, formulations, or criteria used

	Number of Firms Using
Return on investment (ROI) or return on assets (ROA)	6
Cost savings	3
Comparison of R&D expenditures with sales or profits	3
Research/return ratio	1
R&D expenditure/unit sold	1
"Dollarization:" profits/cost of technical personnel	1
Payback of investment	1

7. Kinds of indicators used

	Number of Firms Using
Committee evaluation (judgment)	26
On time, on budget performance	25
Goal achievement	10
New products, patents	5
Product/process improvement	5
Innovative breakthroughs	3
Compliance with regulations	2
Complaint reduction	1
Other	4

Figure 7-24. Methods of R&D evaluation used by 30 large industrial firms.(*Some use both.) (*Source*: 1980 Survey by IASTA, Inc. Eleven firms were interviewed by telephone and field visits. Data on an additional 19 companies were obtained from the open literature and IASTA's technology management files.)

see whether they are on the right track and have a reasonable chance of meeting formal objectives or informal expectations. The sometimes tedious distinctions made in this chapter are aimed at guiding the technology manager and his managers through this complex situation so that they will indeed use appropriate criteria and their associated measures at the various stages of evaluating technology projects—before, during, and after.

	Number of Companies*
1. *Related to Effect on Sales Volume or Revenue*	19
• Increased business	
• Increased output without increasing investment	
• Share of the market	
• Percent of products from research	
• Consumer acceptance	
• Effect of new products on old product sales	
• New customers	
2. *Related to Effect on Savings in Materials, Labor, or Other Costs*	17
• Royalty payments saved	
• Use of by-products, wastes, idle facilities or personnel, or less profitably employed facilities	
• Reduction of product line	
• Closer control of manufacturing quality	
• Better process yields	
3. *Related to Effect on Profits*	13
• Profit on research versus nonresearch products	
• Profit and loss analysis for whole R&D effort	
• Payoff time on projects	
• Percent return on investment	
4. *Related to Time and Cost of the Technical Solution*	28
• Frequent reestimates of time and cost	
• Progress on project or program phases	
• Actual versus budgeted expenses	
• Actual versus scheduled progress	
• Proportion of budget spent versus progress	
5. *Related to Customer Satisfaction*	10
• Number and nature of complaints	
• Broadening of product line	
6. *Related to Information Output*	17
• Number of valuable ideas	
• Percentage of ideas from inside laboratory	
• Learning about new processes and materials	
• Sources of new ideas	
• Training individuals	
• Development of specifications	
• Evaluation of information output to application groups	
• Information developed for sales	
• Repeat requests for work	
7. *Related to Success of Technical Solution*	16
• Number of problems successfully handled	
• Number of patents written up, applied for, or granted	
• Number and nature of project failures	

Figure 7-25. Criteria used to judge progress and/or results of research and development work by 37 laboratories. (*Indicates use by a company in at least one of its laboratories.) *Source:* Albert H. Rubenstein, "Setting Criteria for R&D," *Harvard Business Review*, Jan-Feb 1957, pp. 95-104.

7.5.7   What Can Be Done with the Results of Evaluation?

Some organizations engage in "scorekeeping" routinely for comparison and other purposes (incentives, bonuses, rationale for changes in people and budgets). Aside from that, a major purpose of after-the-fact evaluation is long-term corrective action in the technology management process itself.

Effective evaluation should focus on, in addition to scorekeeping, careful analyses of *what* actually happened and *why* it happened. It should identify and evaluate the impacts of barriers and facilitators on the success/failure of the project or program. This can provide leverage for corrective action in technology management so that future projects/programs will benefit. A succession of projects that suffered from poor pilot plant support, for example, can point the way toward overhaul of the testing and scale-up system for new products or processes. This might involve changes in facilities, personnel, procedures (e.g., the priority system for getting onto pilot facilities), or resources available for that vital phase of the innovation process. Repeated glitches in market research, technological forecasting, pricesetting for new products, and implementation of new manufacturing methods can be identified in a series of careful "after" evaluations that go beyond the mere counting of dollars spent and dollars earned. On the positive side, specific facilitators that contributed to the success of particular projects might be generalized and more widely applied to other projects in the future.

Of course, many of the barriers and facilitators should have been detected *during* a project and corrective action (control) should have taken place in time to affect the course of *that* project. Ideally, that is what the "during" procedures discussed earlier are for—to prevent or correct bad things that hinder or destroy a project and to enhance good things that help it. But the world is not perfect. In the heat of actually doing a project, especially under resource and time pressure, the depth of analysis often does not disclose *general* barriers or problems which have *generic* solutions or facilitators which can be more broadly applied. Project personnel tend to think of their problems and projects as being unique or special and will often overlook solutions found by others elsewhere in the organization. It is the job of higher-level managers and their staff analysts to identify recurring *general* patterns that can be observed and understood in post hoc evaluations. Their longer-run (and calmer) view can contribute to continual improvement in the effectiveness of the whole technology enterprise in the firm.

7.6   THE TECHNOLOGY/INNOVATION (T/I) AUDIT

Many managers are sick of audits. They have been subjected to financial, accounting, environmental, energy, facilities, safety, liability, personnel, and other kinds of audits of what they have been doing and how they have been doing it. Their time and that of their scarce staff specialists have been taken up in preparing for, defending, and responding to audits. So why add another one to the list?

Periodic or occasional audits of operations are good mechanisms for focusing the attention of managers on aspects of their operations that they often take for granted.

Financial and accounting audits are almost in a class by themselves because of the legal consequences for the management of a public firm of not being able to account for its stewardship at least in monetary terms. Many other audits—personnel, equipment, and so on—are analogs of the financial and accounting audits and may have other than legal motives. A few other audits, such as safety and environmental and energy (in times of scarcity), may also have legal implications in addition to cost implications for the firm (insurance, liability, fuel, and other costs).

Using an auditlike approach, one of our colleagues at Northwestern—Phil Kotler—developed a *marketing* audit that provides a basis for an external analysis by marketing specialists or a self-analysis by the managers responsible for the marketing function and their managers, including the CEO. Inspired by that, we have developed and successfully applied an audit of the *technology* and *innovation* capabilities and performance of the firm in a format that can be used for self-analysis or for an outside analysis by specialists in technology and innovation management.

Figure 7-26 displays the seven major elements of this T/I audit checklist. Figure 7-27 provides specific checklists for each of these seven major elements and a rating scale for indicating where the unit being audited stands compared to its competitors. The comparison can also be made with where the unit stood at some previous period, where it would like to be at some future time, or all of the above. The depth and breadth of the analysis depend on why the audit is being done, the resources available to do it, and how the output is likely to be used.

Several specific events or continuing situations should trigger an audit of this type—as an overall evaluation of the technology capabilities and performance of a unit in the firm, which can be a strategic business unit (SBU), operating division, product group, or the whole firm. A necessary condition is that the unit being evaluated actually have some products out in the market over a period of enough years so that they and their technology content and performance can be compared with the competition.

Somewhat truncated T/I audits can be done for start-ups or single projects with a single product recently introduced on the market. However, in such cases the

---

1. Innovation track record compared to competitors in company's markets and fields of technology
2. Capability of company's R&D/Innovation staff
3. Strength of R&D/Innovation organization and planning
4. Idea generation, flow, and utilization
5. Time to commercialization or application of R&D results
6. Costs and benefits of R&D/Innovation projects
7. Relevance to and impact on operating units' interests, problems, and opportunities

---

Figure 7-26. Major elements of T/I audit checklist.

	Much Below Average	Below Average	Average or About the Same	Higher Than Average	Much Higher Than Average
1. Track Record					
Introduction of new products					
Cost of production					
Use of latest production equipment					
Economic success of new products					
Getting into market first with new products					
Staying power of new products in the market (resistance to substitution or obsolescence)					
Warranty record					
Contribution of new products to total sales and profits					
2. Capability of Staff					
Total size of R&D staff					
Number or percentage of professionals					
Number or percentage of advanced degree people					
Diversity of professionals with respect to disciplines					
Work experience of professionals in industry					
Work experience of professionals in your specific product lines					
Contribution to state of the art (papers, publications, presentations, awards)					
External scientific and technical reputation					
Approaches and offers received from other organizations (competitors, universities, others)					
3. Strength of R&D/Innovation Organization					
Flexibility of organization in responding to external threats and new developments (e.g., competitors' actions)					
Ability to anticipate trends and launch preemptive projects or programs					
Ability to handle severe workloads and/or loss of key personnel					

Figure 7-27. Checklist for the T/I audit. The checklist is used for comparison with competitors or other R&D organizations. A scale can also be developed for degree of satisfaction of the rater with respect to these factors.

	Much Below Average	Below Average	Average or About the Same	Higher Than Average	Much Higher Than Average
Ability to continue important longer-term projects and also respond to internal and customer needs for assistance					
Shortness and reliability of communication channels for ideas, decisions, information flow					

4. *Idea Generation and Flow*

Number of ideas generated related to current products and business					
Number of ideas generated related to potential future products and businesses					
Quality of ideas					
Speed of investigation and disposal (acceptance or rejection) of new ideas					
Time lags from idea generation to implementation or introduction of new/improved product or process to factory or market					
"Robustness" of ideas (how well they hold up under scrutiny)					
Cost effectiveness of ideas					
Degree to which ideas incorporate the state of the art					
Patentability or potential for proprietary position as a result of the ideas					
Breadth of base in the organization from which ideas come (is it a one-man show)					
Mechanisms for dealing with new ideas in terms of preliminary analysis and response-to-originator time lags					
Relation of the ideas generated to the main business and technology areas of the company					

5. *Time to Commercialize*

Total time from start of work on a project to introduction to market					
Time from introduction to full-scale marketing					
Time to get reasonably up the learning curve in terms of cost, quality, and output rate					

Figure 7-27. *(Continued)*

	Much Below Average	Below Average	Average or About the Same	Higher Than Average	Much Higher Than Average
Mechanisms for assuring technology transfer from R&D to manufacturing and marketing					
Barriers and delays in the transfer process					
Ability to make and meet milestone estimates					
6. *Costs/Benefits of R&D Projects*					
Total cost of R&D relative to total life-cycle revenue of particular projects					
Cost relative to prior projects (is it going up or down)					
Contribution to profits					
Profitability relative to existing product lines or processes					
General payoff of total R&D program (for whole company, for division, or by product line or business)					
Time to amortize R&D/innovation costs for process improvements					
7. *Relations with and Impacts on Operating Units (Product Divisions, SBUs, etc.)*					
Responsiveness to operating unit needs					
Anticipation of operating unit needs					
Reputation in the minds of operating division people					
Contributions to problem-solving and crisis-averting in operating units					
Degree to which operating units seek out help of R&D					
Transfer of people from R&D to operating units (requests from operating units for such people)					
Joint projects with operating units					
Funding of projects and programs by operating units					
Protection of operating units from technological surprises					
Preparation to provide operating units with "next generation" of technology					
Personal relations with plant personnel					

Figure 7-27. *(Continued)*

elements will have to be changed to provide a useful audit output. Some special conditions and events that might trigger a T/I audit include the following:

1. *Appointment of a New Top Manager—CEO, Division Manager, or Group Executive.* If the new top manager is from outside the company or industry, he may be quite unfamiliar with the products made by his new organization (except perhaps by general reputation) and technology on which the products and their manufacturing processes are based. If he is from within the industry, he may have some ideas for improvements, but he also may have some misconceptions about his new company's technology and its technical people. Reputations can sometimes be very informative and accurate; but at other times they can be severe distortions and can do the people being evaluated an injustice.

Initiating a T/I audit, along with other types of audits or surveys, can provide a quick overview or a detailed analysis of where the strengths and weaknesses are and what kinds of actions and decisions may be needed to remedy the weaknesses. If the new manager has to find out about needed improvements opportunistically— such as when a project blows up or a customer cancels a long-standing contract— he may have lost precious time and leverage on the problem.

It is also easier to pull off an audit in the early days of a new administration; such actions are expected by the organization's members and are less likely to be resented and resisted, in the spirit of "forget the past, let's see what we have to do to make this outfit better."

Even if the new top manager is from inside the corporation, he may be unfamiliar, or not familiar enough, with the capabilities and limitations of his new technology, people, and program. A new division manager may have come from a distant division or group and have had very little contact with his new division until he took it over. If he is the new CEO, he may have come from a functional area of the company—finance, sales, or legal affairs—and may not have had much direct contact with technology matters. He may in fact have a stereotypic or biased view of technology in the company, developed in his previous positions or career path, where technology was viewed as a constraint or problem area in his work.

Whether he comes from inside or outside, the new top manager of a company or an operating division may need a quick fix on his assets—which ones he can count heavily on in his game plan and which have to be improved or watched carefully to make sure that they help rather than hinder him. One of his key assets is technology—the people, the products, the manufacturing base, the capabilities that will strongly influence his performance in both the short and long run.

We have also encountered the desire to make such an audit by a new CEO or division manager who moved into the position from the number two spot in that organization and is in fact intimately familiar with its technology programs and capabilities. This is a function of the "honeymoon" period for a new manager. He may have observed and chafed under situations in his previous position where he could see that technology was being neglected, constrained, and not realizing its full potential. The former head person in the organization was either not able or not willing to see this or was busy on other things more important to him than technology.

The new manager sees a brief window of opportunity in which he can get away with exposing the weaknesses of his technology programs without carrying the onus of having caused or allowed it to happen. We have conducted T/I audits in such circumstances and realized that we were not digging up anything significant that the new manager was not aware of. The ritual of the audit, however, provided him with leverage in getting funding and other resources from his top managers or the board.

Such "new broom" analyses can be anxiety-producing for the people in technology. However, if conducted in an open, inquiring manner, without the threat of censure or retribution for past failings, they can launch a new administration in the right direction with respect to technology.

2. *Following a Major Product or Process Failure or Setback in the Market, Connected with Technology.* Such occasions for an audit are much more tense than are the cases of the new manager wanting a general feeling about his technology assets. The crisis that precipitates an audit generally has some explicit or implicit finger-pointing and blame attached to it. Auditors may dig for bad decisions, sloppy thinking, or other derelictions of duty. Or they may be looking for a scapegoat, no matter who was to blame, if anyone. Some of these audits are conducted in a vengeful or inquisitorial atmosphere, and most likely by outside consultants. The technology management group—the CTO or divisional technology manager— is often left out of the plans for the audit and often finds himself participating only as an interviewee or subject of the audit. This can generate the same kind of reaction as must be felt by the bank manager who suddenly finds the auditors in his office without prior notice. These crisis audits of technology are seldom completely without warning or prior indicators of trouble, however. When a succession of projects are in trouble and when the company's or division's products or market shares have been taking a beating for a while, one specific failure can trigger a reaction and an outside audit. By then it is too late for the technology management people to do much except to wait and hope that they and their programs survive.

3. *In Connection with a Merger or Acquisition.* Such audits are often part of the general analysis which potential or actual new owners or partners make to "see what we've got or what we're getting." These may or may not be "friendly" audits, but they are typically done by outsiders, either from the staff of the new owner/partner or from an outside consulting group. When they are done in a hurry, as part of the negotiations or assessment *prior* to closing the deal, they can be very superficial. Many CTOs and divisional technology managers have complained that they really do not get a chance to "tell their story" to these assessors (or assassins, as some of them put it). If the deal goes through, the impressionistic look at technology in the acquired unit may become part of the perception of the new top managers or owners and may have severe repercussions for the technology program.

Comments in the report like "a lot of fat," "not really on target," "no real connection with the marketplace," or "not our kind of guys" can strongly influence the fate of the technology program for years to come. Initial impressions may be very hard to change or dispel, no matter how superficial their evidentiary basis.

One direct consequence of such an acquisition-driven audit is evident in the post-acquisition moves to combine or eliminate CRLs, other labs, and other technology groups. If the CRL in the acquired company gets a "bad report" on the preacquisition or immediate postacquisiton audit—no matter how superficial the analysis may be—it is a likely candidate for elimination, cutback, or merger into the new company's CRL. We have followed half a dozen cases of this type and can see the virtue of having in hand a relatively objective assessment of the CRL made *prior* to the crisis, which provides a fair picture of strengths and weaknesses and a *program for improvement or maintenance of effectiveness*. For this and other reasons, related to good management in general, I strongly recommend periodic self-assessments or T/I audits.

4. *As a Routine Feature of Good Technology Management*. Being caught by a "new broom" or "crisis-induced" audit can be prevented and such outside inquiries can be preempted by technology managers and *their* managers. Periodic quick checks on "how are we doing versus competitors and what could we be doing better" can help reduce or eliminate the need for such crisis management. A well-run technology program, much as a well-run individual project, should have a built-in monitoring and assessment procedure that can help management anticipate and, where possible, prevent problems and bad situations from developing. Few technology programs remain static, despite my comments about stagnation in divisional technology programs made earlier. Such stagnation is, in itself, a product of change in the interests, skill levels, and capabilities of the staff and the collective technology capabilities they represent.

Deterioration of skills in both an absolute sense and relative to competition or the state of the art can be a gradual process and pass unnoticed, especially to people too directly involved.

That is, the CTO, the divisional technology manager, and their staffs may be part of the aging and obsolescence process that occurs so gradually that there may be no point at which it appears critical *from the inside*. However, from the outside, other members of the firm may be increasingly aware that "their R&D people or engineers are not keeping up," "the competition always seems to be ahead of us in new technology," or "those guys are not as sharp as they used to be and seem to be going downhill from there." This view is sometimes the partial motivation in scaling back or even eliminating CRLs. Even where cost-cutting and other considerations dominate the decision, a jaundiced view of the CRL's capabilities helps justify and make such actions seem "right."

For open-minded and forward-looking technology managers and *their* mangers, the aging process is a natural one and cannot be stopped completely. But they may be willing to take a periodic reading to see how far it may have progressed and where possible improvements can be made. Even if aging is not a major issue, periodic examinations of the technology program and its people and facilities is warranted in the spirit of continual self-examination and improvement. The T/I audit was especially designed for such self-appraisal and works bests in that kind of atmosphere. It can be useful in crisis mangement as well, but in such circumstances, the people being appraised may become defensive and less forthcoming than is desirable for an objective appraisal, diagnosis, and prescription for improvement.

7.6.1 How Detailed and Quantitative Should the T/I Audit Be?

The very term "audit" evokes an image of columns of figures in painful detail. This is far from the intention or the nature of the T/I audit. The R&D/innovation process and the management of technology programs does not allow for such precision and quantification, aside from questions of costs. The procedure recommended here is somewhat deeper than a surface skimming but much less thorough than a deep probing, except when weak spots are found. Perhaps better analogies than accounting audits might be medical exams where the first pass involves general issues of level of health, and where soft spots and problem areas are identified for further probing (pardon the painful metaphor).

Not only might the cost and time required for a detailed quantitative technology audit be unattractive to most managers, but they would probably be bored stiff by the results. In a recent audit of a large operating division, which was doing very well in terms of current operations, our "full" audit identified less than half a dozen areas that needed further attention and improvement. In that sense, the most welcome results of a T/I audit, like that of a medical exam, are "You're doing fine. See you next year."

In a certain specific area, such as the technology budget and expenditure pattern, the distribution of skills, and the "hit ratio" for projects, some quantification and some level of detail are desirable, to provide a basis for informed diagnosis and treatment. Where the technology management maintains or can easily pull such data together, the audit can be done faster, more easily, more credibly, and with less strain and pain than otherwise.

7.7 THE PROJECT PORTFOLIO AND TECHNOLOGY STRATEGIES

Using the categories in Figure 7-28, we can clearly distinguish very conservative technology strategies and portfolios and less conservative ones.

The strategy components in the upper part of Figure 7-28 can be further grouped into three broad strategy categories:

- *Maintenance*, which primarily includes components 1, 2, and 3
- *Expansion*, adding components 4, 5, and 6
- *Exploratory* (or some other designation), which adds components 7, 8, and/or 9

Combinations of these strategy components can form patterns, reflecting the technology policy and perceptions of managers of the company and/or a division. Firms or divisions following "pure" maintenance strategies for their products, man-

Overall Strategies That Describe the General Intent of the Program

1. Service on current materials, processes, and applications (M, P, A). Note: "Applications" include products and services.
2. Minor improvements one at a time on current M, P, A.
3. Continual minor improvements on current M, P, A.
4. Major improvements on current M, P, A, one at a time.
5. Intentional departures from current M, P, A, one at a time.
6. Attempts to meet a future market mission.
7. Coverage of a technical field of *current* interest.
8. Coverage of a technical field of *potential* interest.
9. Search for knowledge for its own sake.

Specific Intentions for Individual Projects or Programs

1. *Work Supporting Current Operations*
   (a) Customer service on current product
   (b) Minor improvement of current product
   (c) Major improvement of current product
   (d) Factory service of current process
   (e) Minor improvement of current process
   (f) Major improvement of current process
2. *Work Leading to Expansion of Present Product Line*
   (a) Work on new product not currently made here
   (b) Work on radically new product not currently made anywhere
   (c) Applied work leading toward an idea related to a new product
   (d) Work on new process not currently used here
   (e) Applied work leading toward an idea related to a new process
   (f) Translation of a research discovery into the prototype of a new product or process or part thereof
3. *Work Not Yet Connected to Any Product or Process*
   (a) Exploratory work in a field of current interest
   (b) Exploratory work in a field of potential interest
   (c) Work producing only knowledge for its own sake

Figure 7-28. Some categories for describing the technology program (portfolio) in a firm, division, lab, or other unit.

ufacturing processes, and materials will confine themselves primarily to "service" work and periodic or episodic minor improvements, driven by external forces in the market.

Figure 7-29 displays some of these alternative technology strategies, from very conservative to much less conservative. The latter involve significant departures from what the unit may be familiar with and feel comfortable with.

Strategy component 6—attempts to meet a future market mission (Figure 7-30)—falls just short of the "exploratory" strategy components and often entails substantial expenditures, risks, and time lags. It is not for the faint-hearted or

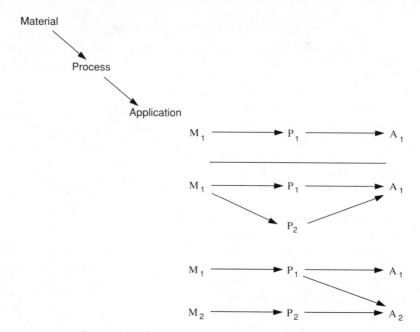

Figure 7-29. Some conservative diversification strategies.

nose-to-the-ground division manager or even, it appears, for many top managers in companies without a strong tradition of leapfrogging their own and competitors' products and technologies.

Figure 7-31 illustrates shifting technology strategy patterns as reflected in the case study of a firm that opted for reducing its percentage of maintenance work and increasing its percentage of expansion and exploratory work in the portfolio. This case, collected at time "now," reflects what had happened in the previous 3 years and what the technology and top managers believed (or intended) about the future. This pattern of actual and intended shifts toward more "outreaching" kinds of technology portfolios was encountered frequently in firms that had earlier been in a relatively stable situation. They had been supplying a relatively constant market via

1. This is what we would like to be doing in n years.
2. This is what we will need to know about:
   Materials
   Processes
   Applications
3. This is what we currently know.
4. This is a time schedule for learning and getting from "3" to "2".

Figure 7-30. A radical diversification strategy (future market mission).

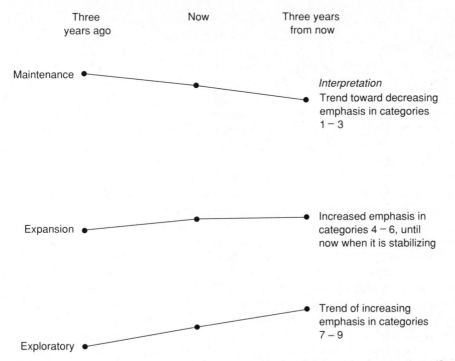

Figure 7-31. An illustrative pattern of shifting program (portfolio) emphasis over time. (See Figure 7-28 for category descriptions.)

a relatively unchanging technology, based on the same materials and manufacturing methods they had used for years. Due to one of the many factors triggering change—challenge from the marketplace, product substitution by competitors, new entries into "their" market, need for cost reduction, a new division manager, or combinations of these factors—they began to expand and modify their technology strategy.

Many of these firms reexamined their underlying technology—materials, processes, product designs—and embarked on new programs of expansion and, in some cases, exploration. However, some of the same kinds of factors, plus impatience with the rate at which results were being achieved (recall the discussion in Chapter 2 on how long it takes new efforts to get organized), then led to some cutting back and reversal of the kind of pattern shown in Figure 7-31. In many cases, another change in divisional management led to the view that "money and time have been wasted in forays outside our traditional and current lines of business and we must get back to our main path."

Presenting such patterns as shown in Figure 7-31 suggests that the firms represented may indeed have a "technology policy" or "portfolio strategy." Many firms have a technology strategy often laced with euphemisms such as "second to none" and "leading in technology." However, at the divisional level, it is difficult to identify such a pattern clearly and to trace it through from general statements by divisional management to the actual allocation of the technology budget and people to the categories in Figure 7-28. This may reflect uncertainty or lack of careful thought relative to the emphasis that is or should be placed on the various categories in their technology portfolio. Frequently, it reflects lack of perception of the importance of even having such a clear plan for divisional technology or reluctance to spend a lot of time and energy thinking about it. As discussed in connection with the idea pool and the factors inhibiting the generation and flow of ideas (Chapter 6), such management attitudes can have a profound impact on the projects undertaken and their results.

This situation can have significant impacts on the technology portfolio and the possible outcomes of it for many years, due to the time-lagged nature of responding to changes in portfolio emphasis. It may in fact reflect a desire to change the essential "mission" of the divisional lab and other technology groups in the division.

Suppose, for example, that a program of work in category "7" is initiated as the result of an intended shift in the lab's mission. As indicated in Section 2.11, the divisional lab may not have the personnel, the facilities, the equipment, or the shared experience to undertake this work immediately. Several months or years may be required to build up a capability in this new area. Once geared up for it, several more years may be required to achieve any useful results. Even intermediate ones, such as a more profound knowledge of the materials which the company uses and might use, takes time. If, during this period, the usual turnover of division managers and his management group occurs, the assigned mission of the lab may be changed again, to fit *his* (the new DM's) vision of what the division needs in the technology area. As a result, the "new" effort may have to be written off as lost and the people hired or trained to work on it may have to be redirected or terminated.

This is not a plea for rigidity or sanctity of the technology portfolio—either in continuing along its traditional path or doggedly pursuing a new direction no matter what happens. It is a plea for better understanding of the *frequency–response characteristics* of such portfolios when subjected to changes. It is not generally cheap, easy, or quick to change direction. A high frequency of changes in portfolios (as compared to individual projects) can make for a very unstable and unproductive platform from which to launch new and improved products and manufacturing processes.

8
COMMERCIALIZATION OF TECHNOLOGY

8.1 INTRODUCTION AND CHAPTER OVERVIEW

This topic—getting new and improved products out the door and new or improved processes into the plant—warrants a book of its own. Indeed, under headings such as Technology Transfer, Product Development, and The Innovation Process, several dozen books have been or are being written. In addition, several trade and professional journals concentrate on this aspect of technology management and the number of conferences and seminars on the subject is huge and growing.

In this chapter I have selected several aspects of the overall subject of "commercialization" for special attention in the book's treatment of the overall process of managing technology or R&D/innovation. These subtopics have been selected because they are crucial, I believe, to successful culmination of the technology management process—that is, "getting innovations out of the lab and into the factory or marketplace." The topics are:

1. Effectiveness of the R&D/production interface.

2. The role of imbedded technology.

3. Technical entrepreneurship in the firm.

The first two topics have been the foci of a long-term research study in Northwestern's Program of Research on the Management of Research, Development, and Innovation (POMRAD). The third is currently (1987–1988) the focus of a study of its own. In addition, all three have been involved in numerous consulting engagements by my colleagues and me over several decades. Aspects of both studies have been and are being reported in the open professional literature and in Ph.D. dissertations and working papers. Those channels carry most of the description of the theory and methodology underlying the studies and the research results in statistical terms. In this chapter I present some of the major ideas that gave rise to the studies themselves and that reflect the kinds of issues with which practicing managers have to deal in attempting to "get technology out the door."

8.2   EFFECTIVENESS OF THE R&D/PRODUCTION INTERFACE

8.2.1   Why Study or Be Concerned About This Subject?

From a very practical point of view—that of the division manager, top manager, or the manager of technology—this interface is one of the most critical in the entire R&D/innovation process. It is not the only important interface, of course, since the transitions or linkages or barriers between R&D and marketing may also serve to make or break a new or improved product or to increase its transition time and cost to a prohibitive level. We have written elsewhere about the R&D/-marketing interface as have other researchers and consultants. I have selected the R&D/production interface for special treatment in this chapter on commercialization because this interface is not as obvious or familiar to most managers without direct production experience or experience in attempting to make the interface effective, from whatever vantage point: R&D, production, or general management.

From a research viewpoint, it is only very recently that academic researchers have begun to focus on this topic and make it more visible and less mysterious to a broad audience of industrial managers and other academic researchers and students.

Practicing *managers* should know more about this topic because of its importance in the overall success of new and improved product projects, and *academics* are intrigued by its complexity—the myriad factors that can influence the rate and effectiveness with which projects pass through (or fail to pass through) this interface. Both groups—practitioners and academic researchers—may also find the topic fascinating in a larger management context, since a careful analysis of the barriers and facilitators affecting transition through the interface can also yield more general insights into factors that influence the effectiveness of the firm itself, whether it is in a "new products" phase or merely pursuing business as usual. Some of the subtopics we dealt with in our multiyear study and the half dozen Ph.D.

dissertations (see Appendix B) that arose from it suggest more general aspects of the topic:

- Technology strategy and make or buy decisions
- The role of organizational politics in project management
- Technology networking between operating divisions
- The role of the pilot plant in the innovation process
- Technical skill maintenance and renewal
- Conflict and cooperation between organizational units
- Organizational design for "temporary" activities or functions
- Personality differences and commitment to organizational goals

Other subissues and more general aspects of organizational behavior were investigated and are mentioned below under the discussions of issues, barriers and facilitators, and other variables.

8.2.2 What Are Some of the Key Issues Affecting or Affected by the R&D/Production Interface?

The overriding issue is: "What factors influence the effectiveness of transition through the R&D/production interface?" A large number of subissues follows from that—based on the individual barriers and facilitators to effective transition.

In the early stages of the interface study by a 10-person team in POMRAD at Northwestern University, we surveyed the literature—both practical and theoretical (although it soon became clear that many of the so-called practical articles—of the "how-to-do-it" variety—were much more theoretical than some of the academic studies). From this literature search, from our own experience (the research team embodied more than 100 years of practical experience in technology management), and from the previous research of POMRAD and other research groups, we assembled a list of more than 100 factors—potential barriers and facilitators—which appeared to have more than a trivial effect on successful transition of projects through the interface. (Actually, as illustrated below, there are multiple interfaces involved in the transition of a project from lab to factory.) The research process or "paradigm" described earlier in Chapter 6, in connection with the idea flow studies, was applied. This huge number of variables of factors was essentially "unresearchable" due to the rules of inference and sample size. The list was reduced to nine "macrovariables," which are listed on the left-hand side of Figure 8-1. Additional variables appeared during the course of the overall study and the individual dissertations, but this set of factors—including both barriers and facilitators to transition—remained the principal focus of the research team over a period of more than 3 years.

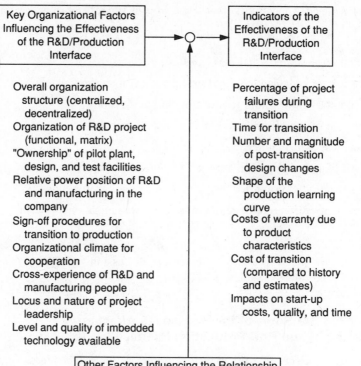

Key Organizational Factors Influencing the Effectiveness of the R&D/Production Interface	Indicators of the Effectiveness of the R&D/Production Interface
Overall organization structure (centralized, decentralized)	Percentage of project failures during transition
Organization of R&D project (functional, matrix)	Time for transition
"Ownership" of pilot plant, design, and test facilities	Number and magnitude of post-transition design changes
Relative power position of R&D and manufacturing in the company	Shape of the production learning curve
Sign-off procedures for transition to production	Costs of warranty due to product characteristics
Organizational climate for cooperation	Cost of transition (compared to history and estimates)
Cross-experience of R&D and manufacturing people	Impacts on start-up costs, quality, and time
Locus and nature of project leadership	
Level and quality of imbedded technology available	

Other Factors Influencing the Relationship

Specific field of technology
Size of company or unit (e.g., operating division)
Importance of project to the company
Priority assigned to project

Figure 8-1. A preliminary flow model of the R&D/production interface. (Source: Research by POMRAD/Northwestern and team members' experience on the R&D/production interface.)

Following the research paradigm, we identified a number of key questions, including those in Figure 8-2. These questions, reflecting the kinds of concerns that technology managers and their managers had expressed to us in a series of preliminary research interviews and in many of our consulting engagements, led us to develop a set of "potentially testable propositions," such as those shown in Figure 8-3. Many of these propositions were tested during the course of the major study and the dissertations. They are reported in some detail in the technical literature and in the technical reports of the research project (mind-numbing statistics and all). Some of the findings and their applications are presented below, in the context of practical application to managing this important interface. Before that, the next section captures some of the research questions and their derived propositions in the form of a pair of conceptual flow models.

1. How can effectiveness of the R&D/production interface be assessed?

2. What specific indicators can be used for such assessment?

3. To what extent can quantitative measures be developed for some of these indicators?

4. What factors (barriers and facilitators) appear to have significant influence on these indicators and on the effectiveness of the R&D/production interface?

5. What roles do imbedded technology capabilities play in the R&D/production interface and related stages in the R&D/innovation process?

6. How could the significant influence factors and indicators be built into a monitoring and evaluation system for improving the R&D/production interface?

7. What organizational design features (structural, procedural) and management practices can be identified as guides to effective and ineffective ways of functioning at the interface?

8. What kinds of subactivities or skills in the organization (or potentially available to it) are relevant and important at the interface? This can include a number of specialties in engineering which are not covered in most engineering curricula and which are not recognized in descriptions of individual skills and organizational capabilities in surveys of R&D or the innovation process (e.g., the annual surveys by the National Science Foundation).

9. What is the makeup of the technical support groups in terms of:

    (a) size—full-time people, full-time equivalents, or consultants?

    (b) funding—budgetary provisions as a distinct entity or as part of another activity such as process engineering; nature and amount of resources allocated?

    (c) skills of individual members and subgroups related to the R&D/innovation process?

    (d) levels of training existing and needed for effective contributions to innovation projects (beyond routine activities which do not relate to new and improved products and processes)?

10. How do imbedded technology groups (ITGs) interface with mainstream R&D/innovation activities: organizational arrangements, supervision, transfer of people and/or of funds, division of responsibilities, and resolution of conflicts?

11. What transfers actually occur or are attempted and blocked at the interface, such as: ideas, information, skills, special knowhow, hardware and software items?

12. How is interface effectiveness affected by design change or engineering change order (ECO) procedures?

13. What policy implications are there in connection with the R&D/production interface for R&D management, production management, new product planning, general management, the federal government, engineering schools, and other educational and training institutions?

14. What are the relations between manufacturing engineering and other groups (e.g., product design) at the interface which can significantly affect interface effectiveness?

15. What role does the interface play in integrating the diverse technical activities and criteria that are involved in product and process development—for example, design for cost, function, manufacturability, reliability, and maintainability?

Figure 8-2. Some research questions about the R&D/production interface.

Note: The positive forms of the statements given below, derived from some of the research questions in Figure 8-2, do not necessarily reflect the true state of affairs pertaining to the R&D/production interface. They are given here as illustrative of the kind of potentially testable propositions that we expected to emerge from the study, based on a combination of (a) the state of the art as reflected in the literature of past research and practice, (b) the empirical findings of the pilot field studies, and (c) our own previous and current experience with and insight into the R&D/production interface phenomenon, deriving from the combined experience of the POMRAD team members and our advisors in industry.

1. Effectiveness of the R&D/production interface depends on a combination of technical, organizational, economic, and individual factors.

2. There is more than one way to operate the interface effectively—that is, there are various combinations of necessary and sufficient conditions for success.

3. Organizations and individual managers with access to a wide variety of imbedded technology capabilities experience more effective interface operations, in addition to improvement of the productivity of the managers themselves.

4. Conflict can serve as both a barrier and facilitator to effective transfer through the interface.

5. Interpersonal relations and trust, built up over long periods of time, are important factors in effective interface operations.

6. Early mutual involvement of production managers and technical personnel in new projects improves interface productivity and effectiveness through providing early information on (a) needs and constraints from manufacturing and (b) technological possibilities from R&D.

7. Continued involvement of R&D past the official "hand-off" stage to production improves interface productivity and effectiveness.

Figure 8-3. Some very preliminary propositions.

Before even that, however, I would like to illustrate some of the potential complexity of this transition process when even a moderate level of technical complexity is embedded in the product-process and where the organization itself has a fairly complex layering of functions, specialties, and responsibilities for product and process development. Figures 8-4 through 8-6 were developed in the very early stages of the research when we wanted to get a feeling as to who the "key players" were in the R&D/production interface. That is, we were looking for an "inventory" of individuals, groups, specialties, and functions which might have significant impacts on how fast, how economically, and how technically effective particular transitions might be.

These three figures were developed and are presented for descriptive and impact purposes only, not for analytic purposes, since they lack precision in a key dimension—level of involvement. *Conceptually,* that dimension is fairly straightforward: How deeply is a particular group or functional specialty committed to the project? *Operationally,* this would imply the assignment of a given level of personnel in a particular time pattern—for example, 2 man-months during a 2-week crisis period for the project or 1 man-year stretched over the multiyear life of the project. Furthermore, it might measure the "depth" of that commitment in terms of how *important* the project was to that group, what percentage of its total resources during that period was devoted to it, and whether they assigned their best people

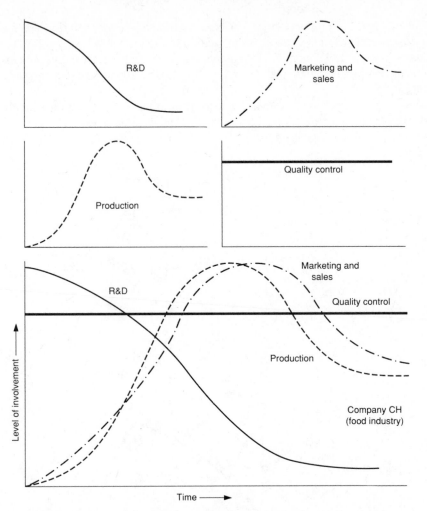

Figure 8-4. Example of patterns of involvement in food company project.

to it. For the rigorous, proposition-testing phases of the research, such operational measures were necessary and they were developed. For this discussion, however, I primarily want to communicate an impression of the potential complexity of this common and necessary transition phase in the life of any project.

The simplest of the three patterns shown here (dozens were developed for the scores of projects analyzed in the formal study)—Figure 8-4—seems fairly straightforward and classical. R&D starts the project and then tapers off and almost phases out as the project is passed on, in sequence, to production and marketing. Because this is a food industry product and because this firm has a strong commitment to the oversight of all new products by the quality control function, the horizontal line indicates that a quality control person or team (depending on the size of the project) is in it from the beginning and maintains its high level of involvement throughout the product's life cycle. The tapering off of R&D is not complete for this product, since (1) continuous minor improvements are made as the product matures and (2) some factory and customer service by R&D and some level of R&D input to improving the product or product line emerging from the project is maintained—perhaps as little as a fraction of a man-year per year.

Figure 8-5 takes a jump in complexity in terms of the number of groups involved and their patterns of involvement. Some, like engineering/drafting, phase in and out cleanly, with a formal sign-off and hand-over to another function. Others, like

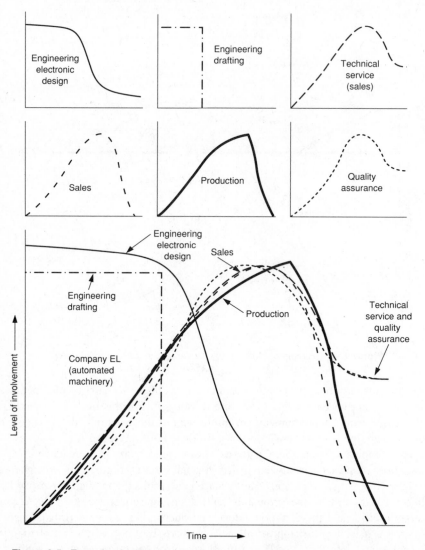

Figure 8-5. Example of patterns of involvement in an automated machinery project.

production and quality assurance move to a peak of involvement and then taper off or phase out. Again, engineering electronic design maintains a continuing level of involvement throughout the life of the product, as a backup to technical service, quality assurance, and production.

Finally, in Figure 8-6, the situation goes wild. In this electrical controls company, where systems have many electronic components and subsystems, groups appear to be (and are) falling over each other to make their contributions to the project. It is not clear, at any time in the project life, who is really in charge of

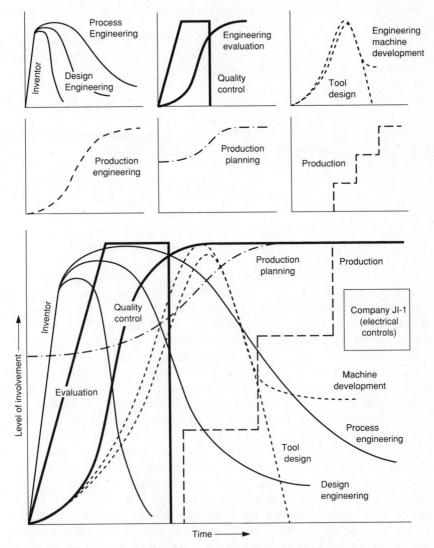

Figure 8-6. Example of patterns of involvement in an electrical controls project.

the project and who has final responsibility for goofs, changes, or policy on the design configurations of the overall system and their execution.

An interesting point in this project is that the original inventor phased out very quickly and the formal organization took over the design, development, and other phases of the project. Another feature of this project—which causes difficulties in many such projects in many companies—is the relatively early phase-out of design engineering. The formal hand-off to process engineering is murky, at best, and it is never clear, as the project moves through the factory and into the market, who is ultimately responsible for its effective design and execution, let alone dealing with complaints or requests from dissatisfied customers or an unfriendly market in general.

Again, let me hasten to caution the reader that these diagrams, especially the third one, are more impressionistic than analytical. However, they are presented to emphasize the need for careful planning of the transition phases in the project life cycle, the role of diverse skills and functions, and the need for clear criteria and procedures for hand-off, sign-off, and evaluation of the readiness of the product/system for production and marketing.

General managers without significant experience in production, let alone specific experience with the R&D/production transition process, often tend to ignore or blissfully "delegate" this important phase of technology management without maintaining sufficient oversight to be able to anticipate or even deal with blockages, crises, or major disasters that can significantly delay or even kill a project. All managers—division or corporate—cannot be thoroughly versed in this area, along with the other skills and experience they need for general management. But they can and should have a policy and set of procedures for oversight of the R&D/production interface and the transition of projects through it. One example that was repeated in many of our sample field sites was the role of the pilot plant as a barrier, as well as a facilitator, to the transition of projects to factory and market. In one dissertation, the student, Martin Ginn, "sat astride" the pilot plant for over a year, observing the transition of chemical projects through the pilot plant and "scale-up" process. It was clear that top management believed they "had better things to do" than worry about the conflicts and delays due to the organization, location, and power relations of the pilot plant vis-a-vis R&D and other downstream functions, such as process engineering. If there is a major point in presenting our approach to investigating this key phase of technology management and the life cycle of projects, it is that management cannot afford to take the transition for granted and assume that "things will go fine."

This is not a plan for "micromanagement" from the top of the division or the corporation. Certainly, the operational aspects of project transition are the business of the functional specialists and lower-level technology managers, especially in the highly decentralized company. However, the role of oversight, planning, monitoring, evaluation, control, and blowing the whistle, when necessary, has to be a company-wide responsibility, including middle and top management. Too many projects grind to a halt or come limping out of the R&D/production interface with cost, time, and technical performance of the products/processes being developed below par, beyond target, and inappropriate for full-scale production or marketing. Management complacency about this transition phase can lead to unacceptable delays, costs, and missed technical targets in terms of product performance and specifications. In many cases, these shortfalls cannot be made up by a crisis-driven "fix," since the market's "window of opportunity" may have closed by the time a transition mess is straightened out, blame is assigned, and "heads roll."

8.2.3 A Flow Model

Figure 8-7 captures the main factors influencing the effectiveness of the R&D/production interface—gathered into the macrofactors mentioned earlier. It also divides the outcomes of the interface process into two levels: (1) the immediate and intermediate indicators of effectiveness of the *transition itself* and (2) downstream indicators of *impact*, in terms of the success of the project and impacts on the product line, division, or firm, relating to economics and market position. The main thrust of this conceptual flow model is that the effectiveness of the transition of a project through the R&D/production interface is not an end in itself. It can only be *evaluated* (see Chapter 7) on how it influences the general fortunes of the firm in the marketplace. The fact that marketplace impacts may be many years away from

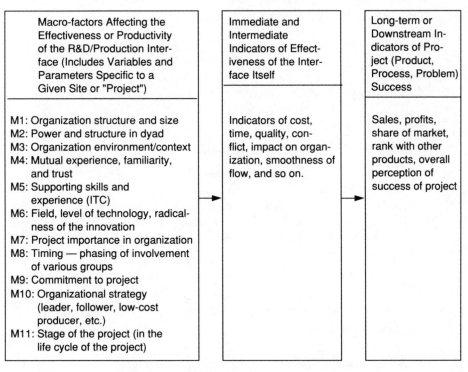

Macro-factors Affecting the Effectiveness or Productivity of the R&D/Production Interface (Includes Variables and Parameters Specific to a Given Site or "Project")	Immediate and Intermediate Indicators of Effectiveness of the Interface Itself	Long-term or Downstream Indicators of Project (Product, Process, Problem) Success
M1: Organization structure and size M2: Power and structure in dyad M3: Organization environment/context M4: Mutual experience, familiarity, and trust M5: Supporting skills and experience (ITC) M6: Field, level of technology, radicalness of the innovation M7: Project importance in organization M8: Timing — phasing of involvement of various groups M9: Commitment to project M10: Organizational strategy (leader, follower, low-cost producer, etc.) M11: Stage of the project (in the life cycle of the project)	Indicators of cost, time, quality, conflict, impact on organization, smoothness of flow, and so on.	Sales, profits, share of market, rank with other products, overall perception of success of project

Figure 8-7. Three-stage conceptual flow model.

the transition we are focusing on makes it important that some of the immediate and intermediate indicators of transition effectiveness be used continuously in a *monitoring* mode to make sure that the process is occurring in a manner likely to impact downstream economic and market indicators favorably. From the conceptual model, we were able to develop a set of potentially testable propositions of the "if...then..." variety to test with the actual data from the field studies. Such propositions derived from Figure 8-3 include:

- If organizations and individual managers have access to a wide variety of imbedded technology capabilities (ITCs), then they will experience more effective interface operations, in addition to improvement of the productivity of the managers themselves.
- If conflict is task oriented and not person oriented, then it can serve as a facilitator to effective transfer through the interface.
- If interpersonal relations and trust are built up over long periods of time, then they can be important factors in effective interface operations.
- If production managers and technical personnel are involved in early planning of new projects, then interface productivity and effectiveness can be improved through providing early information on needs and constraints from manufacturing and technological possibilities from R&D.
- If R&D involvement continues beyond the official hand-off stage to production, then the interface effectiveness improves.

One of the outcomes of the overall study and the related substudies was a set of propositions in the form of a "propositional inventory." The propositions are presented in sets, supported by literature focused on particular aspects of the R&D/production interface and related aspects of the overall R&D/innovation process.

8.2.4   Some Findings From the Research

Some of the general findings[*] were the following:

1. *Technical issues* did not clearly differentiate successful from less successful or "problematic" projects at the R&D/production interface. Organizational and management issues appeared to be more influential, including:

- Cutting of project resources to unreasonable levels.
- Organizational barriers such as several divisions or outside vendors involved.
- Lack of a clearly defined project manager.
- Unrealistic schedules.

---

[*]For more detail, see the final report of the project (Rubenstein et al., 1984).

- Lack of clear and consistent specifications and design details.
- Resistance from sales.
- Lack of in-house know-how, coupled with reluctance to seek outside assistance.

2. *Four major dimensions of technology strategy* were derived statistically, describing the organization's technology policy: analytical strategist, technical innovator, marketeer, and service-driven organization. Most of the organizations in our sample tried to balance these four approaches to technology strategy, with one of them being the "driving force. " Analytical-type organizations seemed to pursue simultaneously "low-cost" and "niche" strategies.

3. *Role conflict* among project team members is more closely tied to the overall level of political behavior in the organization than with politics at the specific project level.

4. *Project-level political behavior* is inversely related to (a) confidence in other team members and (b) project team cohesion.

5. *Projects that were more successful* at the R&D/production interface, in contrast to those that were problematical, had the following characteristics:

- Higher levels of individual commitment to the project by team members.
- Lower levels of project specific politics.
- Lower levels of radicalness/newness of the technology.

8.2.5 Some More Specific Findings Related to the Concept of Imbedded Technology Capability

(See the next section in this chapter for the background on this important concept.)

1. The use of imbedded technology capabilities (ITCs) external to the project group is more effective for long-term and "hands-on" relations with project teams than hit-and-run or quick fixes.

2. Technical assistance is more helpful when the project team has the capability and sophistication to absorb and utilize the assistance offered by the ITC group.

3. Project teams tend to shy away from seeking technical help outside the team. The main reasons are competition, unknown capabilities of external ITC groups, and fear of revealing proprietary information.

4. ITC groups are sought out primarily to resolve nontrivial problems and the ITC groups tend to continue working with the project team until the problem is solved.

5. Project teams tend to seek external ITC help, based on personal experience of team members with members of particular ITC groups and recommendations from the project team's superiors, rather than general reputation of the ITC groups or their presumed expertise.

6. Project teams look first inside the organization for technical assistance, rather than go outside the organization to consultants or other firms.

7. In projects that passed through the R&D/production interface successfully, manufacturing was consulted early in the project.

8. The need for external technical assistance depends on the stages of the project during which the technical problems were encountered and the technical complexity of the project.

9. It was not possible to differentiate successful from problematic projects (with respect to the R&D/production interface) on the basis of presence or absence of ITC or the effective/ineffective use of ITCs. This may be because of the mobilization of all available resources when a project is in trouble and the consequent effective use of ITCs in bailing the project out.

8.2.6 Some Recommendations and Implications for Management Arising From the Research

The following relate to general issues of managing the R&D/production transition and project management in general.

1. *Levels of commitment* of project team members to the project can be increased by the following:
   - Making sure that a project manager is appointed who can effectively coordinate the activities of all the groups and individuals involved.
   - Making sure that the project has reasonably clear-cut, yet flexible, goals and objectives.
   - Assembling a team for the project with each individual having well-defined responsibilities.

2. *Organizational politics* can be lowered by:
   - Involving all the individuals and groups that may be needed on a project as early as possible, preferably during the evaluation or selection phase of project formation.
   - Getting all the involved groups to agree or at least concur with the goals and objectives of the project.
   - Fostering an atmosphere on the project that does not unduly penalize individuals who make mistakes while taking calculated risks on behalf of completing or improving the project.

3. *Measures can be taken to "hasten slowly,"* despite the external factors influencing radicalness/newness of technology involved in a project. They include:
   - Avoiding rushing technologically complex or very new projects; the time and cost of firefighting, subsequent technical problems, and trying to "fix" a "broken" project may cost more than a less hasty approach.
   - Breaking up the project into several stages that can be done in parallel or sequence for very new or frontier technologies.

- Giving team members and their friends (e.g., downstream functions such as production, marketing, service, engineering) time to think through the design early in the life of the project.
- Introducing the results of the project onto the shop floor or market in well-planned stages, allowing for early debugging of problems that might be anticipated.

Delays, excessive costs, technical deficiencies in products and processes, and conflict between individuals and groups involved at the interface can all reduce the effectiveness of the R&D/innovation process and hence the productivity and competitiveness of the firm.

The findings of the main study and related studies point to a number of other factors that can also have important influences on the effectiveness of the interface and the overall R&D/innovation process. These are factors which may not be consciously recognized by managers and policymakers involved in technological innovation but which can be improved or, at least, prevented from interfering with the transition of projects toward the market or factory implementation.

They are mentioned here as guidelines to managers and policymakers seeking to improve the R&D/innovation process and the specific effectiveness of the R&D/production interface. The implications are that if careful attention is paid to such factors in designing and managing organizations, increases in effectiveness are likely. Additional factors emerging from this study in which such improvements can be made are given next.

4. *Technology Strategy*. A much better job needs to be done in both formulating technology strategy for the firm and in making sure that it is promulgated throughout the organization, understood, used as a basis for operating decisions involving technology, and subject to continuous monitoring and, as necessary, revision as conditions both inside and outside the firm change. Unless such actions are taken, specific decisions made during new product or process development may be inconsistent with top management intentions or objectives. Also, specific actions taken during the R&D/production interface (e.g., obtaining the "best available" assistance in resolving technical problems on projects) may be less effective than they might be.

5. *Organizational Politics*. Although this is a subject that is often discussed with a "giggle," it is a fact of organizational life that can have both positive and negative effects on the success of R&D/innovation projects and the general interpersonal atmosphere of the firm. We do not recommend attempts to either "formalize" or "eradicate" politics from the R&D/innovation process. That would be both futile and probably inappropriate. It is important, however, that management be aware of the role of such informal decisionmaking and power usage in the organization as it affects the R&D/innovation process and as it presents barriers or facilitators to the transition of projects through the R&D/production interface. Gentle support of those *facilitating* political processes and gentle attempts to mitigate some of

the *barriers* created or increased by politics can have important consequences for effectiveness.

6. *Organizational Factors Versus Personality.* The behavioral patterns of certain individuals (secrecy, procrastination, combativeness, avoidance of conflict, etc.) can have significant influences on the progress of a particular project through the interface at a particular time. However, some of these individual behaviors and the personal qualities they reflect may be difficult to change, short of replacing the individual. This may be highly disadvantageous to the project, if he has important technical skills and experience that the project needs.

An important step prior to considering such replacement (or milder forms of attempting to change individuals, such as punishment, counseling, or training) would be to examine some of the factors that have emerged as significant for the R&D/production interface to see if they are "causing" or exacerbating the situation and combining with personal characteristics to interfere with progress at the interface. A continuous scan of such factors, to see that they are not at critical levels, could help identify barriers to smooth and effective transition through the interface and help reduce the frequent emphasis on "blame placing" (Ginn, 1983) in favor of problem-sensing and problem-solving.

This "problem" or "barrier" focus should help managers identify *recurring* situations or events at the interface which might yield to more general and lasting solutions than the crisis-reaction or quick-fix mode in which many organizations operate. Such broader and longer-term views can lead to organization and/or policy changes that can prevent or ameliorate some of the crises that are common at the R&D/production interface.

7. *Cooperation Versus Conflict.* Enlisting the energies of individuals and groups involved at the interface in cooperative efforts to improve the design and operation of the interface can divert some of these energies away from conflict and competition into more productive channels.

8. *Institutionalizing the Experience Gained.* Many new project managers seem to be fated to repeat the same or similar mistakes that have already been experienced in the organization before (e.g., offending particular production people, overlooking important sources of information, not dealing appropriately with purchasing). Although many people claim they are too overcommitted to do so, we strongly recommend a formal "debriefing" of project managers as a project transits through the interface and after the transition has been completed. This input to "organizational memory" should include events and situations, good and bad, which affected the transition and the eventual success of the project—technical and economic.

9. *Technical Complexity and Newness.* Although only a few of the sample products involved *radically* new or breakthrough technology (in this sense they were highly typical of most industrial R&D projects), many of the products and their related production processes had a number of sources of technical complexity that affected the transition. Such projects need to be more carefully planned, organized, and monitored than projects that are likely to contain no major technical surprises or areas with which project personnel are unfamiliar. This suggests the

need for a special "module" in the project planning procedure which identifies and flags aspects of the project which are new to the people involved, the organization, the industry, or the field in general. These special aspects should be monitored via a watch list that gives early warning and signals need of a solution mode for unusual problems as they occur or, better yet, are *about* to occur.

10. *Unrealistic Requirements.* A number of the projects in the sample ran into trouble trying to meet unrealistic expectations and constraints imposed by management. Some of these involved time schedules that did not permit an orderly conduct of the project and touching all the necessary bases to make sure that problems would not occur downstream of the interface (e.g., scale-up, introduction to market, service, maintainability, life cycle cost). Others involved what turned out to be unattainable cost limitations for the project. Again, as in item 9 above, special attention should be paid, in the *planning* of a project, to constraints (costs, time, specifications) that may cause difficulty or may, in the end, abort the project or cause it to fail technically or economically.

11. *Phasing in the Key Functions.* As Figures 8-4 through 8-6 indicate, there is wide variation in the pattern by which functions such as R&D, production, marketing, and supporting functions are involved in a given project. Early involvement of some functions is vital to success at the interface and of the entire project. It is recommended that, at the minimum, the functional groups who will eventually be involved in the transition and the whole project life cycle be informed of the general plan and schedule for the project, even if their time and interest commitment in the early stages is very small. Likewise, it is recommended that the formal cutoff of involvement of certain upstream groups, such as research and design, not be as abrupt as it is in many cases. Continued interest by such groups in the product as it goes up the learning curve in the factory and as it is being used in the market can help keep a product sold and help make it economically viable.

12. *Individual Commitment to Project.* Many of the events that occur at the interface (and during the entire project life cycle) are reflections, good and bad, of individual interest in and commitment to the project. Management should design the project team and provide incentives to stimulate, enhance, and maintain high commitment and to take advantage of the extra effort and care that accompany high commitment. In some cases, recognition of individual contributions may be adequate. In others, more tangible rewards may be required to foster a sustained, high level of commitment to the project, in the face of other interests and pressures.

13. *Matching of Project Personnel to Technology Strategy and Type of Project.* Although there were no statistical findings in the study to support this recommendation, it was clear that some project teams in the sample sites were "over their heads" technically in attempting to develop radically new or complex products that were beyond their previous experience. One of the issues involved is that of "key communicators" (Knapp, 1984). This issue implies the need for people who are continually in touch with the leading edge and state of the art in certain technical fields of current or potential interest to the company. In turn, this relates to whether the company is pursuing a predominantly "pioneering" or "defensive" strategy with

respect to the technology base of its products and manufacturing processes. The hiring, maintaining, and continual upgrading of such people as "key communicators" and the "resident genius" may be crucial to the company's ability to sustain a pioneering strategy in the face of strong technological competition from other firms and countries. Many project personnel we encountered in the study (and in U.S. industry in general) do not have the time to read the technical literature or to keep up on the leading edge of relevant technologies by other means.

Such considerations should be incorporated into the company's recruiting, training, and continuing human resources maintenance program as a major concern of top management.

14. *Project Manager System.* Some projects turned out to be "orphans," with no one responsible for pulling the whole project together and making sure that the R&D/production transition occurred effectively. Although a project manager system may not fit into all technical organizations, it is an organizational device that should be *considered* by all managers who are concerned about interface effectiveness. There are many forms of this, from completely separate departments with a cadre of project manager type of people waiting for their next project assignment, to a less formal system where people in functional departments are assigned lead responsibility for a given project, even if the responsibility passes from one specific individual to another as the project moves through the R&D/innovation process.

Finally, the following are recommendations and implications related to the specific aspect of the study dealing with *imbedded technology capabilities* (ITCs) in the firm and their role in the R&D/production interface.

1. *Increased Flexibility.* Project managers should be allowed more latitude in seeking out and "contracting for" technical assistance from ITC groups (ITGs) inside the company as well as outside. This can be done by minimizing formal accounting and other constraints that may inhibit the approach to and use of the services of such groups. In a number of sites it "was not worth the hassle" to try to get such groups to work on a project to which they had not been officially assigned and budgeted. We found very few cases, for example, of cross-divisional or cross-site cooperation, when the project team and the ITG had different reporting relations to upper management.

2. *Skills and Capabilities Inventory.* Such a reference system ought to be established and maintained for general and specific areas of expertise and experience that might be useful for project managers. To be useful, it has to be kept up-to-date and be "realistic" in terms of accessibility and ease of use of the services of such groups or individuals.

3. *Recognize and Reward ITCs.* Individuals and groups with important imbedded technology capabilities (ITCs) should be encouraged to maintain and improve their technical skills and to make them available to the rest of the organization. Many ITC individuals and groups, by virtue of training, experience, or personal

interests, have special knowledge and techniques that are considered a "side issue" to their regular assignments and they have no incentive for letting project managers know about them or making them available to project teams.

4. *Ad Hoc Task Groups.* A number of projects in our sample were "bailed out" by ad hoc groups set up between regular team members and ITC. This should be encouraged, where appropriate, rather than inhibited because of formal organizational or budgetary constraints. Ad hoc groups may contribute well beyond the "quick fix" and may be effectively integrated into the project team for extended periods. This can be beneficial to the organization if it does not tie them into one project and make them unavailable to other project teams that may need them. The use of "crisis" or "panic button" teams that arrive on the scene too late to prevent or repair damage to the project is no substitute for systematically forming ad hoc teams, based on early warning of difficulties (See Section 4.4 on "Design Review").

5. *Encourage Contact with ITCs Even When Not Needed.* A major finding was that project managers tend to seek help from outside people and groups that they *know*, preferably inside their immediate organization, but outside it, if necessary. A routine method of familiarizing project team personnel with ITG personnel, so that they can get to know each other, can improve the chances that such people will be called upon to help when problems arise.

6. *Make Better Use of Academic Consultants as ITGs.* Many initiatives are under way to improve working relations between universities and industrial firms (see Chapter 6). Most of them, however, focus on collaborative research or teaching and other *institutional* relations. Although many individual faculty members do consult for industry in the areas of their *research* specialties, there is much additional technical talent available in university groups that could serve as external ITGs for specific company projects. This would require (a) easing the information barriers—that is, letting the involved parties become better acquainted, since, as with internal groups, *familiarity* is a major criterion for seeking out a group; (b) guaranteeing protection of proprietary information, which is usually part of regular consulting arrangements; (c) providing financial incentives for faculty that are on the commercial/consulting scale rather than the "community service" scale, at which some of these relations are currently taking place.

7. *Retain Scarce ITGs and ITCs.* A number of incentives might be developed for retaining individuals and groups with valuable ITCs to remain active in the company beyond the usual age of retirement, either as full-time "captive" employees or as occasional consultants. Typically, this is done on an ad hoc basis by individual firms and leaves a lot of talent unused as the result of early or normal retirements. This pool of ITCs should be viewed as a corporate and national resource and more effective use made of it. Take care to retain important ITC groups in the corporate network in the face of reductions in force, mergers, acquisitions, reorganizations, and other changes.

8. *Make the skills and technical capabilities* of ITC groups in the company available to it from outside sources *known* to project team managers and project

team members. A directory of such people and groups would be useful, if in sufficient detail and kept up-to-date.

9. *Facilitate the access* of project team members to ITCs via multiple communication means—for example, telecommunication, mail, travel, and seminars.

10. *Keep the ITGs aware of company policies* on technology issues and changes in those policies (e.g., "first in the market," "quality regardless of cost or delay").

11. *Consider the ITGs as important investments* and continuously upgrade their technical capabilities and quality.

12. *Allow project leaders more latitude* in seeking out and contracting for technical assistance from ITC groups both within the company and outside it. Encourage cross-divisional and cross-site use of ITCs and sharing or networking of technological capabilities (See Chapter 4).

13. *Establish a national referral service* for ITCs across some industries and technical fields.

The findings of this study suggest that there are significant gains to be made in productivity and competitiveness via improving this major aspect of the R&D/innovation process—the R&D/production interface.

8.3 THE ROLE OF IMBEDDED TECHNOLOGY (IT) IN THE COMMERCIALIZATION PROCESS*

8.3.1 What Is Imbedded Technology and How Does It Relate to the Overall R&D/Innovation (R&D/I) Process and Commercialization of Technology?

In most discussions of technology transfer, the innovation process, productivity, and other technology-related subjects, there is generally an area of uncertainty and concern which revolves around the issue of "the secret of success" or the "reasons for failure." Comparisons among countries, industries, sectors, and individual firms involve attempts to identify and explain these "secret ingredients" of technological superiority or lag. Certainly, differences in level of research and development (R&D) or science and technology (S&T)—roughly equivalent terms— are used to account for some of the differences among the most advanced countries (industries, firms, sectors) and those least advanced.

The problem is that the usual indicators of R&D or S&T depend heavily on numbers such as expenditures on R&D, numbers of engineers and scientists, and the size of laboratories. Such numbers do not explain enough of the differences in technological capabilities and performance to provide guidance for policymakers and technology managers.

*Adapted from the author's paper in *Research and Innovation: Developing A Dynamic Nation*, Vol. 3, 12/29/80; Joint Economic Committee, Congress of the U.S. Special Studies on Economic Change, Vol. 3.

There are also contradictions. Some countries, industries, sectors, individual firms, and major parts of firms (e.g., operating divisions) appear to be spending a reasonable sum on R&D and appear to be employing adequate numbers of scientists and engineers (in some cases excessive numbers, according to critics). Despite this, however, entities with smaller R&D budgets and smaller numbers of people appear to be more capable of "getting products out the door," achieving technological breakthroughs, producing high-quality and reliable products and services, and performing other technologically related activities in a superior manner.

Many of the discussions mentioned above at conferences, in classrooms, in board meetings, in the press, and in common parlance, refer to other factors and not to R&D or S&T. Among these other factors is one that has been kicking around under various labels since well before the days when "Yankee ingenuity" was recognized as a force in world trade and the development of this country. Although there are many terms used for this phenomenon, I have chosen the term "imbedded technology" (IT) as a convenient hook on which to hang the ideas in this section. No attempt is made at an elegant or rigorous definition here and it is not clear that one is feasible or that one would satisfy all interested parties. Imbedded technology, as I am using the term, involves several concepts, very loosely defined:

1. Specific knowledge that (a) is embodied in materials, products, processes, procedures, and systems; and (b) has accrued or appeared in a gradual, nonbreakthrough manner.

2. Ideas for or knowledge of how to make improvements in materials, products, processes, procedures, and systems which may not have been specifically incorporated, but which may be available "on the shelf."

3. The variety of individual technical skills which are not readily classified or even described, but which involve accumulated experience on how to do things, what works, and what does not work.

4. The aggregate clusters of individual skills which make up organizational capability—a first-rate design group, a savvy start-up crew, a clever and innovative methods department.

Many other terms have also been used to describe the range of phenomena included in this broad definition (Hahn and Doscher, 1977):

Technological infrastructure
Systemic technology
Technological diffusion
Microtechnology transfers
Evolving technology

In some senses, imbedded technology (IT) is a residual when the formal R&D components of the overall R&D/innovation process are removed. One test for "stand-alone" IT is the appearance of innovation and technological capability in the absence of formal R&D or S&T institutions and roles. Examples of manufacturing-related ITs are given in Section 8.3.5.

8.3.2 A Flow Model of the R&D/Innovation Process and the Role of IT in R&D/Innovation

Imbedded technology (IT) plays an important role in the transitions between stages in the overall R&D/innovation process (see Figure 8-8). It involves many of the design, engineering, and production innovations which are added on to or used to replace the original product or process features and specifications that emerge from the upstream or R&D stages of the overall process. Many specifications and features turn out to be nonfeasible, noneconomic, unsafe, too time consuming, difficult to maintain, produce, or operate, and unmarketable. At this point, I do not wish to join the attack, which comes from some quarters, on the impracticability of much work done by scientists and engineers in or for industry. My theme is that science and formal engineering alone are far from sufficient to turn out useful products, services, and processes. For this reason (and in many cases, out of time sequence), the IT infrastructure has been developed and is maintained. Examples of the kinds of skills and activities covered by IT include tool and die making, computer programming, and methods and standards. Additional examples are given later. This section is concerned with the ubiquitous role of IT and all it implies at every stage of the R&D/innovation process, from the making of a drawing or model, to the tricks of the trade in adapting a product or process, to the specific needs and barriers of the market for which it is intended.

8.3.3 Relation of IT to Institution Building, Capability Development, and Productive Infrastructure

Three examples from three quite different situations are given here to illustrate the impact of IT on a wide range of technological situations: new high-technology small companies in the United States, research institutes in developing countries (LDCs), and large multinational companies (MNCs). All these situational examples are composites from our research and consulting experience on all phases of the R&D/innovation process over the past 35 years. All three reflect the importance of IT components of institution building and capability development for R&D/I, often overlooked by policymakers, financiers, and nontechnical management people.

1. New high-technology firms grew explosively on both coasts of the United States and at several isolated inland locations in the United States after World War II. Most of them, in the early postwar days, arose from university or government laboratories and represented applications or continuation of the science and technology (S&T) that their founders had been conducting under the umbrella of the lab or university. As long as such small start-up companies continued their "studies and prototype development" and contract R&D and made "one of a kind,"

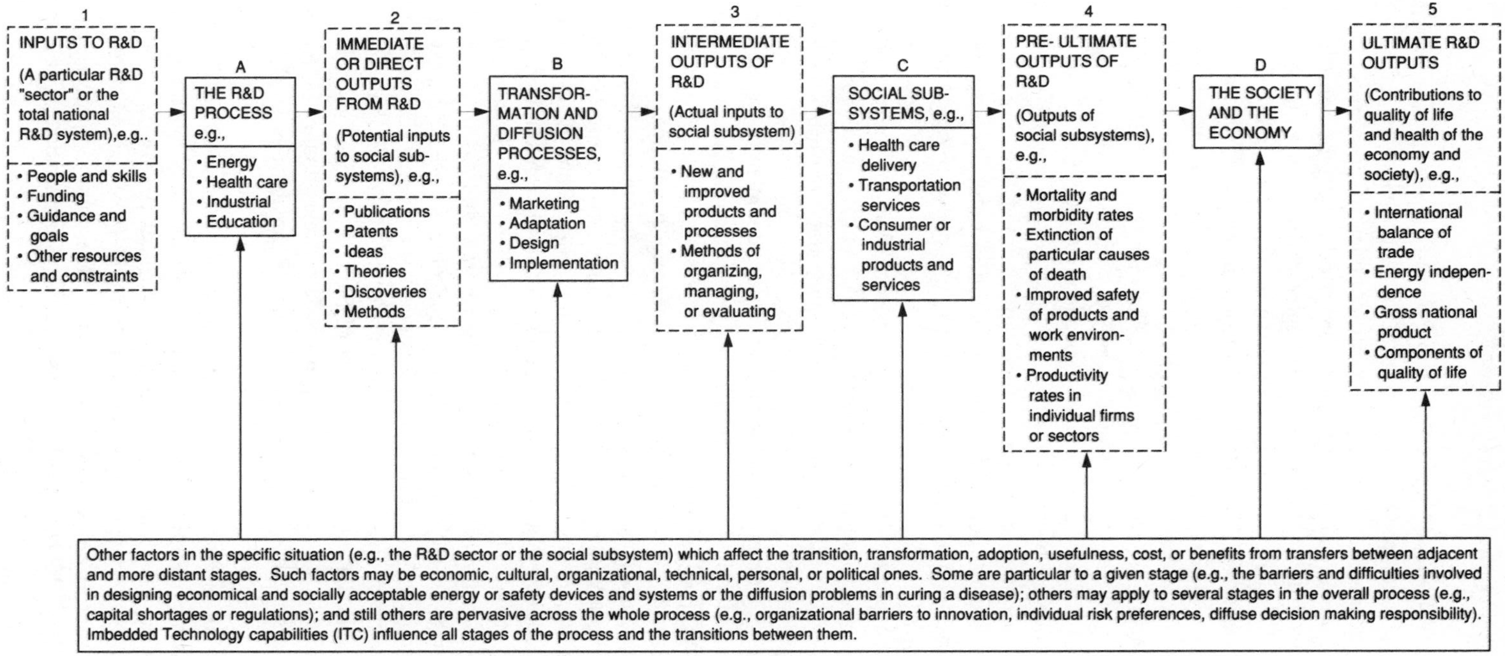
1
INPUTS TO R&D
(A particular R&D "sector" or the total national R&D system),e.g..
• People and skills
• Funding
• Guidance and goals
• Other resources and constraints
A
THE R&D PROCESS e.g.,
• Energy
• Health care
• Industrial
• Education
2
IMMEDIATE OR DIRECT OUTPUTS FROM R&D
(Potential inputs to social sub-systems), e.g.,
• Publications
• Patents
• Ideas
• Theories
• Discoveries
• Methods
B
TRANSFOR-MATION AND DIFFUSION PROCESSES, e.g.,
• Marketing
• Adaptation
• Design
• Implementation
3
INTERMEDIATE OUTPUTS OF R&D
(Actual inputs to social subsystem)
• New and improved products and processes
• Methods of organizing, managing, or evaluating
C
SOCIAL SUB-SYSTEMS, e.g.,
• Health care delivery
• Transportation services
• Consumer or industrial products and services
4
PRE- ULTIMATE OUTPUTS OF R&D
(Outputs of social subsystems), e.g.,
• Mortality and morbidity rates
• Extinction of particular causes of death
• Improved safety of products and work environ-ments
• Productivity rates in individual firms or sectors
D
THE SOCIETY AND THE ECONOMY
5
ULTIMATE R&D OUTPUTS
(Contributions to quality of life and health of the economy and society), e.g.,
• International balance of trade
• Energy indepen-dence
• Gross national product
• Components of quality of life
Other factors in the specific situation (e.g., the R&D sector or the social subsystem) which affect the transition, transformation, adoption, usefulness, cost, or benefits from transfers between adjacent and more distant stages. Such factors may be economic, cultural, organizational, technical, personal, or political ones. Some are particular to a given stage (e.g., the barriers and difficulties involved in designing economical and socially acceptable energy or safety devices and systems or the diffusion problems in curing a disease); others may apply to several stages in the overall process (e.g., capital shortages or regulations); and still others are pervasive across the whole process (e.g., organizational barriers to innovation, individual risk preferences, diffuse decision making responsibility). Imbedded Technology capabilities (ITC) influence all stages of the process and the transitions between them.

Figure 8-8. A preliminary conceptual model of the linkages between the R&D process and social systems.

highly specialized and complex instruments and equipment, they performed well (although few made profits early on, let alone paid dividends).

When, however, they branched out into commercial products or processes and attempted to go into quantity production under cost and time constraints, many of them came to a grinding halt or went through formidable gyrations and changes in organization, personnel, financing, and ownership.

Much of what was lacking was a sufficient level and quality of IT. In the simplest view, most of these technical entrepreneurs and their friends, who also came from noncommercial environments, were lacking in most skills required to put out a competitive, practical product. The most serious shortcoming for many of them was complete lack of training in, experience with, or understanding of the production process and the many special functions that are involved in producing a technology-based product or process. In the late 1970s, with many of the lessons about production technology learned by their entrepreneurial "ancestors," many of todays' founders of such firms still do not have the requisite skills and knowledge to manufacture properly and many of them still do not recognize this gap.

I have been personally affected by this situation for the past three decades as a director of a small business investment company and, for many years, a prime advocate on the board of high-technology start-ups. Although there are notable exceptions, the typical technology start-up company, which spins off from a university or government lab, is almost completely lacking in many of the necessary individual skills and company capabilities (IT), unless they are fortunate or foresighted enough to have included some partners or employees with previous industrial manufacturing and engineering experience. One of the consequences of this gap in company capabilities is the offsetting in the marketplace, both domestic and international, of the highly innovative nature of their products (the raison d'être of such companies) by high costs, long delays in production, and product features that often fail in service or in the marketplace. In addition to lack of some basic IT capabilities, many of these new high-technology firms suffer from the common weaknesses of many new and small businesses—inadequate marketing, financing, and general management skills (Rubenstein, 1958).

Several programs of the National Science Foundation (NSF, 1978) and other agencies have been directed toward this category of small high-technology firms and it is possible that remedies for some of their IT deficiencies are forthcoming. One danger is that the typical partner arranged by matchmaker federal agencies is a university group that often lacks the same IT capabilities as the firm seeking help!

2. Research institutes (RIs) in developing countries (LDCs) play a unique role. Many of them are the only relatively sophisticated source of technology for industry or other sectors. They can (although not always successfully) serve their clients as "windows to the technological world," providers of technical assistance, and sources of scientific and technical information. The common problem, however, is that many of these RIs (the majority, in some views) are no better off in terms of the

range of capabilities needed for industrial innovation and production than the new high-technology firms discussed above. Their personnel typically come (with a few notable exceptions) directly from university or government science departments, laboratories, or bureaus. The exceptions are those few whose top management and senior personnel are deliberately recruited from industry to head the RI or which, in entrepreneurial fashion, are established by experienced industrial people.

Our work with dozens of such RIs in Asia, Africa, Latin America, and elsewhere (Jedlicka et al., POMRAD doc. no. 75/7) suggests that most of them do not have the depth of experience or understanding of the range of ITs needed for effective product and process development along the entire R&D/innovation chain. Therefore, the advice many provide their clients is often "amateurish" with respect to modern techniques in engineering, tooling, procurement, marketing, and production methods. Eventually, if not sooner, many industries turn away from them and seek help elsewhere or do without.

There are many reasons for this mismatch between RI and client needs, in addition to the university bias of its personnel, and they have been discussed in countless conferences and surveys. It is not clear that international organizations, which originally set up most of the RIs, or that national agencies (ministries, departments, bureaus), which are responsible for operating them, recognize even today the important role that IT capabilities play in the RI–industry relation. It is also unclear whether they are doing much to remedy the situation.

One approach we have been pursuing to contribute to improving the situation is to recommend broadening and deepening the scope of activities of such RIs beyond the typical "doing of projects" and "providing of technical information." It involves a fuller partnership between RI and client along the entire R&D/innovation process rather than merely at the front end or R&D part of the spectrum (Rubenstein, POMRAD doc. nos. 75/109 and 76/50).

The RI has tremendous potential impact on helping to build the productive infrastructure of individual companies, industries, and the economy of an LDC as a whole. This is particularly important to LDCs, since the typical company is very small or tiny by U.S., European, or Japanese standards. It often does not have a single college graduate or person with deep and broad experience to do the range of activities included under IT, let alone have formal multiperson R&D or engineering departments.

The RI, in theory, can assist in providing the IT help needed by many such firms and help them build in-house capabilities for the range of ITs needed. This can be done with the cooperation of the emerging industry associations and other technical assistance bureaus in some LDCs. Unless most of the RIs remedy their own lack of IT capabilities, however, they can be of little use to their clients.

3. Large multinational companies (MNCs) provide the model for the full range of IT capabilities needed for modern innovation and production. Even here, however, some problems appear. When the outer skin of a huge diversified MNC is peeled away, we often find a collection of relatively small "companies" or operations which, according to decentralization doctrine, are expected to operate in

a fairly autonomous manner except for legal, finance, and some other corporate-wide services. In many of the top 500 U.S. corporations, mergers and acquisitions of autonomous companies created the operating units. The apparent synergism between units that should provide for skill and capability exchange at the level of innovation and production is not very effective (see Chapter 4). That is, within the large corporate framework, many smaller divisions are lacking, at an acceptable level, some basic IT capabilities and others have a surplus. We[*] have been acutely conscious of this in our small business investment company (SBIC) activities in recent years, as we have concentrated on "buy-outs" of divisions and operations of large diversified companies. We found many of these units, when stripped of corporate staff support, sadly lacking in basic elements of IT when they are set up as independent small businesses, with their former managers becoming the new owner/managers. This situation is a far cry from the new high-technology companies discussed above, where most of the personnel are relatively inexperienced in industry. In the case of the "buy-out" that leads to a new "stand-alone" company, the management and other personnel are quite experienced (we would not back them in a new venture if they were not), but frequently they could no longer afford the luxury of specialized IT capabilities which they formerly obtained "free" (as part of the overhead they paid to corporate staff) or paid for on a prorated basis.

Few of these small new companies can afford the highly specialized people in quality control, operations, accounting, management science, market research, law, corporate finance, logistics and transportation, data processing, and other special capabilities they enjoyed when they were part of a large corporation. In instances where the parent corporation displayed the extremes of decentralization and autonomy—allowing the division manager the option of "buying" such specialist services—many of them declined and did without or used fewer highly specialized people to perform the needed functions.

A relatively new problem in relation to IT in MNCs has emerged recently, although symptoms have been observed for many years. This is the steady loss, through retirement and lack of renewal, of many of the skills that make up the range of IT capabilities in a manufacturing firm. This was brought home to me abruptly a few months ago, when one of my clients—a top 100 company—offered early retirement (55 years old and up) with a benefits package that many people felt they could not refuse. The resulting exodus of highly specialized manufacturing and engineering people (both inside and outside formal R&D) was a profound shock and has left a number of critical holes in the range of ITs in several operating divisions and corporate staffs. A rationale might be that it gives younger people the opportunity to advance, and this is true; but it is also true that for many of these

---

[*]Members of the board of directors and management of Narragansett Capital Corporation—one of the largest of the publicly held small business investment companies (SBICs) in the 1970s and 1980s, of which I was a director for 25 years. Narragansett's early investment strategy is described in "How I'm Deconglomerating the Conglomerates" by Royal Little, founder and former chairman of Narragansett in *Fortune*, July 16, 1979.

early retirees (as well as "on schedule" retirees), the company had not trained or recruited sufficient talent to back them up and replace them.

These three disparate examples illustrate the widespread nature of the IT phenomenon and the different circumstances under which lack of adequate IT capabilities can occur or persist. All three relate to the constant institution- and capability-building required to develop and maintain a productive infrastructure to back up the R&D/innovation efforts of a company, an industry, a sector, or a country.

8.3.4 Why Is Imbedded Technology Important to the U.S. Economy?

It Is a Necessary Condition for Successful Technological Innovation.
"Imbedded," "nonobvious," "infrastructure," "know-how" technology is a dominant factor in the U.S. past and current position of technological and commercial leadership in many fields. It represents skills, knowledge, people, and organizational capabilities which support industrial innovation. This is not to say that formal R&D has not been and will not continue to be a major factor in this success and that published and patented technologies are not critical to our leadership position. IT does, however, constitute a necessary condition for this leadership position and without it the economy would be unable to fulfill the potential provided by R&D results, including patents.

As indicated earlier, the complete innovation process involves the application of the capabilities included in IT at all stages of R&D/I. The arts involved in actually designing and producing a product predicted by theory or getting a new patented process to work efficiently and economically are generally not transmitted through textbook instruction or formal algorithms and paradigms. These are learned behaviors which can be transmitted by various means but which require time, personal interaction, and much trial and error.

And herein lies the essence of IT's role in the innovation process. For example, in the coming shake-out in the computer, peripheral and software field, dozens and even hundreds of "marginal" operators will fail, due in part to lack of essential IT skills, if the growth patterns of many other markets for new technologies apply in this case.

For another example, we have recently done a "fish or cut bait" study for one of the hundreds of firms which have plunged into the water treatment market with new devices, systems, materials, and services. Few have made any money and the vast majority will fail or abandon that line of business over the next few years. The failures will not be all or even largely due to lack of clever technology and R&D-based ideas for treating water. The failures will come from an inability to produce and deliver ideas and technology in a temporal/economic mode that the market will accept. Clever means of removing pollutants from water, based on good physics, chemistry, and biology (which many of the current products and services in this market incorporate) do not make a sustainable business if the embodiment of these clever technology ideas is in poorly designed, tooled, engineered, and produced products and services that fail in the factory, the account books, the marketplace, or the user's plant. The specific case investigated was overloaded with R&D talent and very short indeed, of people who embodied necessary IT skills and capabilities.

Although people from particular schools may view such shake-outs as merely one of the necessary side effects of a free market system, they involve tremendous wastes and misapplications of scarce resources—people, time, money, materials, and energy (both physical and psychic). Such losses should be segregated in the national accounts as "losses due to innovation" or perhaps "losses due to inadequate innovation capabilities." In some of our studies of adoption of new technology by U.S. industry (let alone studies of technology adoption by LDCs), the lack of adequate IT related to installing, adapting, using, maintaining, and improving capital equipment has been brought home countless times. For example, the delivery to a user of a new numerically controlled machine tool, once thought by some analysts to be the culmination of the innovation process, is actually only another process stage in getting a new concept and technology from the R&D or engineering laboratory into routine use. Our detailed, "microdynamic" studies of such introductions into users' plants (Ettlie and Rubenstein, POMRAD doc. no. 78/65) clearly illustrate the role of IT in integrating the new equipment or technology into the technoeconomic–social matrix of production in the user firm and the many barriers that must be overcome.

IT's Crucial Role in the Innovation Process Has Not Been Adequately Recognized by Top Managers and Policymakers. The main point in the previous subsection is that R&D alone is not *sufficient* for producing innovations and assuring their application or implementation. The many other factors included under the heading of IT are necessary to each stage of the overall process. If there were a choice between support only for R&D or IT, a good argument could be made that an adequate range of ITs without much formal R&D could keep us going more effectively in the short (and perhaps intermediate) run than increased R&D without broad and deep IT capabilities. In the longer run, of course, our technological barrel, from which many ITs are refilled continually, would empty. Without a vigorous and adequate level of R&D, our productive and competitive systems would eventually fail.

This short-run–long-run threshold varies greatly among industries and sectors. In military and electronic technology the crossover point may occur (and frequently does) in months, while in others the rate of significant technological change is so slow that a company, industry, or country can coast for decades on applications of old technology through clever use of IT, as long as they retain their skilled people and groups.

The main point in this subsection is that many top managers in government and industry and policymakers in all sectors do not recognize this situation or, if they recognize it in abstract, do not feel that they can or should do much about it. It is relatively easy to decide to set up or dismantle R&D programs, laboratories, or projects because they are so distinct and visible. The response and recovery time lags in making these changes have their own penalties, but at least decisionmakers feel that they can manipulate the organizations, projects, and budgets representing

formal R&D programs. They are much less comfortable with the nebulous, ubiquitous, and sometimes arcane entities that embody IT. How many economists, political scientists, legislators, lawyers, or bankers without direct "hands-on" experience
in the factory have any feeling for what is involved in developing capabilities in
such apparently mundane fields as welding, forging, quality control, patternmak-
ing, coating, lubrication, and heat treatment.

Except through literature and other art forms which depict the apprenticeship
process and the years it takes to train a craftsperson, few nonmanufacturing or nonengineering decisionmakers know of the tremendous investment U.S. (and all
other) industry has in such people and how long it takes to build teams needed to
provide the necessary IT parts of the innovation process, as well as the routine production process. These nonglamorous areas receive little top management attention and virtually none from public policymakers except when a brief flurry of
concern is expressed for a specific industry sector that is threatened by foreign
competition.

At the point when such flurries occur, it is often too late to do much about the
technological base of the industry, which may have eroded both at the upstream
(R&D) end of the innovation process as well as the many impact points for IT
all along the chain. As a strong advocate for decades of the need to maintain
and strengthen our formal R&D capabilities, I cannot take the position that IT
capabilities should be developed, maintained, and expanded at the expense of
R&D. However, I have been concerned that too much attention of policymakers and
top managers has been focused on the glamorous front end of the R&D/innovation
process and that a proper balance needs to be achieved to protect our overall
innovative capabilities. This imbalance is starkly illustrated by the situation in
a multi-billion-dollar U.S. manufacturing company where only one member of
corporate top management had any appreciation of the fact that the company's
productivity and economic bases were being eroded over a period of years by
neglect of manufacturing capabilities, facilities, and equipment. Finally, he was
able to convince a new chief executive officer of the need for emergency injection
of capital funds in this direction, after it became clear that the company was losing
its market position and profit margins through neglect of IT and its manifestations
in equipment, facilities, and human resources.

***Effective IT Can Make a Crucial Difference in U.S. Ability to Compete in
Both Domestic and International Markets.***   The United States still maintains
a technological and commercial lead in many high-technology markets such as
aircraft, computers, military hardware, and some scientific instruments. It has
been losing ground in other product lines and markets where high science or
"breakthrough" technology is not common, however, and where product features,
reliability and price dominate market share. In this latter case, clever design, human
engineering, low-cost manufacturing methods and tooling, easy mantainability,
reliability, serviceability, and other "nonscientific" features of a product influence
success in the international competitive market. Because some Japanese products
have features not matched by any U.S. products, and in spite of higher prices,

Americans buy many high-technology products from Japan. Few of these features appear to be the product of high-level R&D, but appear to have the advantages that come from competent engineering and design, after the formal R&D on the product has been done.

Despite their preferences for U.S.-made production equipment, many U.S. manufacturers are turning increasingly to foreign original equipment makers (OEMs) to buy machinery that may not have a significant price advantage but that has reliability, features, and delivery times that they need. Furthermore, some of the purchasers feel that they can depend more on follow-up service, warranty, and spare parts from some foreign OEMs, despite the geographical distances, than they can from some domestic producers.

International technological competitiveness has been a major issue between the United States and other countries for some time, swinging between the extremes of "technological gap," "technological imperialism by U.S.-based MNCs," to "export of jobs" and the "Japanese or Asian menace" to many of our basic industries. I have no intention of offering a single-factor explanation of the deteriorating competitive position of many U.S. products in the world market (including our own domestic market, where U.S. and foreign products meet head-to-head), let alone attribute specific or general declines in our competitive position to superior IT capabilities of our competitors. It is quite possible that erosion of some of our non-R&D technological capabilities are contributing to this situation, however.

One clue to this economic threat to the United States is the situation we find in our work in some of the LDCs where U.S. MNCs have subsidiaries or branches. Although the U.S. system of "technology transfer" has developed to a high degree and although product and process standardization is almost a religion, many of these standards—both in ways of doing things and the resulting products—deviate significantly from the U.S.-side equivalents. This can be largely attributed to the lack of certain IT capabilities in the LDC plants and, by inference, to the lack of sufficient depth in those particular capabilities in the parent organization to provide assistance to the remote plants.

We have also observed the difficulty of transferring certain IT skills internally between domestic operating divisions or between corporate engineering/manufacturing staffs and operating divisions or plants. This is easily understood, once we get beyond the textbook or polemic definitions of technology transfer and begin to understand the tedious, time-consuming, nitty-gritty aspects of actually transferring the skills needed to "move" technology from one place, group, company, or country to another, and the compounding of the transfer problems by cultural, social, and psychological factors. Our work in international technology transfer, for example, has yielded many lists of barriers to what sometimes appear to be simple transactions which both parties to a technology transfer agreement appear to want to consummate.

Such barriers were identified at a meeting of the United Nations Economic Commission for Europe in Geneva, July 1975, whose topic was "The Management of the Transfer of Technology within Industrial Co-operation." An abstract of these issues from papers and discussions presented by representatives of a dozen U.S.

corporations and their West and East European counterparts (Rubenstein, 1976) shows these potential barriers:

1. Cultural differences among sources and recipient countries.
2. Appropriateness to recipient of technology proposed for transfer.
3. Local conditions and recipient characteristics affecting productivity.
4. Extent of training associated with the transfer.
5. People-dependent nature of technology transfer.
6. Transfer depends on recipient's technical competence.
7. Differences between source and recipient in standards, components, and materials.
8. Need to study thoroughly partner's technology and methods of operation.
9. Preference of many Western companies for limited agreements.
10. Is pricing to be based on cost or value?
11. Selling products versus selling technology.
12. Differing concepts of profit.
13. Role of capital costs in the transfer.
14. Need for mutual benefits to both partners.
15. Buy back and barter versus cash payments.
16. Protection of source's R&D investment over a long period.
17. Role of intermediate organizations such as agents and trading companies.
18. Absence of common commercial traditions.
19. Theory versus practice in technology transfer.
20. Transfer is more complex than just licensing.
21. Unbundling of software from hardware.
22. Start-up technology is needed and often not written down and is people dependent.
23. Ability of recipient to shop around for alternative source companies.
24. Means of payment for technology.
25. Rights of recipient to relicense third parties.
26. Differences in internal structure of source and recipient countries (organizational, political, and economic).

8.3.5 Some Examples of Imbedded Technology (IT)

Supporting Infrastructure for R&D. One traditional supporting technology in the "wet processes" (chemistry, biology) is often listed as glass blowing. In the mechanical industries, model making or drawing (creative drawing, that is, rather than mere drafting and detailing) are key ITs. In electronics, basic supporting ITs vary, but breadboarding, maskmaking, wiring, and instrument modification have played prominent roles. Of course, some of these and many other ITs have been made obsolete (at least theoretically and for the future) by machines that perform many of the tasks that skilled artisans and technicians performed traditionally, and by computer systems that can "outdraw" and "outvisualize" the artists who often went under the unromantic job titles of "draftsmen."

Such IT capabilities in the R&D process are only part of the story, however, and too much focusing on them can distract attention from the many other not-so-obvious skills and capabilities that are often shared by "professionals" (university graduates or, in some industries, advanced degree people) and technicians. Some examples include crystal growing, cryogenics, heat treating, welding, making biological cultures quickly and "cleanly," and removing as much of the "noise" as possible from the pictures achieved through electron microscopy by almost fanatic maintenance and cleanliness of the equipment.

In some labs in some companies, in some industries, the work of technicians in performing some of the tasks implied by these IT specialties and the products of these ITs, are considered more important for the day-to-day maintenance of the products and processes for which R&D is responsible, than academic theories or scientific experimentation.

This does not imply that all functions that are performed by technicians (e.g., animal care, cleaning up experimental equipment, taking routine data) are to be included in the concept of IT and focused on as critical to our innovative capabilities as a nation (although some people might take that position). However, many of the technical chores performed in our R&D laboratories, which are performed by technicians, are not the product of university education, are not found in the scientific and technical literature, and are in danger of costing us dearly as the number and quality of skilled practitioners decline. The dropping of a lot of the traditional laboratory "arts" from science and engineering curricula over the past two to three decades suggests that their importance may not be recognized by the academic community. The failure to train adequate numbers and levels of people to perform them in industry suggests that their importance may not be recognized by many of the managers of organizations that do R&D.

Two particular ITs illustrate a wide range of knowledge, skills, capabilities, and behavioral implications. One is a single action done by a single engineer engaged in development of an engine. This example was suggested by an incident encountered recently in a company developing engines where a manager asked one of the engineers "did you torque the engine?" and received a blank stare. The situation involved a fine art, which required a "feeling" for the amount of torque placed on the engine in adjusting it and about which no manual had been written to communicate that feel. Perhaps a parallel is the "learn by doing" technique used in medical school when the preceptor tries to teach the medical student how to listen to a heartbeat.

The other example is at a downstream phase of the R&D process and involves start-up know-how, also involving a whole series of "feelings" and "judgments" as to whether things are "going right" or are likely to fail or explode. Although this is often looked on as part of the manufacturing responsibility, it is a necessary part of successful innovation for new processes and the products they are intended to produce. Many projects (products or processes) which are considered successful by R&D fail miserably in the start-up phase through unexpected costs, delays, material wastage, equipment breakdowns, extended learning time for operators, below-quality production, and other undesirable effects. Some of these problems and failures may be due to poor original design. Others are due to specific mistakes by start-up people (typically engineers, supported by manufacturing people or vice versa). Still others are due to lack of experience with or attention to the specific characteristics of the new product or process being started up.

Manufacturing Technology. Probably, if a "complete" list of important ITs were compiled which might make a major difference in a company's productivity, costs, product quality, international technological competitiveness, and other factors, those related to manufacturing might dominate the list. When we focus on factors contributing to the ultimate success of technological innovations, such ITs in manufacturing might also figure prominently.

A number of manufacturing ITs related to the downstream phases of the overall R&D/innovation process have been mentioned in earlier parts of this section and will be mentioned in following ones. Again, without attempting an enumeration or a taxonomy, here are some examples that appear important in the downstream parts of R&D/I which involve manufacturing and the commercialization of products and processes.

Automation
Coating
Computer-aided manufacturing
Control systems
Corrosion
Cost control
Design review
Forging
Foundry techniques
Heat treating
Heating and ventilating
Inventory control
Lubrication
Machine design
Materials handling
Materials management
Metrology
Nondestructive testing
Packaging
Plant construction

Plastic molding

Pollution control

Quality control

Reliability

Robotics

Standards development

Tool and die making

Welding

In many plants and companies, some of these ITs are imbedded in the experience or skills of one individual. Others are represented by a formal group or an informal aggregation of individual skills into an organizational capability. Some of them are embodied in college graduates, such as scientists (rare, except in some fields) or engineers who work in individual plants or corporate manufacturing staffs. And still others, perhaps the majority, involve technicians (nongraduates) either working alone or supporting engineers or scientists.

One situation, not uncommon in R&D-performing companies, is that the particular individuals or small groups which embody these kinds of ITs are the subject of intense competition among individual plants, R&D, and other corporate staffs. A recent shift of half a dozen control metallurgists from corporate R&D to plant engineering in a local company left R&D short of such talent and did not fully satisfy either the needs of the plants who received them (because the former R&D metallurgists were not considered practical and experienced) or the individuals themselves (because of their removal from their preferred career paths and the low level of use the plant managers made of their technical talents).

Other IT. There are many other skills, capabilities, and bits of know-how that contribute to successful innovation in the industrial firm. Some of them can hardly be called "technological" in the sense that they are based on science or "hard" technology. They are as essential to success in innovation as many of the ITs mentioned under R&D and manufacturing, however. Some of them in fact are management functions, which, when associated with the R&D/innovation process, can become highly specialized.

Among this set of ITs, procurement/acquisition (terminology varies in the field) and contract administration loom large for many programs and projects. Getting the right materials from the right sources at the right price and time and according to requirements can make the difference between successful and unsuccessful innovation. This is particularly true where value added is a small percentage of total product cost and most of the cost represents purchased materials, components, equipment, and services. The not infrequent conflict between R&D and purchasing people can be quite destructive of the time, cost, or quality performance on a project for an innovative product. Therefore, many organizations pay particular attention to recruiting, training, and maintaining an effective procurement staff specializing in support of R&D/innovation. Contract administration and all that it implies for meeting cost and schedule targets is also a key element in successful innovation. In some contexts, contract administration either includes or ties into project management and related functions.

Although the formal, legal aspects of patent management (writing up, searching, filing, appealing, litigation, and licensing) are often handled by legal staff, at least in large companies, there are other elements of patent management which are more subtle and which might fall into the category of IT. These have to do with the intuitive, experienced-based aspects of patent management, such as searching for patentable items, knowing how to encourage people to disclose and file and follow up, sorting out the wheat from the chaff, and other intangibles that are hard to codify and teach in law school.

Although some of these management skills appear to stretch the "technological" connotation of IT, they figure prominently in the actual transfer of technology among countries and companies and even within companies. For example, while this section was being written, a group of us were engaged in a series of field experiments on "key communicators in R&D." We visited one of our field sites to discuss experimental treatments for one aspect of their R&D/innovation process — the licensing or sale of technology. This very large company had compiled a list of approximately 100 items of know-how, process technology, and other items which were not proprietary mainstream products, and management had been exploring the possibility of selling, licensing, or otherwise making money on some of this technology. When their explorations yielded the fact that most of the potentially profitable technology exchanges would require substantial involvement of their own people, including managers, for extended periods, they decided to shelve the whole idea.

Few transfers of technology beyond simple provisions of new parts or equipment embodying changes can be completed by "mail" or by "arm's-length" exchanges. They generally require "hands-on" instruction and cooperation in installation, start-up, breaking in, adapting, improving, training for use and maintenance, and going through the other activities that are required for effective and "full" transfer of technology. Given this situation, many of the "softer" ITs play prominent roles in the transfer aspects of the overall R&D/innovation process, whether within a given company or across companies and national boundaries.

8.3.6 What Indicators Can Be Used to Measure and Monitor Imbedded Technology (IT)?

For several years a small group of colleagues and I have been exploring the possibilities and problems of identifying and using indicators for monitoring and evaluating the many stages of the overall R&D/innovation process, guided by the general flow model of Figure 8-8. We have used this to locate and identify barriers and facilitators to the R&D/I process at various stages as it progresses from laboratory to application. (*Note:* Although the "usual" flow of events for a new or significantly improved product or process generally involves the sequence of activities depicted in the model of Figure 8-8—R&D, transformation, implementation—many pieces of technology do not originate in formal R&D programs or laboratories. Also, many of them involve later starting points and cycling back through R&D at later stages of development, when technical problems arise that downstream inventors cannot resolve without formal R&D inputs.)

In our work on R&D/I indicators, particular attention has been paid to mainstream variables and entities involved in the flow of a project or program. In the course of these explorations, however, we have encountered many potential indicators and measures for the less obvious aspects of the R&D/I process which come under the heading of IT. As an illustration of how this "spin-off" of our indicators works, Figure 8-9 presents some examples from a research study of indicators for small business R&D capabilities.

In general, the process of identifying, measuring, and monitoring or evaluating ITs as both capability and output (ideas, knowledge, techniques) indicators for the R&D/innovation process is much more difficult than the kinds of measures of mainstream R&D currently included in the National Science Board's "Science Indicator" series. Aside from the usual problems of surveying large numbers of companies and getting reliable data that already exist, measuring ITs has added complications. Few companies have any records that directly reflect the kind of indicators needed to assess ITs. In addition, few managers think of their non-R&D resources involved in the overall R&D/innovation process in separable terms that would lend themselves to identifying and measuring indicators for IT. For example, of the people and knowledge in the company devoted to welding or

---

Factor:

(1) Marketing (of R&D) sophistication	(a) Familiarity with government bid procedures
	(b) Bid strategy
	(c) Information system for learning about opportunities (e.g., requests for proposals)
(2) Technical information	(a) Technical journals and reference sources subscribed to
	(b) Procedure and facilities for searching for technical information
(3) R&D facilities and equipment	(a) Expenditures and inventory of equipment and instruments (E&I)
	(b) Replacement and updating of E&I
	(c) Ability to maintain and adapt E&I
(4) Track record in performance	(a) On-time and on-cost delivery record
	(b) Returns, allowances, and disputes on specifications and quality of delivered items

---

Figure 8-9. Some illustrative indicators of IT in small business. (*Source:* Rubenstein, Geisler, and Thompson, 1977.)

foundry operations (in many cases hundreds of people and multiple facilities) only a small proportion are relevant to the success of new or significantly improved products or processes which are generally identified as "innovations. " An attempt to identify such R&D/I-related ITs would require a new approach to "human resources accounting"—a function that is growing in industry, but slowly.

Despite these difficulties, efforts should be made to include indicators of IT in the measurement of national resource inputs for R&D/innovation and for assessing the outputs of that process. The rationale is not mere "tidiness" in the sense that our current national accounts are far from complete in representing the R&D/I process. The practical purpose is policymaking aimed at maintaining and improving these valuable technological assets.

8.3.7 Use of IT Indicators by Government Procurement People, Company Management, and Federal Policymakers

Given that a manageable set of indicators for IT can be developed and that credible data can be gathered on them on a regular basis, then what use can be made of them for what purposes? Although, as in the case of many indicators, new uses may emerge after they are developed, there are several immediate needs that come to mind, and three groups of potential users have been selected to illustrate the possibilities. Usage by the third group—federal policymakers (in the legislative and executive branches)—is discussed in more detail in Section 8.3.8 on public policy implications of IT. The three illustrative user groups and possible uses to which they might put IT indicators are the following:

1. *Procurement People.* Recently, there has been increasing interest in the Department of Defense, the largest federal purchaser of technological products and services, in a concept called "past performance," which has been used in procurement for many years in widely varying ways and under different names. It involves an attempt to provide indicators of contractor capabilities in a number of key areas as aids in *selecting* contractors for a particular project or program and *monitoring* performance once a contract has been awarded. Possible applications to these two important functions are given in Figure 8-10.

2. *Company Management.* There are many uses to which an alert company management might put improved indicators of their IT resources. An important one is in maintaining its technological capabilities through recruitment, training, upgrading, and replacement of people with important and scarce IT skills. Another is in maintaining an adequate capability level in critical ITs and forecasting or anticipating threats to that capability from turnover, promotions, cutbacks, transfers, retirements, mergers and acquisitions, obsolescence, or inadequate support (funds, facilities, equipment, supplies, and services).

The kinds of lessons learned by the two companies discussed earlier—the cases of forced abandonment of a technical venture due in part to inadequate IT and the unexpected loss of key IT capabilities through early retirements—might serve as

A. APPLICATION TO SOURCE SELECTION

Such indicators should be useful to procurement and contracting specialists in procurement offices involved in the source selection process in the following ways:

- Suggesting new indicators of past performances as guides to likely future performance.
- Providing measures to use for those indicators that are considered appropriate.
- Providing a systematic framework within which past performance can be evaluated and weighed in with other factors.
- Reconciling formal or intended methods of weighing past performance with actual methods used, especially the wide variance among source selection groups and at different times and places and relating these, in turn, with procurement regulations.
- Where decisionmakers in the source selection procedure are relatively *satisfied* with their current indicators of past performance and the weights assigned to them, providing an opportunity to reexamine their current practice and perhaps augment the indicators they use.
- Where there is general *dissatisfaction* about the current methods of using past performance or where they are hardly used at all, providing a basis for introducing such factors. The use/nonuse of past performance in source selection varies widely among federal agencies and among units within specific agencies.

B. APPLICATION TO CONTRACT MONITORING

The indicators should be useful to program monitors and technical monitors in system program offices in the following ways:

- As a supplement to existing monitoring systems that focus on cost, time, and milestones.
- Where certain weak points, hot spots, or potential flaws have been identified in the source selection process or in investigations of past performances.
- To focus attention on key aspects of contractor present and likely future performances through "exception principle" management or monitoring.
- Where current monitoring systems are inadequate, to substitute an improved one.
- Providing inputs to the next cycle of contracting or monitoring for follow-on or new contracts. In this way, monitoring can provide its own "past performance" data for future source selection decisions.
- To provide realistic guidance as to what can be expected from the contractor, within the formal limits of contract terms.

Figure 8-10. Potential application of indicators for IT to procurement/acquisition of high-technology items. (*Source:* C.W.N. Thompson, A.H. Rubenstein, E. Geisler, 1981.)

motivators to many managers to pay more attention to ITs than they currently do. One technique that made a stir some years ago, but that has made little headway in industry recently, might be revived in the context of IT. This is the "skills inventory" of important technical skills—R&D and other—that are represented by the company's current workforce. The lack of adoption and widespread use of this technique is due to cost and the political fallout from gathering and maintaining such information beyond the superficial level of merely listing (in files or on a computer) education, previous job titles, and, in some cases, formal assignments in the company.

Most of the attempts at such skill inventories did not include the depth of information needed for a true skills inventory even for R&D, let alone for IT and the overall R&D/innovation process. It may be worthwhile for company management to reexamine the costs–benefits of such an information system for forecasting, deploying, and maintaining its technological human resources.

3. *Federal Policymakers.* Although specific public policy implications are discussed in more detail in the next section, brief examples are given here of some possible uses of IT indicators by legislators and executive agency policymakers (in addition to the uses in the procurement/acquisition process). Perhaps one of the major kinds of policy actions which the federal government might take with respect to ITs is in the training area, to be sure that the supply and quality of such national assets are at proper levels. Another key area involves the exportation of know-how and other manifestations of IT through technology transfer and technical assistance agreements. A third might involve specific incentive programs to enhance IT development in industrial firms and throughout the supporting infrastructure (consulting firms and trade associations which have technical assistance and R&D organizations). Finally, the federal government, which is already doing a great deal in the area of dissemination of IT information (notably through mechanisms such as NASA's Technology Utilization Program, described in several of the references in Appendix B), might do even more of this to enhance industry's IT resources.

8.3.8 Some Public Policy Implications of Imbedded Technology and Some Suggestions for Action

Introduction. This section undertakes to explore possible policy options for the federal government with respect to IT and develops some approaches that might lead to specific policy options. Consistent with this objective, Figure 8-11 presents some suggested actions which the legislative and executive branches of the federal government might undertake to enhance the U.S. posture with respect to IT. Although many of these suggestions appear reasonable and feasible to me, each requires careful analysis and experimental evaluation before it is put into practice or recommended for legislative or executive action. The reason for this caution is that many programs intended to provide incentives to the innovation process in the United States and other countries have fallen flat because they were not carefully designed, tested, evaluated, and adapted to the specific circumstances involved (differences between industries, countries, companies, technologies, and people involved in the R&D/I process).

The actions in Figure 8-11 can be grouped into a number of more general categories, which are nothing more than a convenience for discussion purposes, since almost all cut across technical fields and legislative committee or executive agency jurisdictions. In no sense are the items of Figure 8-11 and the categories A–E

- Share the cost of training of technicians and other IT specialists.
- Provide tax credits for labor savings on IT investments.
- Increase sharing of IT capabilities by federal labs.
- Ease licensing and waiver of ITs covered by government-owned patents.
- Increase access to federal laboratories for testing of products and materials by industry.
- Stimulate action on technology transfer by federal agencies which give primarily lip service to it (despite policy directives to participate).
- Include development of IT capabilities and products explicitly as part of federally financed or federally performed R&D.
- Subsidize trade magazines that specialize in IT, in addition to professional journals that specialize in scientific and technical information (STI).
- Expand the technology transfer agent programs that were tried by various agencies (e.g., NSF, NASA, Department of Commerce) and design them so that they will be more self-sustaining and more effective.
- Count IT spin-off as part of the social and economic benefits explicitly when evaluating cost effectiveness of a government-sponsored R&D program.
- Identify and target needs for IT in different fields.
- Use government influence (through regulations, procurement, tax laws, direct support) to encourage investment in IT by industry.
- Provide support for use of retired executives to transfer IT skills and knowledge and to provide formal training.
- Conduct or support technical audits of IT in individual companies and industries on a confidential basis.
- Provide incentives to larger firms to assist small firms (including some of their supplier firms) in developing independent IT capabilities.
- Support research and experimentation on IT.
- Direct more attention to the IT aspect of the R&D/innovation process, in addition to the current heavy focus on the R&D aspects.
- Consider IT skills when establishing immigration preferences.
- Consider IT skills when considering export and technology transfer regulations and individual export contracts.

Figure 8-11. Some suggested government actions to enhance the U.S. position with respect to IT. (*Source*: Studies and ideas from member of the research-on-research community [students of the R&D/innovation process], including the Northwestern group, plus staff members of the National Science Foundation, NASA, Department of Commerce, practitioners of the art of R&D management in the Industrial Research Institute and outside it, and members of the various advisory subcommittees working on two major studies of innovation [Domestic Policy Review and Joint Economic Committee of Congress].)

in Figure 8-12 intended as a complete or comprehensive program of government incentives to innovation. There have been and currently are dozens of such lists being drawn up in connection with the several waves of activities on the status of technological innovation since the early 1970s, and the continuing programs of such organizations as NSF, NASA, Departments of Defense and Commerce, other executive agencies, Office of Technology Assessment (OTA), and the Joint Economic Committee (JEC). Most lists are redundant and brand new ideas are rare.

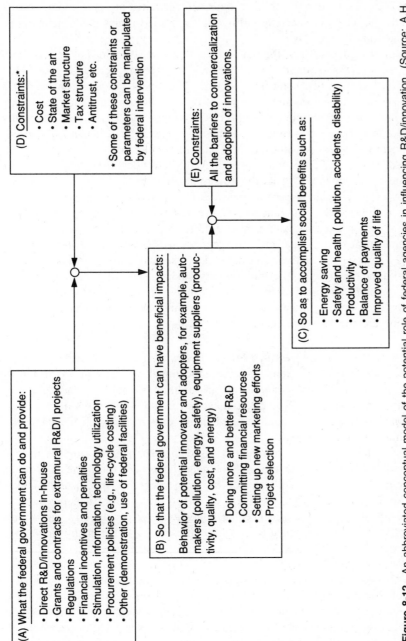

Figure 8-12. An abbreviated conceptual model of the potential role of federal agencies in influencing R&D/innovation. (Source: A.H. Rubenstein and J.E. Ettlie, "Innovation Among Suppliers to Automobile Manufacturers: An Exploratory Study of Barriers and Facilitators," *R&D Management,* Vol. 9, No. 2, pp. 65–76, Feb. 1979.

What is also rare are systematic and credible evaluation and application of some of these ideas—for example, some of the frequently cited incentives to innovation. That is the big gap that needs to be filled in the R&D/I policy area, including that part related to ITs.

Two more figures address the requirements for and difficulties of actually influencing the R&D/innovation process through government action. Figure 8-12 is a simplified version of a general model of government influence on the R&D/I process which we used to guide a series of case studies of innovation in the automotive industry (Rubenstein and Ettlie, 1979). Figure 8-13 is a more generalized version of that model for a series of studies of industrial management responses to government incentives for innovation in half a dozen countries (Rubenstein et al., POMRAD doc. no. 75/95; Watkins et al., POMRAD doc. no. 76/9).

Figure 8-12 emphasizes what the government can possibly do to influence the behavior of firms with respect to innovation. Figure 8-13 focuses on the long, tortuous path between formulation of a government incentive or regulation related to the innovation process and the many informational, perceptual, and behavioral stages that must occur if the incentive/regulation is to have a beneficial impact. It is not an optimistic picture, and deliberately so, since the influence process is far from direct or simple. The multicountry studies guided by this second model focused on the "information–perception–evaluation" stages (boxes 2, 3, and 4) which precede any significant decisions or commitments of resources to actual innovation projects as a consequence of government action.

Government's Role in Training for IT. Below the university level, government (federal, state, and local) dominates the educational and training system in this country. Even at the university level, in addition to direct influence through state universities and colleges, government has a number of ways of influencing educational content and method throughout the system via such mechanisms as research and training grants.

These methods of influence might be used to focus attention on the need for more and better training in the arts included under IT; for the establishment, improvement, and stable maintenance of those programs specifically aimed at particular IT skills; and for provision of direct support for such programs. In addition to attempting to influence the educational sector directly, government agencies might provide direct training facilities through the many government laboratories and other technologically based facilities they operate.

This is not a plea for still another "title X" of an education act addressed to still another special-interest group. Rather, it is a suggestion that the status of our educational and training institutions be examined in the light of IT needs for the future and that influence be brought to bear, through a number of means including direct funding, to assure adequate attention to this necessary and important area of skills. Although it is true that many (perhaps most) individual "skill ITs" need to be imbedded in people's heads and hands over long periods (decades for some skills), an early start in secondary school or early in the college years could point

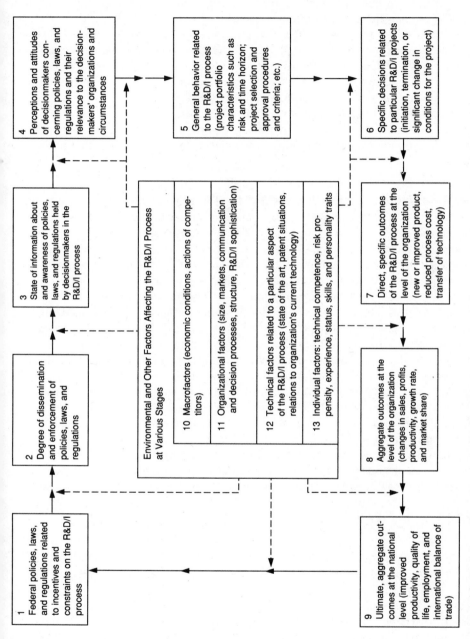

Figure 8-13. A flow model of the potential effects of federal government action and other factors on the R&D/innovation process. (Source: A.H. Rubenstein, Northwestern University, 7/16/73 [revised 8/27/73].)

many people in that direction and provide them with a career focus they might have been unaware of or misinformed about.

Because of the very nature of skill acquisition—the time lags, the role of motivation, the need for practice, and the need for face-to-face, hands-on instruction—I am not suggesting more "orientation courses," which through brief lectures, demonstrations, or films try to catch the imagination of students and then drop them at that point with no way to follow up and get rigorous training. Some critics suggest that such brief and superficial exposures of medical students to nonmainline specialties such as nutrition, gynecology, tropical diseases, or psychiatry can do more harm than good in preparing nonspecialists to cope with serious problems in those areas.

Potential Incentives for Stimulating IT. The general field of incentives for stimulating overall technological innovation has been and is currently being plowed extensively and it is unlikely that many radically new ideas will spring out of the many committees and study papers (including mine) which have been written. The big gap, as mentioned above, is in the many painstaking steps required to implement them, evaluate their impact, and improve them.

Several of the incentives regularly touted for innovation in general—for example, tax reforms, cost sharing, cooperative research, reduced regulation, and direct subsidies—might have an impact on stimulation and development of IT. Several other countries have been offering incentives (often not even identified as such) and programs/facilities for the development of the kinds of IT skills and capabilities listed earlier. Not much is known in this country about the details of such incentives, despite our several surveys of incentives to technological innovation (e.g. Rubenstein et al., POMRAD doc. no. 75/95; Watkins et al., POMRAD doc. no. 76/9; Allen et al., 1978; Pavitt et al., 1976) because of our focus on the mainline aspect of technological innovation, especially the upstream aspects (R&D).

Many people have called for a review of such incentive schemes in general, and attention should be paid to those that might influence ITs. I believe that a much better result may be achieved through some attention to stimulating ITs through incentives as compared to concentrating only on mainline R&D aspects of the R&D/I process.

Role of Government R&D in IT for Industry. As in the case of incentives for innovation and export policy, most attention has been focused on dramatic achievements and mainline R&D products when considering technology transfer, technology utilization, or spin-off from federal R&D programs. Several articles listed in Appendix B describe NASA, DOD, and CERN (the European Nuclear Reactor Center in Geneva) contributions of this nature. They also include additional kinds of outputs, many of which are clearly ITs in the context of this book. The difficulty has been that this attempted transfer has occurred (or been attempted) primarily on *paper*, rather than by *people*. A major theme in this book and other work on technology and skill transfer is that effective transfers occur primarily through people—either actual movement of people from one location to another, or intensive and continuous communication between sources and users. This is particularly true for IT, which is not covered well in textbooks, patents, scientific and technical information (STI), or other formal channels. Although several federal agencies, notably NASA and DOD, have had long-standing policies of encouraging dissemination of such information, these policies have often amounted to lip service and have included little access to people in federal laboratories as compared to paper (technical reports and summaries) and formal conferences and briefings.

My general suggestion is that these and other agencies review the actual behavioral patterns associated with such transfer activities and attempt to shift the emphasis from paper to people. This will require more actual access to federal labs for cooperative work, detailed demonstration, intensive training, and other modes of IT skill transfer through people.

Dissemination of IT Techniques and Information. Finally, improved methods should be sought for the focusing of attention on IT, the extraction and dissemination of information about it, research and experimentation on it, and other forms of communication which are generally used for mainline R&D (conferences, seminars, demonstration projects, and special handbooks and manuals). Figure 8-11 contains some specific suggestions in this direction.

8.3.9 Selected References to Material on Imbedded Technology and Related Subjects

There was no intention of doing a state-of-the-art literature search in connection with this discussion of IT. The items referenced in Appendix B, however, provide further insight into and detail on many of the issues discussed. They are presented here for further reading by policymakers and further analysis by researchers. The sample of material is rather eclectic, although there is a heavy representation of reports on the NASA Technology Utilization Program. Of all programs by federal agencies to provide technological inputs to industry, NASA's has been one of the most intensive and extensive and the best publicized. The small number of references to programs and specific contributions of other agencies should not be taken, however, as a measure of their activity in the field of imbedded technology and related matters, since no statistical sampling was involved. It may be an indication, however, of the low profile that many of them maintain in this field. Included in this list of references are some of the POMRAD[*] papers cited in the text.

---

[*]Program of Research on the Management of Research, Development and Innovation at Northwestern University.

8.4 TECHNICAL ENTREPRENEURSHIP INSIDE THE FIRM

A special approach to commercialization and "getting products out" has recently been the subject of revived interest by many large firms. It is a *revived* interest because, originally, many of the successful components of these firms—operating divisions, products, and whole product lines—were successful in this mode. This approach—technical entrepreneurship inside the firm—is very old and its origins in a particular firm may predate the advent of many of the current managers.

Advocates of reviving or initiating such an approach in the large, decentralized firm view it as a way of bringing into the large firm the vigor, focus, dedication, speed, and enthusiasm with which the small entrepreneurial firm seems to act. However, the rate of success of such "implantation" or "rejuvenation" in most of our very large firms is not very encouraging. Many factors act to inhibit the initiation, development, maintenance, and success of such ventures.

In this section, we look at some of these factors and raise some issues that must be dealt with or at least faced by top management if they want to reap the benefits of this approach to commercialization. I call the action units of this approach V/E or venture/entrepreneurial projects. V/E projects are usually identified a little way downstream from the original research or even advanced development of a technical idea. Unless the R&D project was specifically undertaken to attack a specific urgent and important market threat, need, or opportunity, it is not likely to be identified and organized as a V/E project until some signs of potential commercialization are observed and/or until the need for it becomes very urgent and/or important. Another precondition, discussed in more detail below, is that it does not fit comfortably, or at all, into the firm's existing R&D/innovation or business structure.

Figure 8-14 lists a few of the factors that make it difficult for the existing, "normal" organizational structure and mode of operation to accommodate or encourage technical entrepreneurship or venturing in the large decentralized firm. Figure 8-14 suggests some factors making it difficult and unlikely that individual operating divisions, with certain exceptions, will successfully launch V/E projects.

Barriers to Technology-Based Venturing (TBV) by Individual Operating Divisions:

- Limited charter/mission/territory—for example, "confine yourselves to big blue ones"
- Narrowness of outlook and focus; overspecialization
- Relative impermeability to technology transfer from CRL or outside sources

Barriers to TBV by Corporate Research Lab (CRL)

- Often CRL researchers are amateurs in production, marketing, finance, and/or business in general
- Less consciousness of time and cost criteria for hand-off and start-up
- "Good enough" product specs generally not considered adequate
- Lack of access to markets and factory for introduction of new products

---

Figure 8-14. Barriers to technology-based venturing.

Some exceptions were mentioned earlier, in terms of very large and diversified divisions, high-tech divisions, and divisions led by managers who are themselves very entrepreneurial and technically trained. For most product divisions and their managers, however, their main mission, their traditional operating mode relative to technology projects, and their limited focus and resources make it unlikely for them to foster and nurture V/E projects.

Once a V/E project has developed sufficiently to show strong feasibility and market potential, many division managers will try to capture it and bring it into their divisions, subject to the "price" and other terms of the deal.

The corporate research lab (CRL), at first glance, appears in many companies to be the ideal place for V/E projects to flourish. This is certainly true for the upstream phases of a V/E project. The loose, tolerant, and sometimes leisurely atmosphere of the typical CRL provides the environment for getting such ventures going. However, that very atmosphere and its own intrinsic limitations prevent most CRLs from going very far with such projects and seeing them into successful commercialization (bottom of Figure 8-14).

The typical CRL (see Chapter 2 for a more extensive discussion of its limitations) generally is not very well staffed with people who have the necessary skills to take a technology far down the R&D/innovation life cycle and into the factory or the market. Deep experience with and high skill in such practical arts as production, marketing, finance, procurement, tooling, design-for-production, and design-for-costs are limited, almost by definition, in the CRL and not expected of its members.

Where they do in fact have those skills and try to exercise them, they are frequently viewed as a threat or nuisance by the operating divisions and admonished not to wave those skills around or to impinge on divisional and other staff groups' territories. CRL people are generally less cost conscious and more inclined to resist transferring or commercializing products when they have merely achieved "good enough" or minimally acceptable product or performance specs. Although the myth of CRL people seeking "perfection" may be a bit overdrawn, they do tend to hold onto a product until they have done as well as they can to improve it, including incorporating many good ideas that have little market value and that add to cost and time to produce.

In addition to lack of many of the skills needed for commercialization, most CRLs are organizationally cut off from direct access to plant or market and must negotiate with operating divisions to see their projects carried further along the life cycle. The barriers to this smooth transfer are many, and some are insurmountable in the kind of time frame that characterizes most V/E projects. They can also generate conflict, ill-will, future problems, and territorial struggles which can fester for years.

Such factors, affecting the ability of both operating divisions and CRLs to mount and succeed with V/E projects, makes it desirable and even necessary to have an alternative path for such efforts. Such alternatives *also* pose problems and encounter barriers but, if used with foresight, care, and vigor, can provide at least a temporary or an occasional "entrepreneurial climate" in the large firm.

Figure 8-15 lists some of the issues that must be faced if a serious effort is to be made at launching successful V/E projects in the large firm. Some of these issues are relevant for any large (or even medium-size) firm, whether it is decentralized or not. However, the particular features of technology organization in the decentralized firm, as discussed throughout this book, make such issues particularly relevant to the decentralized firm.

8.4.1 Recognizing the Need for Special Treatment of Some Technology Projects in the Firm

The very idea of "special treatment" sets the teeth of many people, especially division managers, on edge: "All my projects get special treatment; they just have to wait their turn for resources and my attention" or "All our technology projects are urgent and important or we wouldn't be doing them with the limited time and people that we have."

Such responses by division managers and their technology staffs (and by the managers of some CRLs) clearly indicate the need to consider some ways of identifying a limited number of urgent and important projects to be pulled out of the normal technology "queue" and given special attention and facilitating mechanisms.

V/E projects may represent radical departures from the firm's usual technology approaches or fields. They may reflect market opportunities that have a limited "window" in time. Or they may reflect serious short-term or longer-term threats to the firm's or a division's product lines or market positions. We have been involved with all these types of obvious needs for V/E projects but sometimes despair when the management people with the power to launch them and provide the necessary resources are unable or unwilling to recognize that such efforts do indeed require "special treatment."

---

1. Recognizing the need for special treatment of some technology projects in the firm
2. Organizational forms for V/E projects in the firm
3. Evaluating and financing projects in a venture or entrepreneurial (V/E) mode
4. Selecting personnel for V/E projects and phasing them back into the organization
5. V/E projects as a part of the corporate research lab
6. Feeding the V/E project into the operating division
7. The captive venture capital approach
8. The venture or new products division
9. The high-tech spin-off from the firm
10. Incentives and rewards for entrepreneurship and risk-taking
11. Sources of venture capital for high-tech projects

Figure 8-15. Some issues involved in technical entrepreneurship inside and outside the company.

Such projects may not fit into the regular structure or operating modes of the firm. They may not be well served by the usual decision and communication channels or fit with any existing group or function.

The required time perspective (speed, duration, time pattern of resource application) and the funding patterns may not fit the existing environment and may require continual exceptions and negotiations which can further aggravate already strained relations.

When such "nonfit" conditions prevail or can be anticipated, then it is the occasion to consider an alternative approach to overcoming such barriers and tailoring the *environment* to fit the *project*. This, of course, is considered heresy in many conservative and bureaucratic organizations and such special treatment may not even be considered, let alone tried and institutionalized.

If these inhibiting conditions prevail and there is little likelihood of changing them, then it might be better to avoid a new approach that is bound to cause conflict and disruption. If the manager or CEO who has just read this sentence accepts it, then he is certainly not a promising candidate to try to set up a special V/E system in his company. If, on the other hand, he sees the need for and importance of trying to make such a change and is willing to follow through and take the repercussions, then even this set of unpromising conditions might be a fallow ground for planting the seeds of technical entrepreneurship.

Not all important and urgent projects can or should be afforded the kind of special treatment I am advocating here. On the one hand, too much special treatment of too many projects would dilute their impact and the situation may deteriorate into an "all V/E project portfolio" where none is really afforded the necessary special treatment. In addition, most technology projects in the divisions or in the CRL have their own advocates and special conditions necessary for progress. Singling them out for extraordinary treatment may cause more problems than it solves and may also not move the project along faster or increase its chances of commercial success. So this leaves a very small number of potential V/E projects to be selected for special treatment and set up in a V/E mode. Those selected should certainly be "special" in terms of urgency, importance, and need for a tailored environment.

There is another category of project which can be a candidate for setting up as a V/E project and which needs, at least temporarily, the special attention and treatment that V/E projects get or should get. These are "nonventure" projects in the regular R&D/innovation portfolio which are important and promising but which have become hung up or stalled and may die a quiet death or, if they are completed, miss their market or the manufacturing window of opportunity.

They are seldom as urgent as "true" V/E projects when they are initiated and, when the suddenly urgent need for their results in the plant or market are recognized, they are generally found embedded in the general, slow-moving portfolio of a divisional or corporate lab or the downstream development and engineering groups.

Sometimes such projects just need an injection of new leadership, a signal from top management that they are important, or some additional resources. Others need the *full* V/E treatment in order to shake off the cobwebs surrounding them and get them moving along the life cycle path to implementation.

Faced with the division manager's comment that "all our projects are urgent" or the CRL's position that "this project is no more urgent or important than the other ones we are working on," management may hesitate to make too much fuss over a particular project unless it is indeed critical to the firm. If that decision is made, then V/E treatment can be applied and the kinds of special conditions discussed in this chapter can be considered for their relevance and potential impact on the project.

In one situation, we were able to identify a group of important stalled projects in a large company and get them going again by "encasing" them in a V/E environment under the leadership of a half dozen "entrepreneurial" engineers and other specialists recruited from among several thousand technical personnel in the company.

Labeling too many routine projects as "special" can be self-defeating in terms of actual results as well as the devaluing of "true" ventures. This device has to be used selectively and carefully as a means of getting stalled and lagging projects back on the path to successful completion.

8.4.2 Organizational Forms for V/E Projects in the Firm

Once the decision is made to tailor the environment for V/E projects, there is a wide range of organizational forms that can be used. Some are familiar and are used for other kinds of projects—construction, system installation, or reorganization.

The most familiar form is a special project group or task force which, it is understood, has a life limited to the focus of its efforts and is designed to phase itself out when the task is completed or terminated. Some V/E projects start off and even continue as "one-man shows" with only the project leader or manager assigned full time. He is given resources as he needs them—funds, facilities, people on temporary assignment—and he pursues the project within the general framework of the organization.

In many cases, he may have a small permanent cadre of people he has worked with and trusts who may continue with him for the life of the project. Such teams may even stay together after their project is phased out or passed on to other groups for the downstream stages (production, marketing, steady-state maintenance) and pick up another V/E project.

The most common treatment of V/E projects of high visibility and importance is the new organizational unit reporting to a high-level "champion" of new ventures. Often this is the CEO himself, or a board chairman who has given over day-to-day operations to a CEO and/or COO and concentrates on planning and special projects. Many V/E projects fit nicely into such an executive's personal portfolio. He can give it the time and support it needs, along with the prestige of his position to help it through the difficulties of making such projects successful.

A trend in the 1960s and 1970s was to set up a whole new department or division to serve as a generator and incubator of V/E projects. A common designation for such units is "greenhouse. " Many of these are set up entirely within the firm's financial and organizational structure as profit centers or cost centers. Others are set up as somewhat independent subsidiaries with a name such as "Company X Enterprises" or "X Development Company."

Other variations on these general organizational themes are feasible and some work better than others. The organizational form, just as the general organizational deployment of technology in the firm, can certainly affect the progress and ultimate success of a V/E project. However, particular organizational placement and form are seldom sufficient conditions for success and not very often completely necessary for success. There are many ways for V/E projects to succeed or fail and some of the other conditions mentioned below may be more critical to success than these.

8.4.3 Evaluating and Financing Projects in a V/E Mode

One of the most common reasons for failure of a V/E project to be completed or to achieve its full potential is the method of financing. The time scale and pattern of expenditures of most V/E projects do not resemble or fit well into the normal budgetary cycle for technology projects in the firm. The long lead-up to the technology budget (in many firms taking 6–10 months), the compressed and sometimes overlapping V/E expenditure patterns (production may start before R&D is completed), the need for extraordinary expenditures at unforeseen phases of the project (going after a suddenly available new piece of technology on the outside or acquiring a special skill grouping via consulting or recruiting) can wreak havoc with normal, orderly budget development and allocation procedures.

My experience with many high-tech start-ups suggests that, however much the financial backers insist on plans and pro forma financial statements, the unexpected can always be expected to happen. The experienced venture capitalist is used to the entrepreneurs "coming back to the well" many times, after promising that they will "not have to do it again after the last time."

Internal entrepreneurs and V/E project managers may not be as loose in their financial planning as that, due to their corporate backgrounds. But they also must be in a position to take advantage of opportunities in a fast-moving technical field or market and to throw resources into a problem that appears suddenly.

Where the firm is not able to bend its capital budgeting and resource allocation procedures or make its working capital appropriations more flexible than usual, V/E projects can stall or strangle. They may do so for want of timely infusions of funds and assurances that if and when opportunities/problems arise, they can count on swift action to back them up or bail them out.

A representative instance of gross mismatch in funding patterns was encountered in a large firm which had decided, in principle, on a program of acquisitions of small high-tech companies to support its internal new venture activities. Several times, when the head of the V/E group brought acquisition prospects in for consideration by the board of directors, the matter was placed so low on the agenda that several meetings passed before they were even considered, let alone acted on.

As a result, he lost several excellent opportunities for acquisition and his whole program was slowed significantly.

Boards of large conservative companies have difficulty in dealing with small, unusual, sometimes "mysterious" and highly technical projects. They may be intrigued by them and, when they get around to them, spend an inordinate amount of their time on them. But their typical pace of analysis and decisionmaking is frequently mismatched with the needs of fast-moving V/E projects.

A few boards have set up subcommittees to deal with such unusual cases or, as suggested above, delegated their handling to the chairman or even to a senior vice president or another member of top management. This can solve the timing of appropriations but often leaves the venture itself without a clear and continuing spot on the board's long-term agenda.

Not just the timing and procedures for financing V/E projects are different. The whole philosophy of financing and the evaluation of investments have to be different from those used for "regular" technology projects in the firm.

Most ongoing programs of R&D have some interconnection and even if they fail or do not achieve the expected results, there is frequently a residual. This may be in terms of learning by the individuals involved and the firm itself (in the form of "institutional memory") on what works and what does not work and how to distinguish between them for future reference.

In contrast, many V/E projects are "one off" efforts that will either succeed or fail and that will not add much to the *company's* general skills and knowledge, although it can add tremendously to the development of *individual* skills and experience (more on this later).

As stand-alone projects, they are often outside the usual interests and capabilities of the firm in terms of technology, production, or marketing. As a consequence, the "odds of failure" or the "probability of success" has no real statistical meaning. The project will either succeed or fail and it is hard to make a forecast of either outcome. What it is easier to do is recognize that many such projects, since they are new and done on a fast track with a hitherto untried team, are very risky and that "complete" success is a rare event.

On the other hand, the payoff from success may be so high (the "upside" of the investment) that it seems well worth the risk, even though the salvage value ("downside") may be almost nothing. Most capital budgeting committees and financial executives are not used to dealing with such wild swings of investment outcomes. Certainly, the occasional venture into the raw materials market (through prospecting, futures options, or land speculation) or a new distribution system, in an unfamiliar market and under uncertain economic conditions, does not leave them innocent of risk. But the normal sequence of capital investment decisions in the well-run firm has to meet certain criteria and be presented in a familiar form before these groups are comfortable with them. Despite the slick business plans and prospecti for V/E projects which experienced staff people can put together for a funding presentation, the covers enclose a lot of uncertainties and unfamiliar territory for the financial executive and the regular financing procedures.

One philosophy that is a part of the downstream evaluation of such proposals involves a "sunk cost" approach to funding. The decision or "go, no-go" milestones are not burdened by how much has been spent (and whose fault it was), but primarily by the comparative merits of continuing or abandoning the project, based on its likely financial results "from this time forward." This is strong medicine for most people used to life cycle, return on total investment, and other evaluation approaches which weigh in *all* expenditures, even though they have already been made and are unsalvageable (see Chapter 7).

The "one-shot" nature of such projects and their differentiation from the ongoing technology portfolio strongly suggest that there have to be different "pockets" or sources in the firm for them. These sources should be governed by a different investment philosophy, style, and time pattern than the common ones used for routine capital investments and working capital allocations.

Some companies have set up "venture" funds or other mechanisms for dealing with V/E projects. In some of them, however, the traditional criteria and methods of allocation and accounting used for non-V/E projects are applied, thus leading to a "Simon says" atmosphere, where it is not clear to the V/E manager what he can and cannot get away with. Some V/E projects are "forced" on a division manager, making him support the project from funds above his P&L line as "divisional expense." If the project succeeds, the CEO may take away the profits from below the line as "corporate profits." Such projects can be "sure death" even before they get under way.

8.4.4 Selecting Personnel for V/E Projects and Phasing Them Back into the Organization

While financial considerations are very important and can sink a V/E project, properly providing personnel for such projects is the critical factor in their success and failure. It is, of course, a platitude to say that "the right people are needed to make a venture successful. " This is true of many kinds of ventures. In the special case of V/E projects, however, the very reason for setting up a V/E structure and procedure is to allow certain entrepreneurial individuals to succeed against the high odds that such ventures entail.

The characteristics of successful V/E project leaders and their champions or sponsors have been studied and widely discussed. There is not much convergence in the descriptions beyond the usual attributes of "shakers and movers" and "good businesspeople" and "people who can tolerate ambiguity, take risks, take the heat, and roll with the punches." A list of such attributes, beyond a very few basic ones (are they credible, competent, enthusiastic, robust, etc.), is not much help in *finding* the few good people needed to initiate and push such projects through the choppy waters or quicksand (whichever metaphor is preferred) of the large corporation.

Such a list, however, can be helpful in *screening out* possible candidates — people who are likely to fail under pressure or drift aimlessly without strong guidance from someone else. "Crybabies" are also not good bets for this role, nor are a number of other stereotypes. Our experience has been, in searching for these kinds of people in a number of large companies, that they are easily recognizable and well known by their colleagues. Some are also heartily disliked by their less-aggressive or less-entrepreneurial colleagues as too pushy and too hard to get along with. So be it. Nasty people are not necessarily successful entrepreneurs, inside or outside the firm; but nice guys (passive, compliant, self-effacing) are also not prominent among the winners. Without having to belabor the precise definition of what we mean by an "internal entrepreneur," many managers and many technical colleagues can readily suggest candidates. Some of these nominations reflect limitations in the perceptions or experience of the *nominators* who do not really know what is involved in a V/E project and what qualities it requires of its leader and other key personnel.

In any event, convergence is not hard to achieve, but sometimes this procedure yields disappointing results. Many of the nominees have one or more "fatal flaws" as potential V/E project leaders. They may have low technical credibility and would probably not be able to recruit and motivate top-quality people. They may be "damaged merchandise" who have suffered so much rejection and frustration in the company that they have become bitter and are nominated because they seem to talk a good game of entrepreneurship: "If I were running things I would do it this way."

Others lack the ambition and the "hunger" that venture capitalists look for in the people they back: "We don't want to get in bed with a person who doesn't want to get rich, because he is not likely to make us rich either."

The disappointment has come from a yield of only three or four people in some very, very large companies who seem to be suited for the role of internal entrepreneur. The *optimistic* view is that may be enough to get things started and to smoke out others who are not so visible but who have the necessary basic qualifications. The *pessimistic* view is: "My goodness, what has the large company environment done to these people to squeeze out the entrepreneurial spirit and make it so hard to find a few good, high-potential candidates?"

The high-potential V/E project leader or internal entrepreneur generally has, among his attributes, a high level of risk tolerance. He should have a high risk propensity and low risk aversion. In that respect, and others, he may also be a "nonorganizational person," one who does not quite fit or just barely fits into the organizational scene prior to being "discovered" as a potential V/E leader. Or, more likely, he did not "really" fit but was adaptable enough (another interesting entrepreneurial attribute) to keep it from being obvious and to be viewed as a team player, even if a little unusual in his attitudes and behavior. So, such people are not standing out in the company's driveway with an "I am an entrepreneur" sign on their foreheads. But they are readily identifiable with a little effort and some "misses."

There is a lot of mythology about the "need to fail" that characterizes the careers, not the personalities, of such people. That is, many people in the venture capital and entrepreneurship field say that "until or unless a man has one or two failures under his belt, he is not seasoned." Maybe. It depends on the circumstances and consequences of the failure. The "damaged goods" syndrome mentioned earlier can take the heart out of a potential winner and make him too bitter to try again, let alone to "risk all" again. Failures *beyond the control* of the entrepreneur—unforeseeable shifts in market, the economy, or competitors' strategies—may in fact leave him wiser, stronger, and less vulnerable to such events in the future.

Some people even advocate "planned failures" for such people (on a small, noncritical scale, of course) to see how they react and if they *can* recover from adversity. In the capital-raising area, we have encountered many external, as well as internal, entrepreneurs who were repeatedly rebuffed by the money people, but finally were able to obtain financing. One CEO brags that he "automatically turns down his self-promoting entrepreneurs at least a couple of times to see if they have the conviction and guts enough to keep coming back."

We followed the adventures of one CRL member in a French company for several years, listening to his story of rejection each year until his CTO finally said "o.k., go ahead and try it, but stop bothering me."

Persistence through adversity can be an admirable quality for V/E type people, but this is not to be confused with the stubbornness to go down with a sinking ship. Many of the "other shoe" V/E leaders and entrepreneurs we have encountered both inside and outside the large firm know how to take their lumps, allow their "sunk costs" to sink, and go on to the next opportunity. Stubbornly flogging a dead project or idea beyond reasonable limits may be the mark of a determined person, but it is not necessarily proper behavior for a successful V/E leader.

There is much to be said on the issue of the "nonorganizational person" syndrome, which is frequently but not necessarily always characteristic of successful or promising V/E leaders. The very nature of V/E projects suggests (requires) that rules be broken, procedures be violated, well-plowed and acceptable paths not be followed. The reason for setting them up or allowing them to arise and continue is that the regular, standard ways of doing things—getting new ideas and products to market against strong odds—does not work or does not work well enough.

Along with the technology, the V/E leader may have to break new *organizational* ground and, unfortunately for some perceptions of him, "fight the system." If he is a person who does that as a matter of principle, even when there is little at stake and no real need for it, we may be dealing with an organizational "misfit" rather than a potentially successful V/E leader who can use and modify the organizational environment to suit his needs. This misfit or maverick may look and talk like a swashbuckler who can cut through the brush and get things done, but he may fall on his face because he cannot gather and use the resources he needs from other people and groups in the firm.

Especially if he is operating in the style of the "honcho without troops of his own" (see Section 8.4.2 above on organizational forms for V/E projects), he needs the cooperation of others in the organization to do the job. Successful V/E leaders and their cadre are, in a sense, *really* organizational people. They are pushing for a project that will benefit the organization if successful and they are using the organization's resources to help accomplish that single-minded goal.

Playing the renegade, the maverick, or the fool will not help in this highly political task of obtaining and maintaining cooperation. So, it appears that two necessary attributes of a V/E leader at many, if not all, stages of the total project life cycle, are that (1) he knows the organization and knows how and where to obtain support, and (2) he engenders enough credibility and goodwill to obtain that support.

We have worked closely with a number of V/E managers brought in from outside the firm to launch or save a V/E project already under way. Some of them are "quick studies"—attractive people who can garner cooperation and support from other people, even strangers. Many others grind to a halt or stumble, however, in their approaches or reactions to certain people, groups, or functions in the organization. When the whole V/E team is made up of outsiders, no matter how experienced and sophisticated, the potential for such stumbling situations is high and this frequently kills V/E projects.

The dilemma of whether to try to launch V/E projects with all internal people and hope that they can shake off the crust and inhibitions they have accumulated in the past or to go outside for a whole new team has an obvious, and perhaps trivial, solution. This is the combination team of insiders and outsiders. The outside V/E leader, if none can be found inside, can be supported by one or more "old soldiers" from inside who know the ropes and the people and can get things done. Often such team members are cast in the role of "master sergeant" who is a master, among other things, of scrounging resources.

This combination strategy can work if carefully thought through and carefully monitored (not tightly controlled) by the project champion in upper management. If signs of unmanageable conflict or divisiveness appear which will not just go away or subside to a nonsignificant level between the insiders and outsiders on the team, then some action is indicated to reform or shake up the team. Conflicts on goals, styles, career aspirations, and "protection of the firm" can and do occur. They must be anticipated, watched for, and dealt with before they become destructive.

One more comment may be useful, about the "beards and sandals" syndrome of some R&D and entrepreneurial types in the organization. Not all people who wear sandals and beards are creative or born leaders of V/E projects. Some are pretending to be mavericks. Others may have sore feet or tender skin and are so adorned for comfort. Still others may be the real article, whose dress and general appearance are unself-conscious and do reflect the kinds of qualities needed for V/E projects.

The wise project champion will try to look beneath the outward signs for the V/E qualities he needs and try to accept things that might ordinarily annoy him: "I love these guys and I know they are important for the future of the company and its technology, but why can't they wear socks to work and trim those bushy beards?" If they won't or can't, a choice must be made on which of their attributes are more important for the company and the project.

Aside from their outward appearance and general attributes, V/E project leaders and their teams may have to be rewarded with special incentives. More on that below. At this point, it is important to note that the usual compensation system in the company may not be enough to attract and hold such people, either on the project or even in the company. Referring to an earlier discussion of "locals" versus "cosmopolitans" in terms of their loyalties, reference groups, and mobility, I can assert that good V/E people are much more mobile than their colleagues and more likely to be externally oriented in their values and reference groups.

They may seek out kindred spirits in other companies, universities, or the business and financial communities. Many of those who are risk takers and entrepreneurial in nature can be found dabbling in the stock markets and associating with the venture capital community when they are accessible. They may keep close watch on the price of the company's stock and its operating ratios as well as that of competitive and other comparable companies. Keeping them in the organization and motivated may be quite a different task than doing the same for their less-entrepreneurial colleagues.

Once the entrepreneurial fever is caught, such people may no longer be satisfied with routine operations or may no longer be compatible with the kinds of things they were doing before being annointed for a V/E project. As a spin-off of a study we did in the early 1970s on the transfer of NASA technology to industry,[*] we found that a high proportion of the internal people who had picked up pieces of technology from the space program and attempted to commercialize them via V/E projects had left their companies within a 3–5-year period, *whether the project had succeeded or failed*. There are many influences acting on the V/E type person in the large firm which may come together and cause him to leave the firm or be terminated. Some of his personal attributes, which made him a good bet for V/E projects, may be intolerable in the normal routine of the company. He may have accumulated enemies and black marks while attempting to swashbuckle his way to a successful product. He may be disillusioned, bitter, discouraged, or just plain tired of fighting for resources, attention, and approval. The wise V/E project champion will think seriously about these potential consequences of V/E projects and try to anticipate and ameliorate them. If not, then the next time around there will be even fewer V/E leader types from which to choose.

8.4.5 V/E Projects as Part of the Corporate Research Lab (CRL)

As mentioned earlier, a natural starting place for many V/E projects is the CRL. The technical environment, the pace, the skill levels, the curiosity, and other attributes of such an environment encourage dreaming and generating ideas for ventures. A large percentage of technology-based V/E projects do in fact start in the CRL or equivalent groups. The problem is that a lot of them also die there. The key to their success, if they are indeed technically feasible and have a good chance to make it to a successful commercialization stage, is being able to cut them loose as true business ventures. I have discussed a number of barriers to this, including the frequent amateurism of CRL people in routine business functions such as marketing, sales, production, finance, and negotiation. Perhaps the best way

---

[*]Alok Chakrabarti and A. H. Rubenstein, 1972.

to illustrate the barriers to success of CRL-generated V/E projects or ideas for V/E projects is by a very representative detailed case history of one. It is the Miller Chemical Case (see Appendix C), which I researched and wrote for the Harvard Business School case series in the 1950s and is still greeted with "instant recognition" from managers in many companies as reflecting the situation in their own companies. A key aspect of that case is the problem dealt with in the next section.

8.4.6 Feeding the V/E Project into the Operating Division

Wherever the idea for a V/E project arises and wherever it goes during its early upstream phases (typically research or advanced development), it must soon be transferred to an operating environment where the resources and facilities for commercialization exist. As indicated earlier in this chapter, that environment can be "special," in the sense that it is created or maintained specifically for the purpose of incubating new ventures.

No matter how special, however, it can seldom command all the resources, experience, skills, and technology transfer and application mechanisms that exist in a mature, "full-line" operating division. Some "new business" or "venture" departments or divisions do have their own production facilities and marketing or sales groups. For a large diversified company, which experiences a stream of V/E projects arising from different technologies and headed for different markets, such a specialized unit cannot handle more than the midphases of the V/E life cycle and often can only handle a very few of them at a time.

This poses the central theme of the Miller Chemical Case. At what point, how, and with what safeguards can and should a V/E project be transferred to the "real-world" environment of an operating division? This is a critical technology transfer phase and will strongly influence the outcome of the project. The timing and choice of a receiver—a division manager who is willing to continue the project and give it the additional special treatment needed to get it to market—need to be made very carefully and with a great deal of contingency planning.

The many reasons presented in earlier chapters on why operating divisions are not generally good places for radical idea generation and fostering entrepreneurial projects in their early stages should not be taken as counsel to avoid transferring V/E projects at *some* appropriate stage into a divison. The keys are the timing of that transfer, which division is to get it, and what the accompanying conditions are for the transfer.

Do the project leader and his team go with the project? Does he get a favored place in the division's hierarchy and continued special treatment (including an individual parking space and his own private office and secretary)? What are the reporting lines? Will the original corporate champion still be involved? Will the division manager take on the role of champion? And many others. These must be thought through, planned, negotiated, and monitored to increase the chances that the project will indeed get to market and realize its potential. Otherwise, it can grind to a sickening halt or slide into the queue of routine projects waiting for their turn at the pilot plant, the factory, the funding source, the marketing program, and the salesperson's sample cases.

A critical factor in this whole transfer process is "who pays for what." Does the division manager or the corporation absorb the upfront costs that have already been incurred—in other words, does the receiver "buy" the project at some markup or markdown from the costs-to-date? Does special equipment purchased or developed for the project come with it and does it come free or at some cost? Who pays the premium on the salaries of the V/E team, if they are above the going rates for divisional personnel? Neglect of such matters has killed many such transfers.

8.4.7 The Captive Venture Capital Approach

This has been discussed earlier in the context of "make or buy" of technology. It is also relevant here, because it can provide a source and a mechanism for dealing with questions of "who pays" raised in the previous section.

If indeed V/E projects—one or a series of them—are looked on as a form of venture capital investment, then the funding process and the use of the funds should be in a *venture capital* rather than in a *banking* or *accounting* mode.

Whether a project is transferred to an operating division or goes all the way through its life cycle on its own (resulting in formation of a new operating division), the sources and uses of funds have to be dealt with as "special" and not as part of anyone's regular operating or capital budget. This means, contrary to the practice in many firms trying to foster internal entrepreneurship and V/E project mechanisms, that the funds need a special appropriation mechanism and have to be kept segregated from other investment capital and working capital funds. I argued earlier for a special philosophical view of financing V/E projects and this should be reflected in how such funds originate and are managed.

Many V/E projects have stalled or aborted following a lot of fanfare about "heralding a new era of entrepreneurship" for the firm. This is because the V/E manager suddenly finds himself at a critical junction of financing the project with a new (old) set of criteria and procedures for getting the money he may need urgently.

The original funding may have been provided in a true venture capital mode (not reckless, but with few strings), but suddenly he finds that continuation funding must stand in line behind all other capital appropriations in the firm—plants, equipment, distribution networks, institutional advertising, R&D, and so on.

If V/E projects are to have a good chance of success, the mode as well as the level and time pattern of financing must be adapted to their special needs.

8.4.8 The Venture or New Products Division

This kind of "incubator" for V/E projects or certain stages of them is not very common. Nominally, the organization charts of many large companies display departments, divisions, or groups with such designations, including "special projects." Closer examination of their actual portfolio, however, often reveals a collection of "cats and dogs," only some of which may actually be "true" V/E projects. The others may be problem product lines at a low point of their life cycle, small acquisitions, or products or business segments that are phasing out or that management wants to phase out. The managers are often very experienced business executives, but they can be distracted by their non-V/E problem projects and also not really in sympathy with *high-tech* V/E projects.

In such an environment, the V/E project may get "special treatment" alright, but that may consist of benign neglect and lack of the kind of support and push it needs at critical junctures. This is especially true when the original champion, generally at top management levels (if not the CEO himself), gets fed up with having to worry about the project and fight its battles for it. He may decide to hand it over to an experienced manager who does not have a regular, large-scale operating division to run and "can devote time to their special needs." His washing of his hands of the project can be a signal to the new manager that the project is really not that important or urgent and thus the benign neglect or even a little punishment and scaling back may be in order, to show the V/E leader that he is not so special after all.

If a venture or new products division is to accomplish the incubator, tech transfer, and commercialization functions it is intended to, then it must be designed, managed, and monitored accordingly. Most of them are not. We have encountered many V/E projects languishing in supposedly supportive environments of such an organizational unit, where the decisionmaking and response processes have become as bureaucratic and inflexible as they are elsewhere in the firm.

8.4.9 The High-Tech Spin-Off From the Firm

Ultimately, many V/E projects cannot survive and thrive inside the large decentralized firm, no matter what mechanisms are put in place to help them along. They violate the pace and peace of the managers of the firm and exhaust their patience. They may "contaminate" other projects and groups and cause more trouble than they are worth. They may become smothered in the firm and stall or die.

In such cases, a viable alternative to allowing them to die or languish is to spin them out as the closest thing a large firm can tolerate to an "independent" operation. The organizational mechanisms are straightforward, although there may be later complications. The entrepreneurial group may be given special status, such as a leave of absence, where their fringes and seniority rights are maintained for a period of one or more years. Or they may be terminated and allowed to set up on their own with perhaps a reasonable termination settlement from the firm, including a fee, royalty, license, or part ownership arrangement. Some ties with the new venture may be maintained and technical and other support and advice (e.g., legal, financial, labor relations) provided on a less-than-market price basis, or even free. Some people with special skills may even be loaned for a limited period to get the enterprise launched.

Financially and legally, this can be accomplished in a number of ways. At one extreme, the venture can be sold to entrepreneurs and the parent firm cuts all ties with it (except, of course, personal ones which may lead to future involvement). Another extreme is that the new venture is independent in all senses except financial and may in fact be owned by the parent firm as a "silent or absentee owner." In-between, there are a variety of financing modes, including joint venture, limited partnership, exclusive supplier, and semicaptive shop.

Aside from the goal of getting the project off its neck and perhaps providing a better chance for its ultimate success, one of the prime drivers for this kind of spin-off is related to the next section—incentives and rewards for the entrepreneur and his V/E team.

8.4.10 Incentives and Rewards for Entrepreneurship and Risk-Taking

For some V/E people, the excitement of the game is enough to sustain them. For many technical people, V/E projects provide a chance to get out of a rut and get their pulses going again. Some may value the challenge and the risk for their own sakes and view the project as an opportunity to show that they have got the right stuff to succeed, if given a chance and the right conditions.

Despite these psychological and spiritual motives, most people who stick their necks out voluntarily and work harder than they have to want something more material out of their efforts. The typical viewpoint of the typical entrepreneur is "what's in it for me? I don't mind making a lot of money for the company, but I want my share." In the world of outside venture capital and entrepreneurship, this share is generally and literally a "piece of the action" or *equity* in the venture. This ownership permits the entrepreneurs to benefit directly from the outcomes of their efforts in terms of dividends (not common in the early life of high-tech start-ups), capital appreciation, and, ultimately, cashing in through public financing or selling out.

Such a direct ownership position is not feasible (or even possible) in the large public firm, except through the usual mechanisms of stock purchase plans and stock options (for which low-level employees seldom qualify). These mechanisms and even some bonus plans do not reward the entrepreneurs inside the company directly and proportionately *for their own efforts and results.* They are generally filtered through the overall operating results of the large company and bear little relation to their own project's success. Some special bonus plans, accelerated promotion, perqs, fringes, and other mechanisms can come closer to individual reward and incentive systems. Where they are feasible and do not founder on "equity and fairness" grounds relative to other employees, such special recognition—symbolic and material—may serve as a reasonable substitute for direct ownership in equity of the venture, at least initially.

The career paths of V/E leaders and team members can be helped just by being associated with such activities, whether they succeed or fail. If they succeed, better yet! If they do remain inside the company, other opportunities may arise to stand out from the crowd and move onto the fast track. "We need a leader for this new group or project (which may or may not be a V/E project). How about Bob? He seemed to do a reasonable job on that new venture even though the project itself went belly up. His experience could help in this new job."

Or, in some cases, career paths may be slowed or hurt by V/E experience. "Yes, Bob's a good man, but his high-flying year with that new product got him into some bad habits and made him impatient with the usual ways we do things.

Let's keep him on hold for another venture and give this job to someone else who is liable to appreciate it more."

If Bob leaves the company, he can likewise benefit or be hurt by V/E project experience. He can be viewed as a hot shot by another company and hired for his perceived entrepreneurial qualities. Or he can be viewed with suspicion by more conservative recruiters who may see him as a potential boat-rocker and trouble-maker.

We have followed the career paths of over a dozen entrepreneurial-type people for many years, watching them move from company to company. Sometimes they have moved up in status and compensation. Sometimes they have moved laterally, and sometimes they slide back down the ladder they climbed so quickly in an earlier period of their careers. Involvement in V/E projects can provide intrinsic and career rewards, but direct, material rewards in the form of cash and securities appear to motivate most people in the game.

8.4.11   Sources of Venture Capital for High-Tech Projects

Some of the philosophy and mechanisms of financing high-tech V/E projects inside the firm have been discussed in other parts of this chapter. This section contains brief comments on sources of the funds used for such projects.

Given the predispostion to treat such funds differently from the usual capital investment and working capital funds, from where do you get them? For the large firm, the amount required even for a series of such projects does not warrant a public offering or a major borrowing program. If even the blue chip firm were to go into detail with their bankers on the uses of such funds as part of their line of credit or general short-term loan programs, they might get some very unsupportive responses. Even longer-term loans might not be a good source of such funds.

Consistent with the venture capital frame of mind needed for this kind of investment, the funds should come out of operations, profits, and surplus as part of the firm's long-term "renewal" program. Aside from the "one off" short-term opportunity that might give rise to a quick V/E *project*, a continuing V/E *program* should be viewed as an investment in longer-term changes to the firm's technologies, products, and markets. These funds, like those used in a continuing acquisition program, *may* be funded by borrowing but, in my view, *should* be funded by plowing back profits or a portion of cash flow into higher-risk renewal and expansion projects.

Such behavior clearly signals to observers both inside and outside the firm, including stockholders and security analysts, that the firm is seriously using part of its income stream to make sure that it will remain technically viable into the future. In this sense, the V/E funds are akin to corporate R&D investments, beyond the short-term expenses of supporting current products and operations. It can also be viewed as part of its overall renewal investments in people, facilities, and equipment, beyond the usual short-term "maintenance and repair."

Such a strategy can clearly signal the firm's intentions to keep up and even "leapfrog" technically, as well as to exploit new opportunities peripheral to its main businesses.

9

SPECIAL CASES: "NEW" AREAS OF TECHNOLOGY MANAGEMENT

9.1 INTRODUCTION AND CHAPTER OVERVIEW

In this final chapter, two special aspects of technology management in the firm are considered. They are not the only ones that could have been selected for inclusion in this "miscellaneous" chapter. They have been selected for two reasons. The first is that they are "new" in the sense that new threats, needs, and opportunities have arisen or become more prominent in connection with them, almost unnoticed by many managers and policymakers. The other, not independent of the first, is that we—both the POMRAD research group at Northwestern and my own consulting practice—have become heavily involved in them in the past few years. The first special case—the software development process—has been the focus of a multiyear research program sponsored by the National Science Foundation. It has led us into another aspect of "organizing for technology" than most of the research and practice we have done before, as reflected in the first eight chapters of the book. The second—based on an idea from my colleague Don Frey, former chairman of Bell & Howell Company who recently joined the faculty at Northwestern—has taken us into a relatively new area of technology and R&D management in the service industries. In 1986 we established a Center for Information Technology (CIT) to (1) focus research on this area and (2) provide a forum for intensive interaction between academic researchers and practitioners from the service industries around major issues related to information systems in the service sectors: productivity, competitive posture, cost effectiveness, design of information systems, and related topics.

   The two sections in this chapter only skim the surface of these two important topics, raising issues, making suggestions for improved management in these two areas, and urging more management attention to them. The research literature on both topics is relatively new, but beginning to grow as more academic and industrial groups begin to do serious basic and applied research on them. The references in Appendix B should assist interested technology managers and their managers in exploring the field in terms of ideas, the experience of other firms, and a watch list of factors they should be aware of and alert to in terms of their own technology programs.

9.2 SOFTWARE PROJECTS[*]

9.2.1 Introduction

A radical change has been occurring during the past decade or so in the kind of technology projects that make up the portfolios of many firms. The shift has been in the direction of an increasing proportion of the R&D and engineering effort being devoted to the software components of their products and production processes. The availability of cheaper and more versatile microelectronic components, systems, computers, and the software that accompanies them has led to a subtle but significant shift in the content of many firms' technology portfolios.

Initial penetrations were made in the use of electronics in manufacturing processes and in the engineering and design activities that support them. Computer-aided design and manufacturing and the many variations and combinations of those functions have made software development and use commonplace. The past decade has seen a drift or rush (in some companies) into major efforts in developing, acquiring, adapting, and implementing software in both manufacturing processes and products themselves. Electronic sensors and controls, microprocessors, and many novel features have become integral parts of many traditional products either as "add-ons" or as the core of new products around which the functional features are built—appliances, vehicles, watches, toys, typewriters, communication equipment, and so on.

This shift away from primary or exclusive focus on traditional metal-bending, or wet processing, or assembly in their traditional technology areas has caught many firms unprepared for the new technical skills and management procedures and styles that are needed to assure the success of software projects. During the period 1984–1987, members of our POMRAD research group at Northwestern undertook a study of the software development process for manufacturing. The objective was to determine what problems and issues have arisen to date and are likely to arise in the future in this field that is new to most companies' technology programs, except for those specifically in the electronics and computer fields. In addition, we were concerned, as always, about how the effectiveness of the

---

[*]Adapted from "Measuring the Effectiveness of the Process of Developing Computer Software for New Production Systems," Rubenstein, 1985. Reprinted courtesy of the Society of Manufacturing Engineers. Copyright 1985, from *Advanced Systems for Manufacturing*.

"software development life cycle" could be evaluated. Some of the findings from that study are discussed briefly in this section. Further detail is available in the many reports and publications from the study itself. (See Appendix B.)

9.2.2 Some Key Issues in Software Development

Early in the study we identified a number of issues that were already of concern to managers of software projects and *their* managers, including top management of the division or the corporation. Interviews were conducted with developers, users, and vendors of software for manufacturing systems and other knowledgeable people who closely watch the field. Some of these issues are likely to remain or become major policy concerns as the field develops. Others are likely to be overtaken by events as the field progresses and will become moot for some companies. The ones discussed briefly below appear to be of continuing concern.

The Software Development Life Cycle. Many companies have slipped into or been jerked into the computer age in their manufacturing process development without paying much attention to the need for planning and managing the software aspects of their equipment and production system development as a complete organizational and technical process. Refinements, equipment, computer programs, and people have been added in piecemeal fashion as opportunities presented themselves for improvement of manufacturing processes via computer software. Only a few companies we have encountered so far have treated the overall process in terms of its life cycle, as suggested in Figure 9-1, including the far upstream—need recognition—and far downstream—maintainability and updating—phases of the life cycle.

Interaction Between the Computer Scientist and the Design Engineer. Several potentially opposing trends are evident in this area. In some companies, heroic efforts are being made to bring the computer scientist and the design engineer together in a continuing, interactive mode—often as part of a software development *team*. In other companies, efforts are being made to accelerate the development and acquisition of "software development tools" by the computer scientists that can be "handed over" to the design engineers so that they can do their *own* software development. In still other cases, attempts are being made to do both—provide the tools and, at the same time, avoid the gaps between the specialists in the software development aspects and the design engineers who are responsible for new equipment and systems, using or incorporating computer software.

Front-End Versus Downstream Costs. A burgeoning literature on the costs of various stages of the software development process has pulled management's attention toward the *early* stages of software development, in the hopes of reducing the mounting costs of the software versus the hardware components of a product or system. At the same time, disasters, delays, and rising costs of software

408

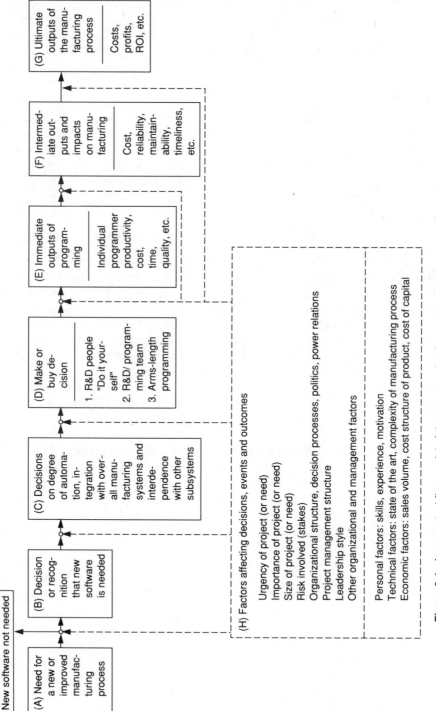

Figure 9-1. A conceptual flow model of the process of software development or acquisition for manufacturing.

maintenance and upgrading in the factory have called attention to the *downstream* costs. Those can add significantly to total life cycle costs for the user—whether the software is developed in-house or purchased from outside.

Does the Engineering Hardware Model of Project Management Work for Software? It is not yet clear in many companies whether the people and "organizational memory" in hardware technology projects can be applied effectively to software development. In many cases, software is being developed by people with little or no background in traditional R&D management or project management. They are making up their own procedures and following their own intuition about how to plan, organize, manage, and evaluate projects. In other cases, the "old soldiers" of traditional project management are being called on to bail out or assist in software development projects and programs that have bogged down or gone astray.

The Super Designer. One view of the future direction of engineering design is that an individual designer, equipped with a modern design workstation, voluminous data bases, and the most advanced design software (CAE, etc.) can "do the job himself." This extreme view, which seems to be a policy target in some organizations, espouses the goals of eliminating interface bottlenecks, fixing responsibility for design, speeding up the process, and other advantages. Critics and skeptics are concerned about possible losses if the process is *too* integrated—losses due to the lack of *intellectual interaction between specialists* who may not be able to reduce all their knowledge and tricks of the trade to formalized data bases or "expert systems."

Make or Buy of Software. A surprising number of large companies we have been in contact with are relying heavily on purchase of either major portions of their manufacturing software and supporting equipment, or "packaged" or "turn key" software for their manufacturing processes. Advantages to such a policy include, in their view, shorter lead times versus developing the software in-house; definite cost estimates they can count on; dispensing with the management problems of finding, hiring, and retaining programmers; taking advantage of the vicarious experience of other companies (even competitors) as represented in the software; and greater responsiveness of vendors. Disadvantages are beginning to surface, however, at the downstream stages when software maintenance and updating are needed to take advantage of new developments and requirements. In addition, the lack of trained, highly competent in-house software people is being recognized as a competitive disadvantage. One factor in the make or buy decision is the difficulty some companies have had in using their existing management information systems (MIS) or information systems (IS) people for developing manufacturing software.

Mr. or Ms. Software for Manufacturing. Our field research methodology required that a major contact in the companies where we interviewed be an individual who has corporate or divisional oversight for the whole software development life cycle. In a number of companies, it is easy to identify such an individual, who is supposed to bring the latest tools and developments in software into the company and "sell" them to the operating units. In others, computer-oriented engineers sometimes perform this function in an informal gatekeeper or key communicator mode—with no official backing. Lack of such a focal point for software development/acquisition is likely to become a barrier to effective exploitation of software technology.

Focus on Quality Assurance. Stimulated by working groups in professional societies and inside individual large companies, a recognition is emerging that quality assurance of the overall software development process is essential. In addition to emphasis on configuration management, software cost management, and life cycle management, quality assurance is rapidly gaining attention as a way of not only controlling technical quality of software but also protecting users' and developers' interests in terms of costs, time, and performance.

The Role of New Tools. As indicated at the beginning of this section, the development and availability of new tools for software development may make some of the above issues moot for, at least, the leading edge companies. We are also interested in seeing how rapidly they are adopted and how effectively they are used by the passive, mainstream user companies.

Other issues are emerging and it would be well for CEOs, CTOs, and their division managers and their staffs to be aware of them before they rise up unexpectedly and seriously interfere with the development/acquisition of software and the effective management of the whole software life cycle.

9.2.3 Some Brief Cases

Some brief cases illustrate some of these issues and other problems that can arise, unseen by upper levels of management who are assuming that "things are going o.k." with major new software or software-dominated technology projects. Primarily, however, they illustrate some of the "facilitators" of effective software development, which can make a difference between success and failure in terms of cost, time, and performance of the product or system.

Case 1. This project was a large-scale (cost over $2 million) in-house software development, involving a real-time operating system for control of a manufacturing system. Approximately 15 programmers were involved for about 18 months. It was considered successful and some of the factors that may account for its success were: (a) tasks were clearly described and several design reviews occurred *early* in the project; (b) the design of every major software component was extensively reviewed by the team *before* coding and testing had begun; (c) the project was developed with programming tools used in previous designs and most programmers were familiar with them; and (d) the project had a relatively high corporate priority.

Case 2. This was a relatively modest effort. The project was successful and the productivity was relatively good. (a) The needs of users were not researched before the start of the project, so some needs were not addressed. (b) The original time estimates were very close to the actual time needed due in part to the fact that a large portion of the work was modification to an existing product. (c) The supervisor was very close to the development process and thus had very accurate information on progress.

Case 3. This project developed software for a staff department. The project team began programming without an internal specification document. External documents were distributed for review by user groups. The project quickly became well behind schedule. "Unexpected" design situations and, especially, interfacing errors started to appear on a regular basis. Hence, the project expense, which was being closely tracked to this point, was soaring. The project ran into a whole series of implementation problems once it was finally deemed ready to be tried. The system had a large number of programming, interface, and logic problems that took months to correct. Some programs had such blatant errors that it is clear that they were not tested with any kind of live data. The project team was pushed to get the project out well over a year behind the initial schedule. The users had yet to sign off until the system tested smoothly for a month. This project was haphazardly put together, rather than approached in a logical and structured manner.

Case 4. This project had already reached the stage of release to manufacturing and had proved itself quite reliable in the field, although customer complaints were still being received. The main pressures to get the product out are off, and all future releases will be to fix problems found out in the field or to add new features, mostly enhancements. This project produced a major software product in a timely fashion and its success was important to the company because it showed the ability and commitment of the team, leading to reliable performance of the software development. The project took about 25 man-years and had complete upper-management support.

Case 5. This project was a development of software for use in control mechanisms for monitoring purposes. The project required much more time to complete than was scheduled. It suffered from interface and user problems due to the short time allowed for the design specification and high-level design phases. Overall, customers were satisfied. The project was oversupervised, and the project team had access to knowledgeable people with problem solution skills.

Case 6. In this project, the project team produced internal documents to be reviewed by local programming management and circulated the documents to the information systems departments in the divisions which would eventually use the system. The project was on schedule with a fair number of program bugs due to new programmers, but few or no interfacing or design errors. The programming errors were easily detected by both peer programmers and programming management and fixed. The system went on to be instantly signed off by the user groups once it was implemented, and minor cosmetic software modifications were easily and swiftly put into the code. The internal documentation also made the local support of the system much easier.

Figure 9-1 summarizes some of the major phases of the "software development/ acquisition life cycle" in terms of decisions involved and levels of output associated with a given project or program. The emphasis in this flow diagram is on the factors affecting decisions, events, and outcomes in the software life cycle. Many of them are familiar to people who have been involved with hardware technology projects. A major difference is the lack of experience and "organizational memory" to draw on or fall back on in planning and executing software projects.

Boxes (A), (B), and (C) are critical to the initial planning and design of a software project, if indeed one is needed. And herein lies a tale. We have encountered many automation and computer-aided manufacturing projects that were undertaken primarily because the "technology was there" and not as the result of careful cost–benefit or even standard investment analysis and business planning. Some degree of false starts, nonsuccessful and abortive projects are, of course, characteristic of the leading edge of many new technologies. Technology people and nontechnical managers want to "try out the idea," "keep up with the Joneses," or "get their feet wet." Even so, the progress in adopting new manufacturing technology that depends heavily on computer software has been slowed in some companies by premature leaps into and inadequate planning for major projects. The chaos caused by failed, or even successful, projects often biases key players from continuing on that path and may block additional attempts for some time.

As indicated above, under "issues," box (D) has become a major management concern in most companies. Those who have had success with their early in-house software projects may be inclined to persist, subject to other considerations (cost, time, available skills). Those whose experience has been less rewarding may be more inclined to "go outside" for their software—either custom-designed or off-the-shelf, again subject to other considerations.

As part of the study of the software development life cycle we carried out, one member of the team did a Ph.D. dissertation focussed on the issue of "make or buy" of software.[*] Some of the factors affecting make or buy for software are analogous to those involved in the make or buy decisions on technology in general (see Chapter 6). Others, however, are peculiar to software technology and require new skills on the part of the decisionmakers and users of software.

As indicated earlier, we were also interested in indicators and measures of the effectiveness of software projects. Boxes (E), (F), and (G) in Figure 9-1 list three general levels of output or impact of a software project or program—immediate, intermediate, and ultimate. These are parallel to the levels of output discussed in Chapter 7 on evaluating projects/programs before, during, and after completion.

In the case of software projects, their performance can be "controlled," using a number of indicators of "how the project is going." One major component

[*] Bruce Buchowicz (1987).

of this has been the productivity of individual programmers. Ratios of an order of magnitude have been reported between marginal or "ordinary" programmers and those who are on the leading edge and, in some cases, in the "genius" category. Speed, of course, is not the only important indicator, because some fast programmers make a lot of mistakes, do not document their work, or do not come up with the most efficient software. This is why "lines of code produced"— the most common measure of progress used traditionally—has been falling out of favor. From the user's point of view, the intermediate outputs and impacts on manufacturing are of most interest—how the software affects the cost, reliability, timeliness, and maintainability of his manufacturing systems and the flow of production.

From the top manager's point of view, of course, the *ultimate* outputs or inputs are the ones that count, and the reasons why he invested in the software in the first place—how it impacts on profits, costs, and ROI.

Figure 9-2 lists some specific indicators used by managers of software projects and *their* managers to evaluate the progress and the outputs/impacts.

Figure 9-3 lists some suggested mechanisms for helping to manage, monitor, and evaluate software projects and to avoid some of the barriers commonly encountered. They come, again, from managers of software projects and *their* managers and the technical and staff management specialists who participate in such projects.

---

1. Extent to which users' needs are met
2. Number of milestones met
3. Reliable reporting on programming progress
4. Project time expenditures (actual versus forecast)
5. Project expense (actual versus forecast)
6. Number of interfacing errors
7. Accountability of tasks
8. Programming errors (both detected early and later)
9. Number of unexpected situations or problems
10. Lines of code per unit of time per programmer
11. Impacts on costs of production equipment
12. Impact on variable production costs
13. Impact on tooling costs
14. Impact on product quality
15. Total system design costs
16. Software reliability
17. Feasibility and cost of software modifications downstream
18. Usability of documentation on progress
19. Maintainability of programs in factory
20. Impact on training of software applications and maintenance personnel
21. Factory downtime due to software

Figure 9-2. Some indicators of effectiveness of software development for production.

1. Assign an individual to watch for and handle problems as they arise (versus after-the-fact crisis management).

2. Insist on clarity of documentation and orientation toward user, with continual updating.

3. Identify realistic milestones, reflecting actual technical situations and problems to be dealt with versus nominal or bookkeeping milestones.

4. Anticipate, flag, and follow up problems likely to be encountered in maintenance of software once released to the factory.

5. Arrange for reviews of the software by other (peer) programmers in addition to administrative review by management.

6. Provide team training, feedback on progress, and access to users during the software development process.

7. Determine resource needs for the whole life cycle of a software development project, allowing for negotiation between departments and flexibility for unexpected events.

8. Improve the release-to-production procedure.

9. Assure that the codes turned over to production are up-to-date.

10. Insist on complete and clear documentation at each stage of the process.

11. Keep records of total costs of the project, including maintenance and updating after release to production.

12. Spend more time on the conceptual designs of software, investing in group sessions among programmers, including those not directly involved in the project.

13. Establish a design review group external to the programmers assigned to the project.

14. Record more accurately the actual time spent by programmers on a specific project.

15. Devote time to testing under various operating conditions.

Figure 9-3. Suggestions for mechanisms to help in managing, monitoring, and evaluating the software development/acquisition process.

Figure 9-4 summarizes some of the factors reported to influence the outcomes discussed above—the three levels of output/impact mentioned earlier.

9.2.4 Implications for Managers of Software Projects and Their Managers

This section gives some suggestions for designing and improving the process of developing/acquiring manufacturing software. Many of them apply to all kinds of software—for office automation, factory automation, supporting systems such as inventory and transportation, and for software incorporated in products and services. Some of the general lessons that may be learned from the experience of companies that have either drifted or plunged into software technology projects include the following:

1. Except for a few company-wide systems, most software development and/or acquisition goes on at the divisional level. This is the case where the centralization–decentralization issue of "who is in charge of software for the company" has been resolved in favor of the division manager and his people. Where it is still centralized or where responsibility is split, care must be taken that compatibility is achieved and

Project Software Related Factors

- Characteristics of project
- Program maintenance
- Program modularity
- Experience of software producer
- Project group solidarity
- Use of shelf programs
- Setting of reasonable goals by software clients/users
- Availability of software development tools

Organization/Structure Related Factors

- Interface between developers and users
- Access and reliability of computers and peripherals
- Clear definition of user's needs
- Number of concrete milestones
- Awareness of problems encountered
- Awareness of methods used to alleviate problems
- Sophistication of user
- Level and awareness of decisions on software acquisition/application
- Influence or importance of user
- Flexibility, resiliency, and motivation of user
- Interface of software application with other corporate information resources

Management Related Factors

- Management support
- Method used to measure software effectiveness
- Frequency and method of performance monitoring

Figure 9-4. Key organizational factors affecting software project outcomes.

maintained between divisional and corporate software—both internally developed and externally acquired. As argued in Section 9.3 on the service industries, there may be pressures for "experimentation" with different software packages and whole systems and for avoiding "standardization" or uniformity at the expense of optimizing particular operations. Where this is the case, it is still important to strive for compatibility between systems in different parts of the company. This is becoming easier as more and more software, systems, and peripherals vendors focus on compatibility and multiple-system applicability of software. Failure to assure compatibility can have many negative impacts throughout the firm.

2. CEOs, division managers, CTOs, and other upper-level management people should continually promote high frequency and intensity of interaction between software developers/acquirers and the users of that software and the systems that incorporate it. Industry is still in the learning stages of software usage and the more opportunity management provides (or insists on) for interaction, the more

likely it is that usable and effective systems (according to the intermediate and ultimate criteria of Figures 9-1 and 9-2) will be achieved. Second-guessing the users (e.g., the factory manager and his people) or stonewalling the designers of software by forcing them to "slide the results under the door and don't bother me with the details" are not good ways of managing software technology projects. The atmosphere for such cooperation has to be initially created and continually supported by top management. In the kinds of centralization–decentralization or division–corporate office conflicts discussed in the early chapters, the most likely path is noncooperation. This will tend to be the mode unless top management, whatever its organization "theory" or ideology is with respect to decentralization, takes affirmative steps to create a cooperative atmosphere. How to do it? Make it known that that is the rule of the game, reward people and groups who do cooperate and punish those who do not. Sounds too simple? In principle it *is* too simple. In practice, however, these are the only kinds of actions top managers can take in a highly decentralized company to influence the key players to cooperate.

3. Recognize and plan around the sad fact that, as one software technology manager said, "application software in manufacturing is never a ready-to-use commodity, even when purchased from a promise-you-anything vendor, or even when delivered to the user by credible and conscientious internal software developers." Expectations of "bang-bang" or "slide in the slot" software are occasionally met, but the norm is far off that mark. Realistic managers must add fudge factors to time and cost, as well as performance estimates provided by vendors and in-house software developers. Otherwise, the potential for disappointment and conflict are high.

4. Preoccupation with the "efficiency" or "productivity" of the *front-end* phases of the software life cycle can lead to neglect of the total life cycle cost or cost effectiveness. The costs of software versus hardware in many manufacturing installations have crossed over, with software becoming the dominant cost component. However, another crossing-over is being experienced in manufacturing systems. The downstream costs of *maintaining* the software, once it is in the plant, are beginning to equal and, in some cases, exceed the cost of developing or acquiring the original software package. Management should focus on the *total life cycle cost* and effectiveness in terms of the ultimate outputs/impacts of a software project and not be distracted into overemphasis on the shrinking front-end costs. Like some other aspects of the software development/acquisition process, front-end costs may take care of themselves as more powerful software development tools become available and many programs "write themselves."

5. In contrast with overconcern about the front end of the software project cycle, management should be very aware of and supportive of the needs of the ultimate users and their managers. Frequently, software projects run out of funds. When implementation is in progress, there are few resources to assure effective implementation. Users may need specialized people, training, and retraining—both technical and management—and more autonomy and risk-taking elbow room in the implementation phases. Users who perceive that they will suffer if the project crashes may be reluctant to identify with it or take ownership of it, without adequate

resources and assurance of indemnity from blame if it does fail for reasons beyond their control.

6. Finally, top management should insist on and support efforts at thorough *planning* of the implementation phases of significant software projects—such as automating a whole product line or plant. They must make it clear that implementation of such projects is a joint responsibility of the technology people and the users, plus all the other supporting groups essential to its success. Clear corporate and divisional technology policies are needed to send signals to all concerned parties that this is indeed a joint responsibility and that the project is very important to the company and/or the division. Apparent indifference or arm's-length management of important software projects from the top may be taken as signals that cooperation, extra individual efforts, and risk taking are not valued, thus leading to the frequent "falling between the chairs" fate of many software projects.

9.3 TECHNOLOGY AND R&D/INNOVATION IN THE SERVICE INDUSTRIES[*]

This section examines the opportunities for and barriers to establishing and conducting in-house R&D by firms in service sectors, such as banking, insurance, transportation, retail trade, law, advertising, publishing, and health care. The focus is on R&D related to information technology and telecommunications in such firms. There is little tradition of in-house R&D in the service industries comparable to that found in the manufacturing industries. However, it is becoming evident that increased internal R&D capability is needed to provide a competitive advantage to service sector firms faced with increasing competition in both their home markets and abroad. Barriers to effective internal R&D are discussed, as well as the important policy and practice of "make or buy" in the field of information and telecommunication technology.

9.3.1 Introduction

R&D in the telecommunication sector has traditionally been dominated by a number of large telecommunication firms. With the breakup of AT&T and the "breakout" of many small- and medium-size firms offering telecommunication equipment and systems, the user firms now have a bewildering array of choices with respect to equipment, systems, and services. The "information system" market also provides many choices of computers, peripheral equipment, software, and entire systems for the office, factory, and other applications.

For firms in the service industries—banking, insurance, transportation, retail trade, professional services, and so on—the choices of source of technology for information and telecommunication systems are even broader. Many of them, especially large ones that are dominant in their fields, appear increasingly willing to consider establishing or expanding their internal capabilities for performing "front-

---

[*]Rubenstein, A.H., "In-House R&D on Information Systems and Telecommunication in the Service Industry Firm," paper prepared for the Symposium on Management of the R&D–Marketing Interface, University of Southern California, February 1988, JAI Press, in preparation.

end" R&D on the systems they need to improve their own operations and their services to customers and to meet increasing competition.

Many service industry firms have traditionally had information and telecommunication systems tailored to their own internal needs and for the information technology components of the services they offered to customers. Increasing competition, both domestic and foreign, emphasizes the need for even more tailored and proprietary information technology. This section focuses on the information technology* imbedded in the *services they offer to customers* and clients.

9.3.2 What Is the Role of Technology in a Service Industry Firm?

Depending on which service sector, there are "production" technologies used for various aspects of preparing and delivering services. Transportation companies use materials-handling equipment, vehicles, and other "hard" technology. Retailers use refrigeration, display cases, and packaging equipment. Even banks and insurance companies employ a limited amount of hardware in their operations.

However, the major thrust of technology in the service industries relates to information processing and communication. Indeed, the essence of many service sectors is the unique or cost-effective way in which they obtain, process, store, retrieve, and transfer information and put it to use in their operations. As a consequence, a walk through a typical service sector firm will seldom include a visit to the R&D lab or the product development and engineering departments that are common in manufacturing firms, even some very small ones.

Their analog—the information systems groups, the data processing groups or departments—and similar activities do not give the impression of being "labs" or even technology groups. Large companies in the service industries do have such groups and they are generally staffed by technically trained people—engineers, computer scientists, systems analysts—and are generally supplemented or even dominated by business analysts and nonengineering systems people. This is understandable, because the business of such service businesses is related to systems and procedures which are not necessarily based on physics, chemistry, and the traditional engineering arts and sciences.

Exceptions, of course, are found in the labs of firms in the information and telecommunication sectors of the service industries, other than the manufacturers of information-handling or communication equipment itself. Software, consulting, and systems firms do employ significant numbers of engineers, operations researchers, mathematicians, economists, and others who are in effect doing R&D on the information systems software they sell or support. Our focus in this discussion, however, is on firms whose primary involvement with information systems and their underlying technology are as users, not as vendors.

Many small *manufacturers*—probably 90% of them—do not have a formal in-house R&D group. They depend heavily or entirely for technology on their

---

*For purposes of this discussion, the term "telecommunication" is included under the general term "information technology" or "information systems," covering optical, voice, electronic, and other means of handling information.

suppliers. They buy technology in "packages" or embedded in products, materials, tools, and equipment. They may even buy their whole manufacturing facility on a package basis, ready to use. However, virtually all large and most medium-size manufacturing firms do maintain some kind of in-house technology capability and activity—R&D, engineering, and so on.

This contrasts sharply with the *service* industries where many of even the largest firms have no such internal technology capability. This leaves them, with very few exceptions, highly dependent on outside vendors for their essential "production" technology—information systems. This may not be a bad situation, since it can provide certain advantages to the service industries firm, such as:

- Providing the opportunity to choose information systems and their components from a wide range of sources, including those whose products are at the leading edge of technology.

- Freeing up the management of the firm for their main tasks of planning, selling or delivering, and managing the services they provide.

- Avoiding costly, long-term research and development projects that may or may not pan out and that take valuable resources to formulate, conduct, monitor, and evaluate.

- Letting their suppliers and competitors take the risks of developing, producing, and commercializing new and improved information technology, some of which may not make it in the market and some of which may be obsolete or at least superceded before it is implemented and "up the learning curve."

For more than 90% of the service industry firms, there is no viable alternative to contracting out or buying technology. They are too small, too specialized, too busy, too new, and/or too vulnerable to the costs, waits, and uncertainties of R&D even to consider organizing and maintaining an in-house technology capability. This discussion then is directed at those larger, more mature, more diversified (with certain exceptions) firms in such service sectors as banking, insurance, publishing, retailing, accounting, law, advertising, and transportation. Why should they even consider setting up and maintaining an in-house technology activity, let alone R&D labs for their information technology?

The overriding general motivation for supporting in-house technology is to gain and keep a *competitive edge*. The case of information technology is no different. If all the firms in a sector have the same information systems and related technology, then how can technology provide a comparative advantage in reducing costs, gaining market share, or maintaining customer acceptance? Advertising and promotion can *claim* to offer a differentiated and superior (from a cost-effectiveness viewpoint) "product"—that is, the service being provided. This may work for a while in many consumer markets where the level, flow, and use of information about such services is "imperfect," as an economist would say. The consumer may not be able to take the trouble to differentiate some services provided by competing insurance companies, banks or stores. In the longer run, however, superior services, measured in whatever ways consumers use, generally elbow out or seriously threaten those that are marginal. Some banks and insurance companies grow, others remain static, and others decline or disappear. Technology is not the main cause in such changes in fortunes, but increasingly, it can and does provide the competitive edge—faster, cheaper, easier to use, more understandable, more flexible, more "friendly."

More sophisticated customers and clients—"user" firms—do competitive analysis and comparative shopping and are not as heavily influenced by advertising and promotion as individual consumers. They are less likely to stick with a supplier who is behind the state of the art and whose service is less cost effective than its competitors (including the in-house services of its customers). Information technology is at the core of the competitive position of most of the service sectors and is therefore a key to survival and prosperity to service firms. The key question for a top manager is whether the firm wants to continue to depend on outsiders for this essential technology base. The issue comes down to a policy and the decisions that flow from it on whether to build an "almost self-sufficient" internal capability for this technology, to depend completely on vendors, or to select a position somewhere in-between.

The choice seems easy—an "in-between" position, of course. Putting that choice into practice is much more difficult. Information technology is a broad, diverse, and rapidly changing field. The time lags in "getting organized" can be extensive, and the difficulty in hiring and retaining top technical people and their managers and the high costs of entry make many of these mixed strategies very risky. It may be very difficult to attract top-level researchers and even development engineers to an R&D activity in a firm that "really is not going to get into R&D on a large scale or in depth." Information technologists who may be attracted to such a situation may be far from "leading edge" people, may be highly mobile, and are very unlikely to have the time to really understand the firm's business and the essential roles of technology in its operations. However, such a group may be better than none. It can follow the literature, analyze and select equipment and systems, find and evaluate vendors, monitor the work of the vendors, and implement the technology when it is delivered. These in fact are major roles of many technology groups in manufacturing companies. However, it is unlikely that such a limited technology capability will provide the competitive edge that information technology is capable of providing in many service sector firms through the development of proprietary radical innovations.

The dominant factors in survival and success for many firms in the service industries, of course, are not necessarily technology as such. They may include the experience, imagination, and energy of their people; the loyalty of their clients or customers; and the specialized command of the "arts" underlying the nature of the business. Top-flight merchandisers, creative advertising people, lawyers with good track records, and bankers who know their customers can make or break the service firm. However, it is becoming increasingly clear that such arts are not enough. They have to be backed up by technology—information technology—in order to make them most effective. The actual realization of creative new services for their clients/customers, beyond the "idea" stage, frequently requires application of fairly sophisticated information technology, preferably of a proprietary nature.

As part of the research program of our new Center for Information Technology at Northwestern University, we are examining a number of service sectors from the viewpoint of: "How do you make money and beat the competition in this sector?" In some sectors, the contribution of information technology or any kind of technology places second, well behind some of the arts and personal talents. For many others, it has become not only indispensible for survival but also the means of growing and increasing their market share.

Examples abound of the critical nature of information technology in the service sectors. Some are: cheaper and better typesetting, transmission, and printing; faster and more effective claim settling; automatic handling of deposits, funds transfers, and other banking services; automated reservation systems; computer-based security transactions and following the markets; improved order pickup and delivery; and faster credit checking.

The future prospects are even more exciting. Predictions about the "paperless office" and other information-age developments have not yet been realized, if they ever will be. But advances in information technology have impressed many top managers with the need to be sure that their firms are in a position to take full advantage of developments as they become available. For those firms in the service industries which clearly see that need for at least some, if not a "complete," in-house capability in information technology, we now turn to the question of how to organize it.

9.3.3 Organizing Information Technology in the Service Industries Firm

A major interest in this section is managing technology in the *decentralized* firm. I primarily address that issue. This is not to say that problems of acquiring, organizing, and effectively using information technology in the *centralized* firm are straightforward. Many of the issues that have to be dealt with in organizing technical activities in the decentralized firm will also apply to the centralized, typically single-line or single-sector firm. The special nature of the diversified, decentralized firm makes the management of technology particularly complex and interesting.

The decentralized but "single-line" service firm—a retail chain, a geographically decentralized bank or insurance company, an accounting or law firm with offices in major cities—has more opportunity for a CRL* type of technology organization than one made up of a collection of different kinds of service businesses—such as banking, insurance, real estate, food services. That is, the "production" of such single-line firms suggests that a pooled effort in technology could help share the costs and provide high opportunity for leveraging the results of research.

In some firms, the management philosophy that led to or maintains the decentralized form of organization also carries with it the constraints that the corporate staffs only have a limited role in "guiding or supporting" the autonomous divisions.

---

*Corporate Research Laboratory.

In such cases, the chances for a successful CRL or corporate technology staff are about the same or perhaps even less than in the case of the decentralized manufacturing company. Many issues encountered in decentralized firms will influence the ability or the willingness of the CEO, his top management team, and the corporate staffs to involve themselves directly in divisional operations, even to "help" them introduce new technology. If the managers of the operating divisions resemble some of the stereotypes I have drawn of division managers (DMs) in many manufacturing companies, they are unlikely to seek, support, or tolerate such "helping" activities from the corporate staff, no matter how expert and how well-intentioned.

We have encountered stubborn resistance to standardization of computer and communication equipment, software, or systems and procedures by many DMs in service industry firms. They frequently prefer to go outside and have the information technology provided by or developed for them by consultants and vendors who are "*their* contractors, working directly for them, and on a very short leash." This is one of the main causes of the rapid proliferation of equipment, software, and systems that are often not compatible and are seldom of uniformly high quality and leading edge technology, due to hasty acquistion and only limited analysis and testing.

Of course, this dedication to their own division's needs, problems, and opportunities can provide fast, focused application of information technology. The downside is that, even from a division's point of view, such applications may be so narrowly focused and short range that new developments and new threats, needs, or opportunities may find the division locked into systems, equipment, and software that are difficult or too expensive to change. "That's o.k.," say some observers. "We can view such individual, decentralized decisions and applications as a sort of 'experiment or test bed.' We can examine the different approaches taken by the different divisions and, eventually, select and mandate the 'best' ones for use company-wide." Easier said than done. The commitment and comfort developed by divisional personnel to "their" information technology may take years and many bitter struggles to replace with better, more standard technology. In addition, the challenge to a DM's autonomy may be far greater in attempts to get him to reverse or abandon his approach to information technology than a mandated cooperative effort in the first place, before he has committed himself and his division to a particular set of information technologies.

The logic of this independent (stubborn?) position by the autonomous DMs of nondiversified but decentralized operations is hard to challenge on grounds of organizational philosophy. If, in accordance with the decentralized doctrine, he is expected to run his own show and to optimize his operations, then he should (must?) have control over the resources needed to do it. Exceptions may be obvious things such as national advertising, legal services, corporate finance, credit functions, and some other functions governed by economies of scale or legal considerations.

The "technology logic" may be much less apparent but can be equally compelling for identical or similar operations in regional offices or different market segments of the same basic market. They share a need for company-wide technology that can provide economies of scale in development, procurement, and even operation and maintenance. There is a need to assure that the development effort is of a critical size that can afford to probe and utilize the supporting science and technology, the adjacent fields that might contribute, and the over-the-horizon research that can help the firm prepare for the future.

This all sounds like an argument for centralized R&D for this kind of firm. It is. The critical mass needed to carry on such a program is best done, in my view, in a central facility or set of facilities supported, in some viable way, by all the potential beneficiaries—the operating divisions. Arguments about above-the-line, below-the-line, fee-for-service, a tax on operations, shared costs, and other financing schemes need to be worked out to avoid the kinds of problems that increasingly plague the CRLs of manufacturing firms (see Chapters, 2, 3, and 4).

9.3.4 What About the Diversified Service Firm?

At first glance, the feasibility of a centralized technology effort in a *decentralized and diversified* service firm may seem even less than for one that is not diversified. "My business is different. How can a corporate information technology group or lab hope to understand and address my special problem along with those of the other divisions who are in quite different businesses?" A deeper look, however, and a careful analysis of the threats, needs, and opportunities faced by individual divisions, with respect to information technology, may yield much more commonality than that first glance did.

The diversified *manufacturing* company's divisions may use drastically different kinds of materials, equipment, and manufacturing processes to make entirely different kinds of products. On close scrutiny, however, many of the "materials" and "processes" involved in different *service* sectors have strikingly similar underlying technologies. Their outward manifestations may look quite different—an automatic teller machine, a typesetting machine, a personal computer, or a telephone. But the "raw material" and the way it is processed have many commonalities. The underlying sciences and technologies may also have much more in common than in the typical diversified manufacturing firm. Small groups of specialists in computer science, system design, communication, artificial intelligence, graphics, and human factors can be used to address a wide range of what appear, at first, to be highly specialized applications.

"Let's not rush into this CRL thing too fast for the service firms just because they seem to have some technology and underlying science in common. Haven't you argued elsewhere[*] that commercialization and implementation are very difficult in general, but especially in the decentralized firm?" Yes, and for that reason, the analogs of these difficulties must be taken into account in organizing information technology as well as "hard" technology.

Central and pooled research and advanced technology have to be supported by divisional application and technology transfer capability, as well as by divisional

[*] See Chapters 2, 4, 6, and 8.

networking, to provide the advantages of synergism and critical mass for the company as a whole. In other words, the issues and choices the division manager and his people face in the service firm are no less complex and challenging than those faced by division managers in manufacturing divisions. A major difference is the temptation to view information technology as less complex, more easily understood and managed, and more easily contracted out than "hard" technology.

Rapid advances are being made in pattern recognition, image processing, expert systems, communication networks, voice/data transmission, copying, data storage and retrieval, word processing, printing, and office automation in general. The advances are too rapid and too discontinuous to accommodate a small, plodding, highly constrained development group with rapidly obsolescing technical people. Unfortunately, this characterizes the divisional efforts mounted by many operating divisions in service companies. If "firsts" are to be achieved in the market and technical superiority attained and maintained, more substantial information technology efforts are needed.

As in the case of large and high-tech manufacturing divisions, which are in essence separate stand-alone companies, some components of diversified, decentralized service firms can indeed support a substantial information technology effort. Many large banks, insurance companies, real estate firms, and others have substantial information system (IS) staffs and groups with related names. Some of them include individuals and subgroups that are pretty well up on the leading edge of technology for their industry. They do indeed design and develop or contract out for equipment, systems, software, and application technology, which puts them out front competitively.

That situation is rare, however, and, in some sectors, becoming even more rare. The early delays, high costs, aggravation, and mistakes associated with many such groups have convinced division management (and some of their top corporate management) that they would be better off buying rather than "making" their information technology. Some DMs may reverse that decision later, as they see their competitive edge and ability to keep trade secrets and move quickly toward implementation hampered by overdependence on outside vendors.

As a group, however, division managers in most service industry firms have not yet faced the full impact of information technology on their division's operations—present and future. They are somewhat behind the manufacturing industries' amount and depth of experience with supporting and using advanced technology. Part of the reason for this is the subtle way in which information science and technology have been developing over the past few decades, on the one hand, and the suddenness with which specific technologies and applications have burst into the service industries in the past decade, on the other hand. Both patterns of development left the DM with inadequate planning, budgets, and staffing to deal with the full potential of information technology for his business. Coupled with this is the almost complete absence of technically trained people as division managers in most service firms. The careers of most of them have been in specialties associated with their own sector—for example, banking, insurance, and transportation—or general sales and service. Many division managers in manufacturing firms arrived at those postions via an engineering or manufacturing path and have had some training and/or experience in dealing with the basic technology of their operations. They may have the narrow, short-term view which I have ascribed to them,[*] but they typically have some familiarity with the technology their division uses. As information technology becomes more visible to top management as a necessary part of operations and even the "key to success," there is likely to be an increase in the number of technically trained people in DM positions in service industry firms. However, the pace is not likely to be fast and the percentage is not likely to be anywhere as great as in the manufacturing industries.

Conceptual, selling, and service skills are still likely to dominate management in the foreseeable future as long as these arts dominate the set of factors affecting success in the market. Given that situation, what is a good strategy for assuring an adequate information technology base for the operating divisions in a service industry firm?

9.3.5 A Strategy for Information Technology in the Service Industry Firm

What's the best way of organizing and "injecting" technology into the typical service industry firm? Very carefully, if there is no or little tradition of the development of technology within the firm or the sector. The first groups of engineers and computer scientists who tried to revolutionize the technologies of their firms in banking, retailing, law, and other service firms ran into stiff active opposition and/or passive resistance from both line managers and staff people in their firms. In the past two decades, the presence and activities of such technical people have become commonplace in the very large service firms. Some banks have hundreds of technically trained people in their systems, data processing, information systems, methods and procedures, and related groups. Full-fledged R&D labs are still very rare, even in the largest firms, however.

Despite this, many of the leading firms in banking, insurance, publishing, transportation, and other service sectors have substantial portfolios, which, at least by analogy, parallel the kind of R&D categories and strategies maintained by manufacturing firms.[†] Most firms in the service industries do not, however, have anything resembling an R&D portfolio that will both protect their current "product (service) lines" and comparative advantages and prepare for future ones, based on the cutting edge of information technology and its underlying sciences. Getting started on building a technology capability or expanding and improving an existing one involves several stages:

1. *Audit the firm's current capabilities* for generating, processing, developing, and implementing the results of technically based ideas for improved services and

---

[*]See Chapters 3 and 4.

[†]See Chapter 7.

supporting technology. At this stage, the T/I audit* can be useful, if adapted to the special circumstances and needs of the service industries firm and the particular sector in which it operates or hopes to operate in the future.

2. *Analyze the idea generation and flow process* to ascertain where good technical ideas for information technology applications have come from in the past and where they seem to be coming from currently. In many service businesses, the "needs" part of many ideas do come from customers as well as from competitors and suppliers—that is, from outside the firm. Care must be taken when setting up an in-house technology capability that these channels for ideas are not cut off or attenuated by the organizational structure (inaccessibility) and red tape or the attitudes of the technical people (arrogance or indifference). This is crucial, because many service firms are based on or make their mark through clever merchandising of services and the continual introduction of new products—that is, services.

In addition to the ideas coming from outside, many of the key ones have historically originated inside the firm from nontechnical people—that is, people without engineering or science training. The idea for a new banking, retail, publishing, or legal service does not require a technical background or the years of technical project work required in hardware R&D. Many such ideas are "flashes" that occur to people throughout the firm, including top management and division management. Some of them are already in a form where the need has been correctly identified and even the technical means of satisfying the need has been conceived in general terms where, in fact, a technical means is required. Many existing services, just as in consumer goods, merely need a new label, a new package, or a twist to make them into a "new" service. Others require deep and prolonged technical developing and testing. Again, in assessing the "idea" capabilities of the firm, the historical and current idea sources should not be cut off, devalued, or allowed to attenuate. A newly graduated computer scientist or even one with a few years of experience is still relatively innocent of the specific opportunities, requirements, and constraints of the firm's special market niche(s) and the perennial issue of: "How do you make money in this business or even stay alive in it?"

3. *Design an open technology system.* Any organizational arrangement for a new or expanded technology group should therefore be able to protect and coopt these valuable internal idea sources as well as outside ones. Leading idea generators and idea "pushers" or champions may be veterans of the firm and/or the industry and may be uninterested in or unavailable to participate as formal or full-time members of the technology group. Channels to them must be kept open, especially when they are members of upper management—corporate or divisional. The need to somehow include both these important internal people and a wide variety of external idea sources in the firm's technology activity mitigates against the traditional R&D lab structures common in manufacturing firms. Setting up or allowing information technology groups or departments in the company to become remotely located, organizationally isolated, inward–looking groups, as many of the R&D labs have become, can be disastrous for information technology in the service firm.

---

*Technology/innovation audit; see Chapter 7.

A further argument for an open organizational structure lies in the area of implementation of information technology. Whatever arguments can be made for the cost effectiveness of having some internal technology capability (making your own technology) versus buying it from outside vendors, a large percentage of the actual equipment and software *will* come from outside. This means that the technology group has to be capable of and available for implementing or helping to implement the technology in the operating divisions and other parts of the firm. Except for some of the very largest firms in a service sector, the total number of R&D people is not likely to approach that of comparable manufacturing firms which have some labs of several thousand people.

Aside from people involved in routine data processing and minor modifications to existing systems, software development, and technical service for existing information systems, the "real" R&D group is likely to be fairly small. This means, going back to the centralization–decentralization issue, that it typically should be a shared, in most cases, corporate group. Major exceptions, as noted before, are for very large divisions whose technology is unique in the company and which can afford a "corporate size and level" technology group.

4. *Negotiate and resolve* (even if only temporarily) *the location and funding basis* for the technology group. At one extreme, this can be resolved in the direction of a corporate staff location and 100% funding from corporate overhead. Another pattern is to have it located centrally (organizationally and/or geographically) but directly funded all or in part by the operating divisions. Another deployment is to have a core group at corporate headquarters and satellite groups colocated with the divisions. Again, the funding pattern can vary between complete corporate and complete divisional funding.

5. *Staff the technology group* with a combination of scientific and/or highly technical and highly experienced, practical types of people. The scientific/technical people are needed to interface with the leading edge of information technology and its underlying disciplines. Many companies have made false starts and poor investments in information and computer systems because, among other reasons, they did not have the in-house capability to evaluate progress in the field, identify appropriate technology, and assess the wares of vendors.

We have recently been involved with a large service firm which committed itself to a major computer system for scheduling its distribution activities. The choice made was the wrong one, in terms of the changing state of the art, but the firm is not in a position financially or politically to abandon or radically change their system, due to the heavy financial and ego investment they have made in software, data bases, and decision rules for their existing system. A brief scrutiny of their system by a specialist in the field, who is at the leading edge of technology, spotted this self-entrapment too late to influence the choice of system and the heavy sunk cost associated with it.

Balancing these state-of-the-art people, the technology group should have, as fully integrated members, operations people experienced in the firm's business(es), needs, and environment and steeped in its technologies of doing business and delivering its services. An economist[*] working on the economics of barge traffic got the "radical" idea of riding barge trains and asking some of the captains "how do you make (or lose) money in this business." Most of the armchair analysts had never thought of doing so.

6. *Integrate the theoretical/technical and the practical group members.* To avoid many of the personnel transition and technology transfer problems encountered in hardware projects, the "practical" people should have the status of "full" members of the technology group, even if they are not full time (some of them may have duties in other functional areas and/or in upper levels of management). This integration can be accomplished at several stages of an actual project or program. The formulation of the project, based on an idea from any source, can be the subject of an initial concept-formation and study design stage.

The "practitioner" members of the technology group are essential at this early stage to help assure the eventual practicality of the information system or service being developed and its successful implementation. They are needed to offset tendencies of many of the leading edge information scientists and engineers. Like their parallels in physics and chemistry, many of the latter tend to shoot for systems that cannot be designed, built, or used in a cost-effective way in the available time frame and within the other real constraints that exist in the firm. The practitioners are also needed to provide insights into the configurations and features which will increase the chances of use and success of any related system, equipment, hardware, or software that is developed.

This integration will not be quick and easy. The theoretical people may be viewed as "blue sky dreamers" (or worse) by the practitioner types. The latter may be viewed as "nay sayers" and worse by the former. Mutual respect, if not similar views, must be fostered for such an integrated group to be developed and to operate effectively. We have had reasonable, but not overwhelming success in getting such teams to listen to each other and develop respect for each others' knowledge and points of view. Stripping away the jargon and the mythology helps and needs to be done early on.

It may still be necessary to ego gratification for the group as a whole to project an air of mystery toward others; but inside the group this can be very difficult to live with. "But why do you need to maintain separate data bases for inventory and for distribution?" "Because the inventory guys are 'us' and the distribution guys are 'them'." "Do you mean it's due to organizational politics rather than economics?" "Yes, you finally got it."

Following the integrated conceptual stage, it is even more important that the theoretical and practical people work together and agree on the subsequent stages of the project. This includes the software development; equipment selection, development, and procurement; system design; "bells and whistles" decisions on what features to give the system; and, most importantly, the implementation stages.

Few events will blow such a relation more than one of the practitioner members finding out, too late, that the original concept has been drastically changed without

---

[*]Art Hurter at Northwestern University, who is also an engineer.

notifying or consulting them. The changes may be for good reasons—sudden advances in the state of the art, a new feature that becomes available, an idea from one of the team members, technical limitations of the approach originally agreed on. But the *surprise* involved in the change, especially when it occurs or is discovered very far downstream in the project life cycle, can be very destructive for the technology team itself and, possibly, for the success of the installation.

The threat to success of the project may come through false steps and irreversible actions taken in preparation for what was thought to be the features of the system as originally conceived. In general, except for some revolutionary information systems (many of which have failed in application) most information technology projects have a short lead time or life cycle and often involve overlapping of the R&D and the implementation phases.

Financial commitments must be made, contracts let, organizational changes put into motion, people trained, people hired, and people psychologically prepared for the impending changes. If the technical surprises make some or many of these anticipatory moves incorrect or obsolete before the system is even developed, then blood will (and frequently does) flow and heads will roll. Worse than that, the level of trust and cooperation may be severely damaged for the foreseeable future.

Continued integration of theoretical and practical people, especially when the latter may not be full-time members of the R&D team, can be a time-consuming and frustrating task. "These guys are always asking dumb questions that any computer science student could answer for himself." "Those guys get very defensive when we ask simple questions to make sure they are on the right track and the system design will be useful and cost effective."

It takes time and patience. The payoff can be high from sincere and continuing efforts at building integrated teams. The cost of not doing so can mean failure of specific projects and a general setback of the information technology program.

9.3.6 What About a Back Room Group Focusing on the Underlying Science and Very Advanced Technology?

By all means, such a group is needed for any substantial size information technology effort in a large, leading company. And, by all means, they should not be isolated and protected against the operating divisions as so many CRLs are in manufacturing companies.

Looking at the many specialties in the broad field of "information science and technology," it may seem that a substantial group is needed to cover "all" of them. That is not necessarily so. A group of half a dozen or less broad-based computer science and information technology people can manage to be aware of what is happening across the whole range of information technology to the extent of knowing where to probe further and where to seek deeper professional expertise from vendors, consultants, colleagues in other companies, or university faculty. In a sense, I am conceiving of this "back room" group as "scouts" for the firm. But, like their analog in the military, scouts who keep the information to themselves and become "information sinks" will be of little use to the firm and its technology programs.

Now we require the concept of "key communicators." Members of the back room group need to be able and willing to act as key communicators within the firm as well as gleaners of information from outside. (See Chapter 4.)

We have encountered many such people attempting to fill this dual role at many professional society meetings. Some perform the role effectively for their companies. Others use their company positions primarily as a platform for making themselves visible in the technical communities outside and for playing the "research game" without actually doing the research on behalf of their companies. We have heard many presentations that are clearly uncoupled from the realities of the very firms who "sent" these specialists to the meetings.

This is an unsatisfactory situation and such behavior must be noted and acted on to strengthen the back room group and make it more a part of the firm's integrated information technology program. Continual briefings, trip reports, and orientation meetings may appear time wasting. But time well spent in such internal communication can pay off in more cost-effective information technology R&D for the service firm until it has developed a critical mass and sophistication that makes such intense educational efforts less necessary. Some level of such communication will always be required, however.

The small, back room group must be visible and mobile in the company. Beyond protecting a certain percentage of their time for solitary and contemplative research activities, a significant proportion of their time should be spent interacting with the potential users of their results — in the operating divisions. But this seems difficult, if not infeasible, for large decentralized companies where the operating divisions may be widely scattered geographically, both in the U.S. and abroad. Should much of their time be spent on planes, even assuming that they are welcome in the operating divisions? Fortunately, advancing telecommunication technology itself reduces the necessity for that and makes it feasible for them to "travel" widely without leaving home.

A trend that is yet to develop in industry can be copied from universities and the military. This is the nascent set of computer networks that allow individual researchers scattered around the world to work together as research teams. Of course, telephones and, more recently, electronic mail have made this more and more possible (recall that before modern telecommunication was invented, there was research collaboration between scientists in different countries via mail). The technology is now available for a significant increase in the level, pace, and intensity of such "teleresearch" and it is especially appropriate for people doing R&D in information technology!

As to the political and organizational aspects of such a centralized group supporting and stimulating the divisional technology groups, new attitudes and patterns of collaboration have to be worked out within the divisional framework. Information technology *is* different from R&D in the physical sciences and new mechanisms have to be developed and tried if some of the old patterns of isolation and failure in R&D are to be kept from carrying over to information technology in the service industries.

APPENDIX A

COMPANIES PROVIDING DATA AND IDEAS FOR THIS BOOK

This is a list of companies participating in the POMRAD/IASTA studies of R&D in decentralized companies, networking of divisional technology, technology planning, idea flow, project and program evaluation, R&D/production interface, technical entrepreneurship, software development, R&D in the service industries, and various consulting engagements.

Abbott Laboratories
ACF Industries
Admiral
Alco Products
Alcoa
Allegheny International
Allen Bradley
Allied
Allis–Chalmers
Aluminum Company of America
American Bosch
American Brake Shoe
American Can
American Cyanamid
American Enka

American Hospital Supply
American Marietta
American Metal Products
American Metals
American Standard
American Sugar
American Viscose
Amerida
AMF
Amoco
Amphenol
Amsted Industries
Anaconda
Arden Farms
Armco

Armour
Armstrong Rubber
Armstrong World Industries
Ashland Oil
AT&T Technologies
Atlantic Richfield
Avco
Babcock & Wilcox
Bankers Life and Casualty
Barber Colman
BASF Wyandotte
Baxter Travenol Laboratories
Beatrice Foods
Bell & Howell
Bendix
Bethlehem Steel
Borden
Borg–Warner
Brunswick
Bucyrus–Erie
Budd
Burlington Northern
Burroughs
California Packing
Campbell Soup
Carnation
Case (J. I.)
Caterpillar Tractor
Celanese
Chemetron
Chrysler
CIMLINK
Cities Service
Coca-Cola
Colgate–Palmolive
Collins Radio
Computer Vision
Container Corporation
Continental Group

Continental Illinois Bank
Continental Oil
Copperweld
Corning Glass Works
CPC International
Crane
Crown Zellerbach
Crucible Steel
Cudahy
Dart & Kraft
Dayton Rubber
Dean Foods
DEC
Deere
Dewey & Almy
Dexter–Midland
Donnelley (R. R.) & Sons
Dow Chemical
Du Pont
Eastman Kodak
Eaton
Ekco
Electric AutoLite
Emerson Electric
Esmark
Exxon
Fairbanks–Morse
Fairchild Industries
Federal Telecommunication
Ferro
Fiberglass
Firestone Tire & Rubber
First National Bank of Chicago
FMC
Foremost Dairies
Foxboro
Garrett
GATX
General Dynamics

General Electric
General Foods
General Mills
General Motors
General Precision
General Tire & Rubber
Genesco
Gerber Products
Gillette
Glidden
Goodrich (B. F.)
Goodyear Tire & Rubber
Gould
Grace (W. R.)
Granite City Steel
Grinnell
GTE
Gulf Oil
Hagan
Heinz (H. J.)
Helene Curtis
Hercules
Honeywell
Hormel
Hughes Aircraft
Hydrosystems
Hygrade Foods
IBM
IMC
Inland Steel
Intel
International Harvester
International Shoe
Johnson Controls
Johnson & Johnson
Jones & Laughlin
Joselyn
Kaiser Aluminum & Chemical
Kaiser Steel

Kearney and Traecker
Kellogg
Kennecott
Kimberly–Clark
Koppers
Kraft
Lever Brothers
Libbey–Owens Ford
Libby, McNeill & Libby
Lilly (Eli)
Link Belt
Liquid Carbonic
Liquid Air
Lockheed
Lukens Steel
Magnavox
Manville
Maremont
Martin Marietta
Matheson
McGraw–Edison
McLouth Steel
Mead
Merck
Minnesota Mining &
   Manufacturing (3M)
Monsanto
Morrell
Motorola
Nabisco Brands
Nalco Chemical
National Dairy
National Distillers & Chemical
National Lead
National Steel
National Supply
NCR
Norton
Olin

Omark Industries
Otis Elevator
Outboard Marine
Parke–Davis
Pfizer
Phelps Dodge
Philco
Phillips Petroleum
Phillips (Eintoven)
Pillsbury
Pittsburgh Steel
PPG Industries
Procter & Gamble
Pullman
Quaker Oats
Ralston Purina
Rath Packing
Rayoneer
Raytheon
RCA
Republic Steel
Revere Copper & Brass
Rexall
Rexnord
Richardson Electronics
Ritter Midwest
Rockwell International
Rohm & Haas
S&C Electric
Sara Lee
Sargent Welch
Sharon Steel
Schenley
Scott Paper
Scovill
Seagram
Searle (G. D.)
Security Pacific

Shell Chemicals
Shell Oil
Signal Companies
Signode
Siemens/AG
Sinclair Oil
Singer
Skil
Smith (A. O.)
Sperry
Sprague
Square D
St. Regis
Standard Brands
Standard Oil of California
Standard Oil (Ohio)
Standard Oil (Indiana)
Stauffer Chemical
Stewart Warner
Sun
Sunbeam
Swift
Sylvania
Teletype (AT&T)
Tenneco
Tidewater Oil
Timken
TRW
U.S. Gypsum
U.S. Steel
Union Oil
Union Carbide
Uniroyal
Unisys
US Robotics
Wagner Electric
Wang
Warren Oil

Westinghouse Air Brake
Westinghouse Electric
Weyerhauser
Wheeling–Pittsburgh Steel
Whirlpool

Wilson Foods
Xerox
Youngstown Steel
Zenith

APPENDIX B
SELECTED BIBLIOGRAPHY*

B.1   TRENDS IN MANAGEMENT OF TECHNOLOGY/R&D

Rubenstein, A. H., "Trends in Technology Management," *IEEE Transactions on Engineering Management*, November (1985).

B.2   DECENTRALIZATION OF THE FIRM

Ahlbrandt, Roger S. Jr. and Andrew R. Blair, "What It Takes for Large Organizations to Be Innovative," *Research Management*, **29**(2), 34–37, March/April (1986).

Altman, Fred D., *Performance Evaluation of Division Managers in Large Decentralized Industrial Organizations*, M.S. thesis, Northwestern University, 1963.

Anonymous, "Information Management: Centralization vs. Decentralization," *Information and Records Management*, **16**(3), 22–25, March (1982).

Anonymous, "Decentralizing Information Systems," *Chemical Week*, **134**(21), 26–27, May (1984).

Baker, H. and R. R. France, *Centralization and Decentralization in Industrial Relations*, Princeton University Press, Princeton, NJ, 1954.

---

* Two kinds of citations are included in this bibliography. The first is to publications, reports, dissertations, and selected working papers of members of POMRAD—Northwestern's Program of Research on the Management of Research, Development and Innovation and IASTA—International Applied Science and Technology Associates (my consulting group). The second is to the open literature, including research publications of other research groups and articles by management practitioners and observers of the technology management scene. *Note:* Section headings do not match the chapter titles.

Barnard, C. I., "A Definition of Authority," in *The Functions of the Executive*, Harvard University Press, Cambridge, MA, 1938. Reprinted in R. K. Merton, A. P. Gray, B. Hockey, and H. C. Selvin (Eds.), *Reader in Bureaucracy*, Free Press, New York, 1952, pp.180–185.

Blau, P. M., *The Dynamics of Bureaucracy*, University of Chicago Press, Chicago, 1955.

Cordiner, R. J., *New Frontiers for Professional Managers*, McGraw-Hill, New York, 1956.

Dean, J., "Profit Performance Measurement of Division Managers," *The Controller*, **25**, 423–449, September (1957).

Dearden, J., "Problems in Decentralized Profit Responsibility," *Harvard Business Review*, **38**(3), 79 (1960).

Dror, Israel and A. H. Rubenstein, "Top Management Roles in R&D Projects," *R&D Management*, **14**(1), 37–46 (1984).

Drucker, P. F., *Concept of a Corporation*, John Day, 1946.

Gouldner, A. W., *Patterns of Industrial Bureaucracy*, Free Press, New York, 1954.

I.B.M., "A New Pattern for Progress," *IBM Business Machines*, December (1956).

Kruisinga, H. J., *The Balance Between Centralization and Decentralization in Managerial Control*, H. E. Stenfert Kroese, NV-Leiden, 1954.

March, J. G. and H. A. Simon with the collaboration of Harold Guetzkow, *Organizations*, Wiley, New York, 1958.

Paolillo, J. and W. B. Brown, "How Organization Factors Affect R&D Innovation," *Research Management*, **21**, 12–15 (1978).

Radnor, Michael, "The Control of Research and Development by Top Managers of Large Decentralized Companies," unpublished Ph.D. dissertation, Northwestern University, 1964.

Ridgway, V. F., "Dysfunctional Consequences of Performance Measurements," *Administrative Science Quarterly*, **1**, 2 (1956).

Rubenstein, A. H. and C. J. Haberstroh (Eds.), *Some Theories of Organization*, Irwin-Dorsey, Homewood, IL, 1960.

Rubenstein, A. H. and M. Radnor, "Top Management's Role in Research Planning in Large Decentralized Companies. " Presented at the Conference of International Federation of Operations Research Societies: Oslo, Norway, July 1963. Mimeo May 1962.

Schlie, Theodore W., *Government Science Organizations and Federal–Provincial Relations in Canada*, M.S. thesis, Northwestern University, 1969.

Schlie, Theodore W., *The Regional Effectiveness of Regional Research and Development Institutes in East Africa*, Ph.D. dissertation, Northwestern University, 1973.

Siegel, Gilbert B., "The Personnel Function: Measuring Decentralization and Its Impact," *Public Personnel Management* **12**(1), 101–114, Spring (1983).

Sim, A. B., "Decentralized Management of Subsidiaries and Their Performance–A Comparative Study of American, British, and Japanese Subsidiaries in Malaysia," *Management International Review (Germany)*, **17**(2), 45–51 (1977).

Simon, H. A., H. Guetzkow, G. Kozmetsky, and G. Tyndall, *Centralization vs. Decentralization in Organizing the Controller's Department*, Controllership Foundation, 1954.

Thompson, J. D., "Authority and Power in Identical Organizations," *American Journal of Sociology*, **62**, 290–301 (1956).

Weaver, A. G., "The Progress of Decentralization at the International Business Machines Corporation," unpublished Master's thesis, Massachusetts Institute of Technology, 1957.

Weber, H., "The Essentials of Bureaucratic Organization: An Ideal Type Construction," in *The Theory of Social and Economic Organization* (trans. A. M. Henderson and T. Parsons), Oxford University Press, Oxford, 1947. Reprinted in R. K. Merton et al. (Eds.), *Reader in Bureaucracy*, Free Press, New York, 1952.

B.3 NETWORKING AND KEY COMMUNICATORS

Aldrich, H. and D. Herker, "Boundary Spanning Roles and Organization Structure," *Academy of Management Review*, **2**, 217–239 (1977).

Allen, T. J., "Communication Networks in R&D Laboratories," *R&D Management (U.K.)*, **1**(2), 14–21 (1970).

Allais, P. and P. Rodocanachi, "Research and Its Networks of Communicators," *Research Management* **20**(1), 39–42 (1977).

Barth, R. T., "The Relationship of Inter-Group Organization Climate with Communication and Joint Decision Making Between Task-Interdependent R&D Groups," Ph.D. dissertation, Department of Industrial Engineering and Management Sciences, Northwestern University, August 1970.

Benson, J. Kenneth, "The Interorganizational Network as a Political Economy," *Administrative Science Quarterly*, **20**(3), 229–249 (1975).

Chakrabarti, A. K. and R. D. O'Keefe, "A Study of Key Communicators in Research and Development Laboratories," *Group and Organization Studies*, **2**(3), 336–346, September (1977).

Crane, D., *Invisible Colleges: Diffusion of Knowledge in Scientific Communities*, University of Chicago Press, Chicago, 1972.

Czepiel, A. "Pattern of Interorganizational Communications and the Diffusion of a Major Technological Innovation," *Academy of Management Journal*, **18**, 6–24 (1975).

Dale, B. G., J. L. Burbidge, and M. J. Cottam, "Planning the Introduction of Group Technology," *International Journal of Operations and Production Management (U.K.)*, **4**(1), 34–47 (1984).

Dewhirst, H. D., R. D. Arvey, and E. M. Brown, "Satisfaction and Performance in Research and Development Tasks as Related to Information Accessibility," *IEEE Transactions on Engineering Management*, **EM-25**(3), 58–63, August (1978).

Fischer, W. and B. Rosen, "The Search for the Latent Information Star," *R&D Management (U.K.)*, **12**(2), 61–66 (1982).

Frost, P. and R. Whitley, "Communication Patterns in a Research Laboratory," *R&D Management, (U.K.)*, **1**(2), 71–79, February (1971).

Goldhar, J., L. Bragaw, and J. Schwartz, "Information Flows, Management Styles, and Technological Innovation," *IEEE Transactions on Engineering Management*, **23**, 51–62 (1970).

Herner, S., "Information Gathering Habits of Workers in Pure and Applied Science," *Industrial and Engineering Chemistry* **46**(1), 228–236, January (1954).

Hertz, David B. and A. H. Rubenstein, *Team Research*, Eastern Technical Publications, Cambridge, MA, 1953.

Herzog, A. J., "The 'Gatekeeper' Hypothesis and the International Transfer of Scientific Knowledge," *Journal of Technology Transfer,* **6**(1), 57–72 (1981).

Holland, W. E., "Characteristics of Individuals with High Information Potential in Government Research and Development Organizations," *IEEE Transactions on Engineering Management,* **12**(2), 38–44 (1972).

Jedlicka, Allen D. and A. H. Rubenstein, "Some Observations on the Strategies and Organization of Applied Research Institutes in Developing Countries," Department of Organization Behavior, School of Business Administration, University of Northern Iowa, Cedar Falls, and Department of Industrial Engineering and Management Sciences, Northwestern University, November 1974, Document No. 75/7.

Jolly, J. A. and J. W. Creighton (Eds.), *Technology Transfer in Research and Development,* Naval Postgraduate School, Monterey, CA, 1975.

Katz, R. and M. L. Tushman, "A Longitudinal Study of the Effects of Boundary Spanning Supervisor or Turnover and Promotion in Research and Development," *Academy of Management Journal,* **26**(3), 437–456 (1983).

Katz, R. and M. Tushman, "An Investigation into the Managerial Roles and Career Paths of Gatekeepers and Project Supervisors in a Major R&D Facility," *R&D Management (U.K.),* **11**(3), 103–110 (1981).

Keller, R. T. and W. E. Holland, "Boundary Spanning Activity and Research and Development Management: A Comparative Study," *IEEE Transactions on Engineering Management,* **EM-22**, 130–133 (1975).

Kottenstette, James P. and James E. Freeman, *PATT: Project for the Analysis of Technology Transfer,* Industrial Economics Division, Denver Research Institute, University of Denver, Denver, CO, July 1972.

Lesher, Richard L. and George J. Howick, *Assessing Technology Transfer, Scientific and Technical Information Division,* Office of Technology Utilization, National Aeronautics and Space Administration, Washington, DC, 1966.

McKenney, James L. and F. Warren McFarlan, "The Information Archipelago—Maps and Bridges," *Harvard Business Review,* **60**(5), 102–119, September/October (1982).

Monge, P. R., J. A. Edwards, and K. K. Kirste, "The Determinants of Communicator and Communication Structure in Large Organizations: A Review of Research," in *Communication Yearbook 2,* B. Ruben (Ed.), Transaction Books, New Brunswick, NJ, 1978.

Moor, William C., *An Empirical Study of the Relationship Between Personality Traits of Research and Development Personnel and Dimensions of Information Systems and Sources,* Ph.D. dissertation, Department of Industrial Engineering and Management Sciences, Northwestern University, June 1969.

O'Keefe, R. and K. C. Kirsch, "Preliminary Description of Possible Experimental Treatments in Field Experiments on Key Communicators in the Dissemination and Utilization of Scientific and Technical Information," Program of Research on the Management of Research and Development (POMRAD), Northwestern University, April, 1977, Document No. 77/76.

Persson, O., "Critical Comments on the Gatekeeper Concept in Science and Technology," *R&D Management (U.K.),* **11**(1), 37–40 (1981).

Roberts, K. and C. O'Reilly, "Some Correlations of Communication Roles in Organizations," *Academy of Management Journal,* **22**(1), 42–57 (1979).

Rubenstein, A. H., "Liaison Relations in Research and Development," *IRE Transactions on Engineering Management,* **EM-4**(2), 72–78, June (1957).

Rubenstein, A. H., W. Beal, D. L. Kegan, E. Moore, W. C. Moore, G. Rath, C. W. Thompson, R. Trusewell, and D. J. Werner, "Exploration on the Information-Seeking Behavior of Researchers," in *Communication Among Scientists and Engineers*, C. E. Nelson and D. K. Pollock (Eds.), Heath Lexington Books, Lexington, MA, 1970.

Rubenstein, A. H., R. T. Barth, and C. F. Douds, "Ways to Improve Communications Between R&D Groups," *Research Management*, **14**(6), 49–59, November (1971).

Rubenstein, A. H., R. D. O'Keefe, and C. N. Thompson, *Methods of Field Experiments on Scientific and Technical Information*, Program of Research on the Management of R&D, POMRAD, Technological Institute, Northwestern University, 1978.

Rubenstein, A. H., G. Rath, and R. O'Keefe, "Behavioral Factors Influencing the Adoption of an Experimental Information System," *Hospital Administration*, **18**(4), 27–43, Fall (1973).

Spekman, R. E., "Influences and Information: An Exploratory Investigation of the Boundary Role Person's Basis of Power," *Academy of Management Journal*, **22**(1), 104–117 (1979).

Taylor, R. L., "The Technological Gatekeeper," *R&D Management (U.K.)*, **5**, 239–242 (1975).

Thompson, C. W. N., "Technology Utilization," in *Annual Review of Information Science and Technology*, C. A. Caudra (Ed.), Encyclopedia Britannica, pp. 383–417, Washington, DC, 1975.

Tushman, M. L., "Communication Across Organization Boundaries: Special Boundary Roles in the Innovation Process," *Administrative Science Quarterly*, **22**(4), 581–606, December (1977).

Tushman, M. L. and T. J. Scanlon, "Boundary Spanning Individuals: Their Role in Information Transfer and Their Antecedents," *Academy of Management Journal*, **24**(2), 289–305 (1981).

B.4 TECHNOLOGY PLANNING

Abetti, Pier A., "Technology: A Challenge to Planners," *Planning Review*, **12**(4), 24–27, July (1984).

Birnbaum, P. H., "Strategic Management of Industrial Technology: A Review of the Issues," *IEEE Transactions on Engineering Management*, **EM-31**(4), 186–191, November (1984).

Bisio, Attilio and L. Gastwirt, "R&D Expenditures and Corporate Planning," *Research Management*, **23**, 23–26, January (1980).

Bitondo, Domencio and Alan Frohman, "Linking Technological and Business Planning," *Research Management*, **24**, 19–23, November (1981).

Conley, Patrick, "How Corporate Strategies Are Affecting R&D Today," *Research Management*, **16**, 18–20 (1973).

Cutler, W. G., "R&D Planning—Formulating the Annual Research Program at Whirlpool," *Research Management*, **22**, 23–26, January (1979).

Damaschke, K., "Systems and Methods for Planning Research and Development in Industry," *Working Group Report No. 21*, European Industrial Research Management Association, Paris, 1979.

Denny, Fred I., "Management of an R&D Planning Process Utilizing an Advisory Group," *IEEE Transactions on Engineering Management*, **EM-27**(2), 34–36, May (1980).

Domsch, Michel, "The Organization of Corporate R&D Planning," *Long Range Planning*, **11**(3), 67–74, June (1978).

Ehrbar, A. H., "United Technologies' Master Plan," *Fortune*, **102**(6), 96–112, September (1980).

El Sawy, O. A., "Understanding the Process by Which Chief Executives Identify Strategic Threats and Opportunities," *Academy of Management Proceedings*, 37–41, August (1984).

Ettlie, J. E., "Organizational Policy and Innovation Among Suppliers to the Food Processing Sector," *Academy of Management Journal*, **26**(1), 27–44, March (1983).

Ettlie, J. E. and W. P. Bridges, "Environmental Uncertainty and Organizational Technology Policy," *IEEE Transactions on Engineering Management*, **EM-29**, 2–10 (1982).

Faust, R. E., "Pharmaceutical Research Planning Strategies," *Journal of the Society of Research Administrators*, **7**(3), 35–42, Winter (1976).

Geisler, E., *An Empirical Study of a Proposed System for Monitoring Organizational Change in a Federal R&D Laboratory*, Ph.D. dissertation, Northwestern University, 1979.

Geisler, E. and A. H. Rubenstein, "Long Range Planning of Corporate Research and Development," *Planning Review*, **13** 39–41, March (1985).

Gold, B., "Alternative Strategies for Advancing a Company's Technology," *Research Management*, **18**(4), 24–29, July (1975).

Hambrick, D. C., I. C. MacMillan, and R. R. Barbosa, "Business Unit Strategy and Changes in the Product R&D Budget," *Management Science*, **29**(7), 757–769, July (1983).

Hampel, R. G., "Building the Total Corporate R&D Effort at Alcoa," *Research Management*, **22**(1), 27–30, January (1979).

Hanson, W. T. Jr., "Planning R&D at Eastman Kodak," *Research Management*, **21**(4), 23–25, July (1978).

Hazelrigg, George A. Jr., "Evaluation of Long-Term R&D Programs in the Presence of Market Uncertainties," *Energy Systems and Policy*, **6**(2), 109–134 (1982).

Lathrop, J. W. and K. Chen, "Comprehensive Evaluation of Long-Range Research and Development Strategies," *IEEE Transactions on Systems, Man and Cybernetics*, **SMC-G**(1), 7–17, January (1976).

Mady, Gregory, B. and Burton V. Dean, "Strategic Planning for Investment in R&D Using Decision Analysis and Mathematical Programming," *IEEE Transactions on Engineering Management*, **EM-32**(2), 84–90, May (1985).

McGlauchlin, Lawrence, "Long Range Technical Planning," *Harvard Business Review* **64**(4), 54, July/August (1968).

Mechlin, George F. and Daniel Berg, "Evaluating Research—ROI Is Not Enough," *Harvard Business Review*, **58**, 93–99, September/October (1986).

Miller, T. R., "Planning R&D at Union Carbide," *Research Management*, **21**(1), 31–33, January (1978).

Neubert, Ralph, "Strategic Management the Monsanto Way," *Planning Review*, January (1980).

Ojdana, Edward, "Some R&D Planning Practices of High Technology Firms," The Rand Corporation, Santa Monica, CA, P-5732, October 1976.

Petersen, G. T., "Working R&D into Corporate Strategy," The Conference Board Record, January 1, 1976; condensed version in *Management Review,* **65**(5), 35–38, May (1976).

Quinn, J. B., *Strategies for Change: Logical Incrementalism,* Richard D. Irwin, Homewood, IL, 1980.

Roberts, E. B., "Strategic Management of Technology," paper presented June 21, 1983, conference on *Global Technological Change,* Massachusetts Institute of Technology, Cambridge, MA.

Warren, Kirby, "Perspectives on Planning Trends and Changes," *Risk Management,* **28,** March (1981).

Weil, Edward and Robert Cangemi, "Linking Long Range Research to Strategic Planning," *Research Management,* **26,** 32–39, May/June (1983).

Wetherbe, James and John Montanari, "Zero-Based Budgeting in the Planning Process," *Strategic Management Journal,* **2**(1), 1–4 January (1981).

Young, Earl C., *An Analysis of Selected Strategies for Organizing R&D in Developing Countries with Reference to Policy and Planning Techniques, International Relations, Manpower and Training and Information Requirements,* M.S. thesis, Northwestern University, 1966.

B.5   SOURCES AND FLOW OF IDEAS

Avery, R. W., "Technical Objectives and the Production of Ideas," presented at M.I.T. Industrial Liaison Program Seminar on the Organization of R&D in Decentralized Companies, 1959.

Baker, Norman R., "The Influence of Several Organizational Factors on the Idea Generation and Submission Behavior of Industrial Researchers and Technicians," Ph.D. dissertation, Northwestern University, 1965.

Baker, N. R., Stephen G. Green, and Alden S. Bean, "How Management Can Influence the Generation of Ideas," *Research Management* **28**(6), 35–42, November/December (1985).

Baker, N. R., J. Siegman, and A. H. Rubenstein, "The Effects of Perceived Needs and Means on the Generation of Ideas for Industrial Research and Development Projects," *Transactions on Engineering Management,* **EM-14**(4), 156–163 (1967).

Bolen, Frank, "A Method for Real Time Measurement of the Flow of Ideas in Industrial Research Laboratories," unpublished Master's thesis, Northwestern University, 1963.

Bonge, John Walter, *Perception and Response to Major Change by Purchasing Agents,* Ph.D. dissertation, Northwestern University, 1968.

Bush, G. P. and L. H. Hattery, "Team Work and Creativity in Research," *Administrative Science Quarterly,* **1,** 361-372, December (1956).

Columbia University, *The Flow of Information Among Scientists,* Bureau of Applied Social Research, New York, 1958.

Conway, H. Allan and Norman W. McGuinness, "Idea Generation in Technology-Based Firms," *Journal of Product Innovation Management,* **3**(4), 276–291, December (1986).

Delbecg, Andre L. and Peter K. Mills, "Managerial Practices that Enhance Innovation," *Organizational Dynamics,* **14**(1), 24–34, Summer (1985).

Gershinowitz, H., "Sustaining Creativity Against Organizational Pressures," *Research Management,* **3**(1), Spring (1960).

Geschka, Horst, "Introduction and Use of Idea-Generating Methods," *Research Management,* **21**(3), 25–28, May (1978).

Goldberg, Louis C., *Dimensions in the Evaluation of Technical Ideas in an Industrial Research Laboratory,* M.S. thesis, Northwestern University, 1963.

Goldhar, Joel D. and Louis K. Bragaw, "Information Flows, Management Styles, and Technological Innovation," *IEEE Transactions on Engineering Management,* **23**(1), 51–62, February (1976).

Green, Stephen G., Alden S. Bean, and B. Kay Spavely, "What Happens to Ideas?", *Research Management,* **27**(6), November/December (1984).

Johnson, G., "Outside Experts: Solutions or Fast Talk?", *Industry Week* **218**(6), 63–69, September (1983).

Jones, S. L. and J. E. Arnold, "The Creative Individual in Industrial Research," *IRE Transactions on Engineering Management,* **EM-9**(2), 51–55, June (1962).

Kaplan, N., "Some Organizational Factors Affecting Creativity," *IRE Transactions on Engineering Management,* **EM-7**(1), 24–30, March (1960).

Kramer, S., "The Art of Selling Your R&D Ideas," *Research Management,* **24**(2), 7–8, March (1981).

Lionberger, Herbert F., *Adoption of New Ideas and Practices,* Iowa State University Press, Ames, 1960.

MacLaurin, W. R., "New Products Innovation and Introduction: Their Broad Implications," in A. H. Rubenstein (Ed.), *Coordination, Control and Financing of Industrial Research,* King's Crown Press, New York, 1955, pp.34–41.

Martin, Robert Burton, *Some Factors Associated with the Evaluation of Ideas for Production Changes in Small Companies,* Ph.D. dissertation, Northwestern University, 1967.

Nagpaul, P. S. and S. Pruthi, "Problem Solving and Idea-Generation in R&D: The Role of Informal Communication," *R&D Management,* **9**(3), 147–149 (1979).

Pelz, D. C. and F. Andrews, *Scientists in Organizations: Productive Climates for Research and Development,* Wiley, New York, 1966, and Revised Edition: University of Michigan Press, Ann Arbor, 1976.

Pound, William H., *Communications, Evaluations, and the Flow of Ideas in an Industrial Research and Development Laboratory,* Ph.D. dissertation, Northwestern University, 1966.

Rubenstein, A. H., "Studies of Idea Flow in Research and Development," presented to the New York Chapter, TIMS, November 1963.

Rubenstein, A. H., "Field Studies of Project Selection Behavior in Industrial Laboratories," in B. V. Dean (Ed.), *Operations Research in Research and Development,* Wiley, New York, 1963, pp.189–206.

Rubenstein, A. H. and Robert W. Avery, "Idea Flow in Research and Development," in *Proceedings of the National Electronics Conference,* Vol. XIV, October 1958.

Rubenstein, A. H. and R. C. Hannenberg, "Idea Flow and Project Selection in Several Industrial Research and Development Laboratories," presented at the Conference on Economic and Social Factors in Technological Research and Development, Ohio State University, October 15–17, 1962.

Rubenstein, A. H., J. Siegman, and N. R. Baker, "Control Mechanisms and the Generation and Communication of Ideas," presented at TIMS 1965 American Meeting, San Francisco, February 3–5, 1965.

Smith, J. J., J. E. McKeon, K. L. Hoy, R. L. Boysen, and L. Shechter, "Lessons from 10 Case Studies in Innovation—II," *Research Management*, **27**(6), 12–17, November/December (1984).

Udell, G. G. and D. I. Hawkins, "Corporate Policies and Procedures for Evaluating Unsolicited Product Ideas," *Research Management*, **21**(6), 24–28, November (1978).

Utterback, J. M., *Accuracy of Perception and Enculturation of Researchers in an Industrial Laboratory*, M.S. thesis, Northwestern University, 1965.

Utterback, J. M., "The Process of Innovation: A Study of the Origination and Development of Ideas for New Scientific Instruments," *IEEE Transactions on Engineering Management*, **EM-18**(4), 124–131, November (1971).

Von Hippel, E., "The Dominant Role of Users in the Scientific Instrument Innovation Process," *Research Policy*, **5**, 212–239 (1976).

B.6 MAKE OR BUY OF TECHNOLOGY

Andrews, Geoffrey, N., "Diversification: Government Contractors Can Learn From 'Commercial' Companies," *Research Management*, **26**(6), 26–31, November/December (1983).

Anusklewicz, Todd, *Federal Technology Transfer*. Prepared for National Science Foundation Office of Intergovernmental Science and Research Utilization. George Washington University, Washington, DC, August 1973.

Berge, S. A., "The Make or Buy Decision," *R&D Management*, **8**(1), 39–42 (1977).

Buckhout, W. E., "Chemical Firms Mull Make–Buy Decisions," *Purchasing Management*, 11–16, May 3 (1976).

Butler, R. and M. G. Carney, "Managing Markets: Implications for the Make–Buy Decision," *Journal of Management Studies*, **20**(2), 213–231 (1983).

Center for the Study of Industrial Innovation, *Aspects of Spin-off, A Study of the Concorde and the Advanced Passenger Train on Their Supplier Firms*, London, England, October 1971.

Chakrabarti, A. K., *The Effects of Techno-Economic and Organizational Factors on the Adoption of NASA-Innovations by Commercial Firms in the U.S.*, Ph.D. dissertation, Department of Industrial Engineering and Management Sciences, Northwestern University, June 1972.

Coddington, Dean G., Paul I. Bortz, and James G. Freeman, *PATT: Project for the Analysis of Technology Transfer*, Denver Research Institute, Industrial Economics Division, University of Denver, March 1970.

Culliton, J. W., *Make or Buy: A Consideration of the Problems Fundamental to a Decision Whether to Manufacture or Buy Materials, Accessory Equipment, Fabricating Parts, and Supplies*, Harvard University Graduate School of Business, Boston, 1942.

Dale, B. G. and M. T. Cunningham, "The Importance of Factors Other Than Cost Considerations in Make or Buy Decision," *International Journal of Operations and Production Management*, **4**(3), 43–54 (1984).

Denver Research Institute, *Space Benefits: The Secondary Application of Aerospace Technology in Other Sectors of the Economy,* Program for Transfer Research and Impact Studies, Industrial Economics Division, Denver Research Institute, University of Denver, January 1978.

Federal Council for Science and Technology, *Directory of Federal Technology Transfer,* Committee on Domestic Technology Transfer, Executive Office of the President, Washington, DC, June 1975.

Head, G., "Deciding About Outside Services," *National Underwriter,* **89**, 21–23, July (1985).

Hersey, Irwin, "The Spin-Off from Space," *Engineering Opportunities,* 22–35, February (1968).

Jaucn, L. R. and H. K. Wilson, "A Strategic Perspective for Make or Buy," *Long Range Planning,* **12**, 56–61, December (1979).

Killing, J. P., "Manufacturing Under License," *Business Quarterly,* **42**(4), 22–29, Winter (1977).

Link, A., G. Tassey, and R. W. Zmud, "The Induce Versus Purchase Decision: An Empirical Analysis of Industrial R&D," *Decision Sciences,* **14**, 46–61 (1983).

MacKenzie, I. W. and R. P. Rhystoner, *Commercialization of University Research,* Imperial College, London, 1983.

Martin, J. and C. McClure, "Buying Software Off the Rack," *Harvard Business Review* **61**(6), 32–60, November/December (1983).

McColly, J., *An Investigation of the Factors Affecting the Perceived Impact of Marketplace Events Upon Decision Makers,* Ph.D. dissertation, Northwestern University, 1967.

McQueen, D. H. and J. T. Wallmark, "Spin-Off Companies from Chalmers University of Technology," *Technovation,* **1** (1982).

Monteverde, K. and D. J. Teece, "Supplier Switching Costs and Vertical Integration in the Automobile Industry," *Bell Journal of Economics,* **12**, 206–213 (1982).

National Science Foundation, *Federal Technology Transfer–An Analysis of Current Program Characteristics and Practices.* A report prepared for the Committee on Domestic Technology Transfer, Federal Council for Science and Technology, Office of National R&D Assessment, Washington, DC, December 1975.

Newport, J., "Waiting for Lightning: Will Professor's Research Pay Off for University Patents?", *Fortune,* **111**(8) 105–106, April 15 (1985).

Olson, Walter T., *Making Aerospace Technology Work for the Automotive Industry– Introduction.* Technical paper presented at the 1978 Congress and Exposition of the Society of Automotive Engineers, Detroit, Michigan, February 27-March 3, 1978. NASA Technical Memorandum TM-73870.

Roberts, E. and D. Peters, "Commercial Innovations from University Faculty," *Research Policy,* **10**, 108–126 (1981).

Roe, P. A., "Modelling a Make or Buy Decision at ICI," *Long Range Planning,* **5**(4), 21–26, December (1972).

Rothwell, R., "The Commercialization of University Research," *Physics in Technology,* **13**(6), 249–257, November (1982).

Rubenstein, A. H. and John E. Ettlie, "Innovation Among Suppliers to Automobile Industry: An Exploratory Study of Barriers and Facilitators," *R&D Management,* **9**(2), 65–76, February (1979).

Sen, F., *The Role of R&D in the Acquisition and Implementation of External Technology*, unpublished Ph.D. dissertation, Northwestern University, 1984.

Senkus, M., "Acquiring and Selling Technology—Licensing Sources and Resources," *Research Management*, **22**(3), 22–25, May (1979).

Sharkey, P., "Develop vs. Buy Decisions on DP Software," *Journal of Information Management*, **4**,1–8, Winter (1983).

Stanford Research Institute, *Technology Assessment of Telecommunications/Transportation Interactions, Vol. III, Contributions of Telecommunications to Improved Transportation System Efficiency*, prepared for National Science Foundation, Menlo Park, CA, NTIS PB-262 696 (1977).

Terpstra, Vern, "International Product Policy: The Role of Foreign R&D," *Columbia Journal of World Business*, **12**(4), 24–32, Winter (1977).

B.7 RELATIONS OF INDUSTRY WITH UNIVERSITIES

Abelson, P., "Differing Values in Academia and Industry," *Science*, **217**, 4656, September 17 (1982).

A. B. T. Associates, *Factors Affecting University Spin-Off Firm Establishment*, A.B.T. Associates, Cambridge, MA, 1984.

Azaroff, L., "Industry–University Collaboration. How to Make it Work," *Research Management*, **25**(3), 31–34, May (1982).

Barker, Robert, "Bringing Science into Industry from Universities," *Research Management*, **28**(6), 22–24, November/December (1985).

Battenburg, J., "Forging Links Between Industry and The Academic World," *Journal of the Society of Research Administrators*, **12**, 2, Winter (1981).

Bloch, E., "Some Comments Concerning Industry–University Relationships in the 80's," *Journal of the Society of Research Administrators*, **16**(1), 5–12 Summer (1984).

Bruce, J. and K. Tamaribuchi, "MIT's Industrial Liaison Program," *Journal of the Society of Research Administrators*, **12** 15–16, Winter (1981).

Business Week, "Business and Universities: A New Partnership", **54**, 58–61, December 20 (1982).

Clauser, H., "New University Research Centers Linked to Industry," *Research Management*, **23**(1), p.2, January (1981).

Colton, Robert, "Status Report on the NSF University–Industry Cooperative Research Centers," *Research Management*, **28**(6), 25–31, November/December (1985).

Culliton, B., "Academic and Industry Debate Partnership," *Science*, **219**, 150–151, January 14 (1983).

Cyert, R., "Establishing University–Industry Joint Ventures," *Research Management*, **28**(1), 27–29, January/February (1985).

Dietrich, J. and R. Sen, "Making Industry–University Government Collaboration Work," *Research Management*, **24**(5), 23–25, September (1981).

Fowler, D., "University–Industry Research Relationships," *Research Management*, **27**(1), 35–41, January/February (1984).

Giamatti, B., "The University, Industry and Cooperative Research," *Science*, **218**, 1278–1280, December 24 (1982).

Hise, R., C. Futrell, and D. Snyder, "University Research Centers as a New Product Development Resource," *Research Management,* **23**(3), 25–28, May (1980).

Johnson, E. and L. Tornatsky, "Academia and Industrial Innovation," in G. Gold (Ed.), *New Directions for Experimental Learning: Business and Higher Education–Toward New Alliances,* Jossey Bass, San Francisco, 1981, pp.47–63.

Kiefer, D., "Forging New and Stronger Links Between Universities and Industrial Scientists," *Chemical and Engineering News,* **58**(49) 38–51, December 8 (1980).

Lepkowski, W., "Academic Values Tested by MIT's New Center," *Chemical and Engineering News,* **60**(11), 7–12, March 15 (1982).

National Science Foundation, *Cooperative Science. A National Study of University and Industry Researchers,* Vol. I, November 1984.

Norman, C., "The Growing Corporate Role in University Budgets," *Science,* **220**, 939, February 25 (1983).

Norman, C., "Chip Makers Turn to Academe with Offer of Research Support," *Science,* **216**, 601, May (1982).

Peters, L. and H. Fusfeld, *University–Industry Research Relationships,* Superintendent of Documents, U.S. GPO, Washington, DC, 1983.

Prager, D. and G. Omehn, "Research, Innovation, and University–Industry Linkages," *Science,* **207**, 379–384, January 23 (1980).

Rodman, J., "A Model for Inter-Institutional R&D Administration and Industry/University Relations," *Journal of the Society of Research Administrators,* **13**(3), 39–45, Winter (1982).

Rubenstein, A. H., "Some Observations on the Effectiveness of Federal Civilian-Oriented R&D Programs (FC/R&D)," *Policy Studies Journal,* **5**(2), 217–227, Winter (1976). Also in *Priorities and Efficiency in Federal Research and Development,* Congress of the U.S., Washington, DC, pp.46–64, Document No. 76/1 (1976).

Shepard, Herbert A., "The Value System of a University Research Group," *American Sociological Review,* **19**(4), 456–462, August (1954).

Sirbu, M., R. Treitel, W. Yorsz, and B. Roberts, *The Formation of a Technology Oriented Complex: Lessons from North American and European Experiences,* CPA 76–8, M.I.T., Cambridge, MA, December 1976.

Smith, D., "Contracts on the Campus," *Physics Bulletin,* **28**(12), 559–561, December (1977).

Sparks, Jack, "The Creative Connection: University–Industry Relations," *Research Management,* **28**(6), 19–21, November/December (1985).

Tamaribuchi, K., "Effectively Linking Industry with a University Resource: A Survey of University–Industry Liaison Programs," paper presented at the WPI/NSF Conference on Management of Technological Innovation, May 1983.

U.S. Congress, House, Subcommittee on Science, Research, and Technology of the House Committee on Science and Technology, *Government and Innovation: University–Industry Relations,* Hearings, July 31–August 2, 1979, Government Printing Office, Washington, DC, pp.53–868, 1979.

U.S. Department of Commerce, *A New Climate for Joint Research,* Conference Proceedings, May 13, 1983.

U.S. General Accounting Office, *The Federal Role in Fostering University–Industry Cooperation,* GAO-PAD 83–22, May 25, 1983.

Wade, N., "University and Drug Firm Battle over Billion-Dollar Gene," *Science*, **209**, 1492–1494, September 26 (1980).

Walsh, J., "New R&D Centers Will Test University Ties," *Science*, 227, 150–152, January 11 (1985).

Whiteley, R. and H. Postma, "How National Laboratories Can Supplement Industry's In-House R&D Facilities," *Research Management*, **25**, 31–42, November (1982).

B.8 EVALUATION OF PROJECTS AND PROGRAMS

Allen, D. H., "Project Evaluation," *ChemTech*, **9**(7), 412–417, July (1979).

Andrews, F. M. (Ed.), *Scientific Productivity: The Effectiveness of Research Groups in Six Countries*, Cambridge University Press, New York, 1979.

Augood, D. R., "A Review of R&D Evaluation Methods," *IEEE Transactions on Engineering Management*, **EM-20**(4), 114-120 (1973).

Ayers, Robert U., Jordan Lewis, and Stephen D. Collier, *An International Study of Economic Benefits Attributable to R&D, By Source and Sector of Performance*, (ETIP) 804229, Delta Research Corporation, Arlington, VA, June 1978.

Baker, N. R. and W. H. Pound, "R&D Project Selection: Where We Stand," *IEEE Transactions on Engineering Management*, **EM-11**(4), 124–134, December (1964).

Battelle–Columbus Laboratories, *Indicators of the Output of New Technological Products from Industry*, Research Report to the National Science Foundation, November 1974.

Bedell, Robert J., "Terminating R&D Projects Prematurely," *Research Management*, **26**(4), 32–35, July/August (1983).

Burgess, J. A., "Auditing an Engineering Organization," *Mechanical Engineering*, **101**(9), 22–25, September (1979).

Chambers, A. D., "The Internal Audit of Research and Development," *R&D Management*, **8**(2), 95–99, February (1978).

Collier, D. W., "Measuring the Performance of R&D Departments," *Research Management*, **20**(2), 30–34, March (1977).

Cooley, Stephen C., Jan Hehmeyer, and Patrick J. Sweeney, "Modeling R&D Resource Allocation," *Research Management*, **29**(1), 40–45, January/February (1986).

Cooper, Robert G., "An Empirically Derived New Product Project Selection Model," *IEEE Transactions on Engineering Management*, **EM-28**(3), 54–61, August (1981).

Czajkowski, Anthony F. and Suzanne Jones, "Selecting Interrelated R&D Projects in Space Technology Planning," *IEEE Transactions on Engineering Management*, **EM-33**(1), 17–24, February (1986).

Davy, M. F., "Economic Evaluation of Research Projects–A Procedure," *AACE Transactions*, L.2.1–L.2.4 (1983).

Fox, G. Edward and Norman R. Baker, "Project Selection Decision Making Linked to a Dynamic Environment," *Management Science*, **31**(10), 1272–1285, October (1985).

Garfield, E., "Citations Indexes for Science," *Science*, **122**, 108–111 (1955).

Geisler, E., "Evaluation of Research and Development: A Critical Review," Working Paper 85–1, Northeastern Illinois University, 1985.

Geisler, E. and A. H. Rubenstein, "Methodology Issues in Conducting Evaluation Studies of R&D/Innovation," paper presented at symposium on Management of Technological Innovation, sponsored by NSF and Worcester Polytechnic Institute, Washington, DC, May 1983.

Geisler, E., A. H. Rubenstein, and C. W. N. Thompson, "A Method of Assessing the Technical and Innovative Capabilities of Small Businesses," *Proceedings of the Seventh Annual Department of Defense Procurement/Acquisition Research Symposium*, Hershey, PA, May 31–June 2, 1978, pp. 196-200.

Glaser, M. A., "The Innovation Index," *ChemTech*, **6**(3), 182–185, March (1976).

Glass, E. M., "Methods of Evaluating R&D Organizations," *IEEE Transactions on Engineering Management*, **EM-19**(1), 2–11, February (1972).

Gold, B., G. Rosseger, and M. G. Beylan, *Evaluating Technological Innovations: Methods, Expectations and Findings*, D. C. Heath, Lexington, MA, 1980.

Gold, B., *Explorations in Managerial Economics: Productivity, Costs, Technology and Growth*, London, McMillan, 1971.

Gold, B., "Improving Management's *ex ante* and *ex post* Evaluations of Technological Innovations," paper presented at the symposium on Management of Technological Innovation, sponsored by NSF and WPI, Washington, DC, May 1983.

Goldstein, Paula M. and Howard M. Singer, "A Note on Economic Models for R&D Project Selection in the Presence of Project Interactions," *Management Science*, **32**(10), 1356-1360, October (1986).

Grief, S., "R&D and Patents: An Attempt to Establish a Relationship Between Input and Output on the Basis of German Statistics," Paper No. 1832 submitted to the OECD Science and Technology Indicators Conference, Paris, September 1980, pp. 22–23.

Griliches, Z., "Issues in Assessing the Contribution of Research and Development to Productivity Growth," *The Bell Journal of Economics*, **10**, 96–99, Spring (1979).

Illinois Institute of Technology, *Technology in Retrospect and Critical Events in Science*, (TRACES), Report to the National Science Foundation, 1968.

Jin, Xiao-Yin, Alan L. Porter, Frederick A. Rossini, and E. D. Anderson, "R&D Project Selection and Evaluation: A Microcomputer-Based Approach," *R&D Management (U.K.)*, **17**(4), 227–288, October (1987).

Krawiec, Frank, "Evaluating and Selecting Research Projects by Scoring," *Research Management*, **27**(2), 21–25, March/April (1984).

Lambert, Lee R., "Project Management Evaluation Techniques for Research and Development: A High-Tech Version of the Old Shell Game," *Project Management Journal*, **16**(3), 47–51, August (1985).

Lee, Jinjoo, Sangjin Lee, and Zong-Tae Bae, "R&D Project Selection: Behavior and Practice in a Newly Industrializing Country," *IEEE Transactions on Engineering Management*, **EM-33**(3), 141–147, August (1986).

Lee, Thomas H., John C. Fisher, and Timothy S. Yau, "Is Your R&D on Track?", *Harvard Business Review*, **64**(1), 34–44, January/February (1986).

Levinson, Nanette S., "The Evaluation Cycle: In Research Evaluation Approaches for the Eighties," *IEEE Transactions on Engineering Management*, **EM-30**(3), 119–122, August (1983).

Liberatore, Matthew J., "An Extension of the Analytic Hierarchy Process for Industrial R&D Project Selection and Resource Allocation," *IEEE Transactions on Engineering Management*, **EM-34**(1), 12–18, February (1987).

Maher, P. Michael, *Some Factors Affecting the Adoption of a Management Innovation: An Experiment with the Use of a Computer Based Project Selection Technique in a Research and Development Organization*, Ph.D. dissertation, Northwestern University, 1970.

Mansfield, E. "Rates of Return from Industrial R&D," *American Economic Review*, May (1965).

Mansfield, E., "Industrial R&D: Characteristics, Costs and Diffusion of Results," *American Economic Review*, **59**, 65–71, May (1969).

Mansfield, E., A. Romeo, and S. Wagner, "Foreign Trade and U.S. Research and Development," *The Review of Economic and Statistics*, **61**, 49–57, February (1979).

Martin, B. R. and J. Irvine, "Assessing Basic Research: The Case of the Isaac Newton Telescope," *Social Studies of Science*, **132**, 61–90, February (1983).

Mechlin, G. and D. Berg, "Evaluating Research–ROI Is Not Enough," *Harvard Business Review*, **58**, 93–99, September/October (1980).

Merrifield, D. Bruce, "Selecting Projects for Commercial Success," *Research Management*, **24**(6), 13–18, November (1981).

Mosteller, F., "Innovation and Evaluation," *Science*, **211**(4485), 881–886, February 27 (1981).

National Science Board, *Science Indicators*, 1980, 1982, 1984, Washington, DC.

National Science Foundation, *National Patterns of Science and Technology Resources*, NSF 84–311, Washington, DC, February 1984.

Narin, F., *Evaluative Bibliometrics: The Use of Publication and Citation Analysis in the Evaluations of Scientific Activity*, Computer Horizons Inc., Cherry Hill, NJ, 1976.

Narin, F., M. Carpenter and P. Woolf, "Technological Performance Assessments Based on Patents and Patent Citations," *IEEE Transactions on Engineering Management*, **EM-31**(4), 172-183, November (1984).

Pavitt, K. and L. Soete, "Innovative Activities and Export Shares: Some Comparisons Between Industries and Countries," in K. Pavitt (Ed.), *Technical Innovation and British Economic Performance*, McMillan, London, 1980.

Pearson, A. W., "Planning and Monitoring in Research and Development—A 12-Year Review of Papers in R&D Management," *R&D Management (U.K.)*, **13**(2), 107–116, April (1983).

Pound, William H., "Research Project Selection: Testing a Model in the Field," *IEEE Transactions on Engineering Management*, **EM-11**(1), 16–22, March (1964).

Rubenstein, A. H., "Setting Criteria for R&D," *Harvard Business Review*, **35**, 95–104, January/February (1957).

Rubenstein, A. H. and E. Geisler, "Potential Indicators and Measures of Downstream Outputs of the R&D Innovation Process in Several Sectors of Science and Technology," paper presented to the AAAS meeting, Houston, Texas, August 1978.

Rubenstein, A. H. and E. Geisler, "A Pilot Study on R&D Output Indicators for Selected Programs of the National Bureau of Standards," IASTA Inc., Chicago, November 1979.

Rubenstein, A. H., "R&D Flow Charts Provide Innovation Output Measures," *Industrial Research/Development*, **21**(4), 49–52, April (1979).

Rubenstein, A. H. and E. Geisler, "The Use of Indicators and Measures of the R&D Process in Evaluating Science and Technology Programs," in J. Roessner (Ed.), *Government Policies for Industrial Innovation: Design, Implementation, Evaluation*, St. Martin's Press, New York, 1988, pp.185–203.

Salasin, J. H., *The Evaluation of Federal Research Programs*, MITRE Technical Report, 80W129, June 1980.

Scherer, F. M., "Corporate Inventive Outputs: Profits and Growth," *Journal of Political Economy*, **73**, 290–297, June (1975).

Schmied, H., "A Study of Economic Utility Resulting from CERN Contracts," *IEEE Transactions on Engineering Management*, **EM-24**(4), 125–138, November (1977).

Sherwin, C. W. and R. S. Isenson, *First Interim Report on Project Hindsight*, Office of Director of Defense Research and Engineering, Washington, DC, October 1966.

Sherwin, C. W. and R. S. Isenson, "Project Hindsight," *Science*, **156**, 1571–1577, June 23, 1967.

Silverman, Barry G., "Project Appraisal Methodology: A Multidimensional R&D Benefit/-Cost Assessment Tool," *Management Science*, **27**(7), 802–821, July (1981).

Skolnick, A., "A Structure and Scoring Method for Judging Alternatives," *IEEE Transactions on Engineering Management*, **EM-16**(2), 72–83, May (1969).

Souder, William E. and Tomislav Mandakovic, "R&D Project Selection Models," *Research Management*, **29**(4), 36–42, July/August (1986).

Souder, W., *Project Selection and Economic Appraisal*, Van Nostrand Reinhol, New York, 1983.

Staats, Elmer B., "The General Accounting Office: Appraising Science and Technology Programs in the United States," *Interdisciplinary Science Reviews*, **3**(1), 7–19 (1978).

Stahl, M. and M. Koser, "Weighted Productivity in R&D: Some Associated Individual and Organizational Variables," *IEEE Transactions on Engineering Management*, **EM-25**(1), 20–24, February (1978).

Tansik, David Anthony, *Influences of Organizational Goal Structures on the Selection and Implementation of Management Science Projects*, Ph.D. dissertation, Northwestern University, 1970.

Terleckyj, N. E., *Effects of R&D on the Productivity Growth of Industries: An Exploratory Study*, National Planning Association, Washington, DC, 1974.

Thompson, C. W. N. and G. J. Rath, "The Administrative Experiment: A Special Case of Field Testing or Evaluation," *Human Factors*, **16**, 238–258 (1976).

B.9 R&D/PRODUCTION INTERFACE

Allen, T. G., *Managing the Flow of Technology: Technology Transfer and the Dissemination of Technological Information Within the R&D Organization*, M.I.T. Press, Cambridge, MA, 1977.

Baloff, Nicholas, "Startup Management," *IEEE Transactions on Engineering Management*, **EM-17**(4), 132–141, November (1970).

Barth, Richard T., *The Relationship of Intergroup Organizational Climate with Communication and Joint Decision Making Between Task-Interdependent and R&D Groups*, Ph.D. dissertation, Northwestern University, 1970.

Berty, J. M., "The Changing Role of the Pilot Plant," *Chemical Engineering Progress*, **75**(9), 48–51, September (1979).

Biller, A. D. and E. S. Shanley, "Understanding the Conflicts Between R&D and Other Groups," *Research Management*, **18**(5) 16–21, September (1975).

Davig, William A., *A Study of Inter-Organizational Relationships in the Transfer of Technology in Developing Countries*, Ph.D. dissertation, Northwestern University, 1972.

Dill, D. D. and W. O. Pearson, "Managing the Effectiveness of Project Managers: Implications of a Political Model of Influence," *IEEE Transactions on Engineering Management*, **EM-31**, 138–146, August (1984).

Douds, Charles F., *The Effects of Work-Related Values on Communications Between Research and Development Groups*, Ph.D. dissertation, Northwestern University, 1970.

Eaton, Robert, "R&D is Key to the Future of General Motors Manufacturing," *Research Management*, **29**(2), 15–19, March/April (1986).

Etienne, E. Celse, "Interactions Between Product R&D and Process Technology," *Research Management*, **24**(1), 22–27, January (1981).

Ettlie, J. E., "Technology Transfer from Innovators to Users," *Industrial Engineering*, **5**(6), 16–23, June (1973).

Ettlie, J. E. and A. H. Rubenstein, "Social Learning Theory and the Implementation of Production Innovation," *Decision Sciences*, **22**(4), 648–668, October (1980).

Ford, J. A., "The Suitability of Matrix Management for Development Projects–A Review," *Project Management Quarterly*, March (1982).

Gerwin, D., "Control and Evaluation in the Innovation Process: The Case of Flexible Manufacturing Systems," *IEEE Transactions on Engineering Management*, **EM-28**(3), 62–70, August (1981).

MacLauren, W. R., "New Products Innovation and Introduction: Their Broad Implications," in A. H. Rubenstein (Ed.), *Coordination, Control, and Financing of Industrial Research*, King's Crown Press, New York, 1955, pp. 34–41.

Quinn, James Brian and James A. Mueller, "Transferring Research Results to Operations," *Harvard Business Review*, **41**(1), 49–66 (1963).

Rubenstein, A. H., A. K. Chakrabarti, R. D. O'Keefe, W. E. Souder, and H.C. Young, "Factors Influencing Innovation Success at the Project Level," *Research Management*, **19**(3), 15–20, May (1976).

Rubenstein, A. H. and M. E. Ginn, "Project Management at Significant Interfaces in the R&D/Innovation Process," Chapter 11 of B. Dean (Ed.), *Project Management: Methods and Studies*, North Holland Publishing, Amsterdam, 1985.

Sagal, M. W., "Effective Technology Transfer—From Laboratory to Production," *Mechanical Engineering*, **100**(4), 32–35, April (1978).

Souder, W. E. and A. K. Chakrabarti, "The R&D/Marketing Interface: Results from an Empirical Study of Innovation Projects," *IEEE Transactions on Engineering Management*, **EM-25**(4), 88–93, November (1978).

Souder, W. E., "Promoting an Effective R&D/Marketing Interface," *Research Management*, **23**, 10–15 (1980).

Towill, Denis R., "An Industrial Dynamics Model for Start-Up Management," *IEEE Transactions on Engineering Management*, **EM-20**(2), 44–51, May (1973).

Young, Earl C., *An Analysis of Factors Influencing the Decision to Adopt Changes in Production Technology in Selected Chemical Firms in Mexico and Columbia*, Ph.D. dissertation, Northwestern University, 1971.

Young, H. Clifton, *Marketing and R&D Coupling: The Process of Information Exchange Between Product Managers and Technical Managers Jointly Engaged in Product Development or Redesign*, Ph.D. dissertation, Northwestern University, 1973.

B.10. IMBEDDED TECHNOLOGY CAPABILITY

Allen, T. J., M. Tushman, and D. M. S. Lee, "Technology Transfer as a Function of Position in the Spectrum from Research through Development to Technical Services," *Academy of Management Journal*, **22**(4), 694–708 (1979).

Dalton, G. W. and P. H. Thompson, "Accelerating the Obsolescence of Older Engineers," *Harvard Business Review*, **50**, 57–67, September/October (1971).

Ettlie, John E., *The Impact of New Technologies: Organizational and Individual Learning*, Ph.D. dissertation, Northwestern University, 1972.

Fischer, W., "The Acquisition of Technical Information by R&D Managers for Technical Problem Solving in Nonroutine Contingency Situations," *IEEE Transactions in Engineering Management*, **EM-26**(1), 8–14, February (1979).

Hahn, Walter A. and Susan L. Doscher, chapter in *Industrial Technology Transfer*, Marvin J. Cetron and Harold F. Davidson (Eds.), Noordhoff, Leyden, The Netherlands, 1977, pp. 431–469.

Hetzner, William A., *An Analysis of the Factors Affecting the Implementation of R&D Results by Medium and Small-Scale Industries in the Developing Countries*, Ph.D. dissertation, Northwestern University, 1972.

Hughes Aircraft Co., *R&D Productivity*, 2nd ed., Culver City, CA, 1978.

Layton, G., "Spiraling Personnel Costs a Factor When Determining Whether to Buy or Build Applications Software," *Computerworld*, **SR6-7**, 6–7, January (1985).

National Academy of Sciences, *Research Management and Technical Entrepreneurship: A U.S. Role in Improving Skills in Developing Countries*, Washington, DC, 1978.

National Aeronautics and Space Administration, *Useful Technology from Space Research*, Technology Utilization Program, Washington, DC, 1968.

Roberts, G. W., "Quality Assurance in R&D," *Mechanical Engineering*, **100**(9), 41–45, September (1978).

Rubenstein, A. H., "The Role of Imbedded Technology and the R&D/Innovation Process," Joint Economic Committee, Congress of the United States, *Special Study on Economic Change, Vol. 3, Research and Innovation: Developing a Dynamic Nation*, December 29, 1980.

Rubenstein, A. H., "Designing Organizations for Integrating Technology Exchange Transactions (TETs) in Developing Countries," prepared for Second Caribbean Seminar on Science and Technology Planning, Port of Spain, Trinidad, January 1976, Document No. 75/109.

Rubenstein, A. H., "Technical Information, Technical Assistance, and Technical Transfer– The Need for a Synthesis," *R&D Management*, **6**, 145–150, October (1976).

Saunders, Neal T., *Overview of NASA/OAST Efforts Related to Manufacturing Technology*, presented at the MTAG-76 DOD Tri-service Conference on Manufacturing Technology, Arlington, Texas, November 8–11, 1976. NASA Technical Memorandum TM X-73583.

Schnee, Jerome E., *Government Programs and the Growth of High-Technology Industries*, Graduate School of Business Administraiton, Rutgers University, New Brunswick, NJ, February 1976.

Tesar, Delbert, "Mission-Oriented Research for Light Machinery," *Science*, **201**, 880–887 (1978).

Williamson, M., "Role of the Technical Staff in Product Innovations," *Research/ Development*, **22**(9), 91–95, September (1960).

B.11 TECHNICAL ENTREPRENEURSHIP

Anonymous, "Management Can Deliver Successful New Products If It Has a Blueprint, Entrepreneur, and a Road Map," *Marketing News,* **18** (18), August 20 (1984).

Baty, Gordon, *Entrepreneurship for the Eighties*, Reston Publishing, Reston, VA, 1981.

Benson, George and Joseph Chasen, *The Structure of New Product Organization*, American Management Association, New York, 1976.

Birley, S. and David Norburn, "Small vs. Large Companies: The Entrepreneurial Conondrum," *Proceedings of the Midwest Academy of Management*, 103–114 (1984).

Braden, Patricia, *Technological Entrepreneurship*, Michigan Business Reports, No.62, Ann Arbor, MI, 1977.

Brandt, Steve, *Entrepreneuring*, Addison-Wesley, Reading, MA, 1982.

Brockhaus, Robert H., "Entrepreneurial Research: Are We Playing the Correct Game?", *American Journal of Small Business*, **11**, 43–49, Winter (1987).

Brown, D., "In House Venture Team Broadens Firm's Business," *Business Marketing*, **70,** April 14 (1985).

Burch, John, *Entrepreneurship*, Wiley Series in Management, Wiley, New York, 1986.

Burgelman, R. A., "Corporate Entrepreneurship and Strategic Management: Insights from a Process Study," *Management Science*, **29**(12), 1349–1364, December (1983).

Business Week, "Big Business Tries to Imitate the Entrepreneurial Spirit," **2786**, 84–89, April 13, 1983.

Chicago Tech Connection, "The V-Team: A New Approach to Corporate Growth," 2, 2–3, March/April (1985).

Churchill, N. and V. Lewis, "Entrepreneurship Research: Directions and Methods," in D. L. Section and R. W. Smilor (Eds.), *The Art and Science of Entrepreneurship*, Ballinger, Cambridge, MA, 1986.

Cooper, Arnold and William Dunkelberg, "Entrepreneurial Research: Old Questions, New Answers and Methodological Issues," *American Journal of Small Business*, **11**, 11–23, Winter (1987).

Dean, Burton, "The Project Management Approach in the 'Systematic Management' of Innovative Start-Up Firms," *Journal of Business Venturing*, **1**(2), 149–160 (1986).

Draper, Roger, "Intrepreneuring: Growth Within," *Managing Automation*, **2**, 76–79, June (1987).

Finch, P., "Intrapreneurism: New Hope for New Business," *Business Marketing*, Cover Story, July (1985).

Gordon, R. A., "The Executive and the Owner-Entrepreneur," in *Business Leadership in the Large Corporation*, Brookings Institution, Washington, D.C. (1945).

Hanan, M., *Venture Management*, McGraw-Hill, New York, 1976.

Hornady, John and John Aboud, "Characteristics of Successful Entrepreneurs," *Personnel Psychology*, **24**, 141–153 (1971).

Kent, C., R. Sexton, and K. Vesper (Eds.), *Encyclopaedia of Entrepreneurship*, Prentice-Hall, Englewood Cliffs, NJ, 1982.

Liles, P., *New Business Ventures and the Entrepreneur*, Richard Irwin, Homewood, IL, 1974.

Litvak, I. A. and C. J. Maule, "Some Characteristics of Successful Technical Entrepreneurs in Canada," *IEEE Transactions on Engineering Management*, **EM-20**(3), 62–68 (1973).

Madique, M. A., "Entrepreneurs, Champions, and Technological Innovation," *Sloan Management Review*, **21**(2), 59–76, Winter (1980).

Management Focus, "AT&T Technologies," March 25, 1985.

Pinchot, Gifford, *Intrapreneuring*, Harper & Row, New York, 1985.

Ronstadt, R., J. Hornady, R. Peterson, and K. Vesper (Eds.), *Frontiers of Entrepreneurial Research*, Babson College, Babson Park, MA, 1986.

Rubenstein, A. H., "Problems of Financing and Managing New Research-Based Enterprises in New England. " A study done for the Research Department of the Federal Reserve Bank of Boston, April 1958, Document No. 58/1.

Silver, A. D., *The Entrepreneurial Life*, Wiley, New York, 1983.

Stevenson, H. and David Lumpert, "The Heart of Entrepreneurship," *Harvard Business Review*, **63**, 85–94, March/April (1985).

Venture, "Seven Companies for Seven Products," **8**, 86–88, August (1986).

Watkins, D. S., "Technical Entrepreneurship: A Cross-Atlantic View," *R&D Management*, **3**(2), 65–70 (1973).

Webster, F. A., "A Model for New Venture Initiation," *Academy of Management Review*, **1**(1), 26–37, January (1976).

Welsh, John and Jerry White, *The Entrepreneur's Master Planning Guide*, Prentice-Hall, Englewood Cliffs, NJ, 1983.

B.12 SOFTWARE DEVELOPMENT

Agresti, W. W., "Applying Industrial Engineering to the Software Development Process," in *Productivity: An Urgent Priority, Proceedings of the Twenty-Third IEEE Computer Society International Conference (COMPCON)*, pp. 264–270 (1981).

Basili, V. R. and T. Phillips, "Evaluating and Comparing Software Metrics in the Software Engineering Laboratory," *Performance Evaluation Review*, **10**(1), 95–106 (1981).

Beck, L. and T. Perkins, "A Survey of Software Engineering Practices: Tools, Methods and Results," *IEEE Transactions on Software Engineering*, **SI-9**(5), 541–561, September (1983).

Behrens, C., "Measuring the Productivity of Computer Systems Development Activities with Functions Points," *IEEE Transactions on Software Engineering*, **SI-9**(6), 648–651, November (1983).

Boehm, B. W., "Improving Software Productivity," in *Productivity: An Urgent Priority, Proceedings of the Twenty-Third IEEE Computer Society International Conference (COMPCON)*, pp. 184–193 (1981).

Brooks, F. P. Jr., *The Mythical Man-Month: Essays on Software Engineering*, Addison-Wesley, Reading, MA, 1979.

Brooks, R. E., "Studying Programmer Behavior Experimentally: The Problems of Proper Methodology," *Communications of the ACM,* **23,** 207–213 (1980).

Buckley, F. J. and R. Posten, "Software Quality Assurance," *IEEE Transactions on Software Engineering,* **10**(1), 36–41, January (1984).

Chrysler, E., "Programmer Performance Standards," *Journal of Systems Management,* **29**(2), 18–25 (1978).

Chrysler, E., "Some Basic Determinants of Computer Programming Productivity," *Communications of the ACM,* **21,** 472–483 (1978).

Cooper, J. D. and M. J. Fisher, *Software Quality Management,* Petrocelli, New York, 1979.

Curtis, B., "Substantiating Programmer Variability," *Proceedings of the IEEE,* **69,** 846–847 (1981).

DeMarco, T., *Controlling Software Projects: Management, Measurement and Estimation,* Yourdon Press, New York, 1982.

Dinitto, S. A., "Software Engineering Problems and Progress," *Journal of Electronic Defense,* **9**(8), 41–50, August (1986).

Frank, W. L., *Critical Issues in Software: A Guide to Software Economics, Strategy, and Profitability,* Wiley, New York, 1982.

Fujii, M. S., "A Comparison of Software Assurance Methods," in *Proceedings of the Software Quality and Assurance Workshop,* San Diego, pp. 27–32, November 1978.

Fuss, J., "A Design to Assess the Impact of Software Generator Productivity," unpublished Ph.D. qualifying examination paper, Northwestern University, 1981.

Glass, R. L. and R. A. Noiseux, *Software Maintenance Guidebook,* Prentice-Hall, London, 1981.

Grammas, G. and J. Klein, "Software Productivity as a Strategic Variable," *Interfaces,* **15,** 116–126, May/June (1985).

Gustafson, G. G. and R. J. Kerr, "Some Practical Experiences with a Software Quality Assurance Program," *Communications of the ACM,* **25**(1), 4–12, January (1982).

Howes, N. R. "Managing Software Development Projects for Maximum Productivity," *IEEE Transactions on Software Engineering,* **SE-10,** 27–35, January (1984).

Johnstone, D., "The Motivation to Work—Implications for Project Managers in Charge of Computing Projects," *Australian Computer Journal,* **5**(2), 22–27 (1981).

Jones, T. C., "Measuring Program Quality and Productivity," *IBM Systems Journal,* **17**(1), 39–63 (1978).

Kamijo, Fumihiko, "Software Quality Control—A Project Management View," in *Productivity: An Urgent Priority, Proceedings of the Twenty-Third IEEE Computer Society International Conference (COMPCON),* pp. 54–64, 1981.

King, D., *Current Practices in Software Development: A Guide to Successful Systems,* Yourdon Press, New York, 1984.

Lehman, J. H., "How Software Projects Are Really Managed," *Datamation,* **25**(1), 119–121, 123–124, 129, January (1979).

McClure, C. L., *Managing Software Development and Maintenance,* Van Nostrand Reinhold, New York, 1981.

Metzger, P. W., *Managing a Programming Project,* 2nd ed., Prentice-Hall, Englewood Cliffs, NJ, 1981.

Mizuno, Y., "Software Quality Improvement," in *Proceedings of the IEEE Computer Software and Applications Conference,* pp. 257–262, November 1982.

Miller, R., K. Kovaly, and T. Walter, *The Handbook of Manufacturing Software,* SEAI Technical Publications, Madison, GA, 1985.

Scacchi, W., "Managing Software Engineering Projects: A Social Analysis," *IEEE Transactions on Software Engineering,* **SE-10**, 49–59, January (1984).

Thayer, R. H., A. B. Pyster, and R. C. Wood, "Major Issues in Software Engineering Project Management," *IEEE Transactions on Software Engineering,* **SE-7**, 333–342 (1981).

Voss, C. A., "Determinants of Success in the Development of Applications Software," *Journal of Product Innovation Management,* **2**, 122–129 (1985).

Werner, F., *Critical Issues in Software: A Guide to Software Economics, Strategy, and Profitability,* Wiley, New York, 1983.

Whiting-O'Keefe, P., "Software Engineering Methodology in Japan," in *Proceedings of the COMPSAC 81 IEEE Computer Society's International Computer Software and Applications Conference,* pp. 102–103, 1981.

B.13 R&D IN THE SERVICE INDUSTRIES

Bank Administration Institute, *Banking Issues and Innovations,* Chicago, 1985.

Department of Commerce, "The Service Economy: Opportunity, Threat or Myth?," in *Proceedings of a Workshop on Structural Change,* October 22, 1985.

Eckstein, A. and D. Heiden, "Causes and Consequences of Service Sector Growth: The US Experience," *Growth and Change,* **16**(2), 12–17 (1985).

National Science Foundation, "Technical Employment Growth Accelerates in Selected Nonmanufacturing Industries," in *Science Resources Studies Highlights,* Washington DC, October 17, 1983.

Office of Technology Assessment, *Trade in Services: Exports and Foreign Revenues,* Washington, DC, 1986.

Office of Technology Assessment, *International Competition in Services,* Washington, DC, 1987 (particularly Chap. 4).

Rubenstein, A. H., E. Geisler, and B. Grabowski, "Managing Technology in Firms in the Service Industries," in *Proceedings of the First International Conference on Technology Management,* Miami, 1988.

Rubenstein, A. H., "In House R&D on Information and Telecommunications in the Service Industry Firm," in *Proceedings of the Conference on R&D/Marketing Interface,* USC, February 1988.

Salomon Brothers Inc., *Technology in Banking: A Path to Competitive Advantage,* May 1985.

Schneider, Keith, "Services Hurt by Technology," *The New York Times, Business Day,* Part 1, Section D 25–26, June 29 (1987).

Tatum, C. B., "Examples of Innovation on Engineering and Construction Projects and Implications for the Construction Innovation System," OTA, 1986.

U.S. Department of Commerce, *1982 Census of Service Industries: Miscellaneous Subjects,* Washington, DC, December 1985.

U.S. International Trade Commission, *The Relationship of Exports in Selected U.S. Service Industries to U.S. Merchandise Exports,* 1982.

B.14   STAFF SPECIALIST GROUPS

Baker, Norman R., *Descriptive Model of Several Environmental Factors Affecting Industrial Operations Research-Management Science Activities,* M.S. thesis, Northwestern University, 1963.

Haretog, Curt and Robert A. Rouse, "A Blueprint for the New IS Professional," *Datamation,* **33**(20), 64–69, October (1987).

Heap, John P., "The Role of Management Services in an Innovation Strategy," *Management Services (U.K.),* **31**(5), 12–17, May (1987).

Higgins, J. C. and K. M. Watts, "Some Perspectives on the Use of Management Science Techniques in R&D Management," *R&D Management (U.K.),* **16**(4), 291–296, October (1986).

La Belle, Antoinette and Edward H. Nyce, "Whither the IT Organization?", *Sloan Management Review,* **28**(4), 75–81, Summer (1987).

McGregor, D., "The Role of Staff in Modern Industry," in G. P. Shultz and T. L. Whisler (Eds.), *Management Organization and the Computer,* University of Chicago, Chicago, 1960.

Radnor, M., A. H. Rubenstein, and A. Bean, "Integration and Utilization of Management Science Activities in Organizations," *Operations Research Quarterly* (U.K.), **19**(2), 117–141, (1968).

Rubenstein, A. H., "Integration of Operations Research into the Firm," *Journal of Industrial Engineering,* **15**(5), 421, September–October (1960).

B.15   OTHER READINGS RELATED TO TECHNOLOGY MANAGEMENT

Abernathy, William J. and James M. Utterback, "Patterns of Industrial Innovation," *Technology Review,* **80**(7), June (1978).

Abernathy, W. J., *The Productivity Dilemma: Roadblocks to Innovation in the Automobile Industry,* Johns Hopkins University Press, Baltimore, 1978.

Allen, T. J., D. M. S. Lee, and M. Tushman, "R&D Performance as a Function of Internal Communication, Project Management, and the Nature of Work," *IEEE Transactions on Engineering Management,* **EM-27**, 2–12, February (1980).

Douds, C. F. and A. H. Rubenstein, "Review and Assessment of the Methodology Used to Study the Behavioral Aspects of the Innovation Process," in P. Kelly and J. Kranzberg (Eds.), *Technological Innovation: A Critical Review of Current Knowledge,* San Francisco Press, San Francisco, 1978.

Drucker, Peter, *Innovation and Entrepreneurship,* Harper & Row, New York, 1985.

Geisler, Eliezer, "Artificial Management and the Artificial Manager," *Business Horizons,* **29**(4), 17–21, July (1986).

Gerstenfeld, Art, *Effective Management of Research and Development,* Addison-Wesley, Reading, MA, 1970, pp.41–53.

Keifer, David M., "Winds of Change in Industrial Chemical Research," *Chemical and Engineering News,* **42**, 88–102, March (1964).

Livingston, Robert T. and S. H. Milberg, *Human Relations in Industrial Research Management,* Columbia University Press, New York, 1957.

Pavitt K. and W. Walker, "Government Policies Towards Industrial Innovation: A Review," *Research Policy,* **5**(1), 2–12 January (1976).

Rubenstein, A. H., Charles F. Douds, Horst Geschka, Takeshi Kawase, John P. Miller, Raymond St. Paul, and David Watkins, "Management Perceptions of Government Incentives to Technological Innovation in England, France, West Germany, and Japan," *Research Policy* **6**(4), 324–357, (1977), Document No. 75/95.

Rubinger, Bruce, "Industrial Innovation: Implementing the Policy Agenda," *Sloan Management Review,* **24**, 43–57, Spring (1983).

Twiss, Brian, *Managing Technological Innovation,* 2nd ed., Loggman, London, 1980.

Utterback, J. M. and W. J. Abernathy, "A Dynamic Model of Process and Product Innovation," *Omega,* **3**, 639–656 (1975).

Watkins, David, A. H. Rubenstein, Raymond St. Paul, Barbara Peters-Koehler, and Charles F. Douds, "Innovation Incentive Programmes in Three West European Nations: France, West Germany, and the United Kingdom," in *Technological Innovation, The Experimental R&D Incentives Program,* Denver Research Institute, Boulder, Colorado: Westview Press, Boulder, 1977, pp. 265–289, Document No. 76/9.

Williamson, Merritt A., "High Hopes and Hard Facts in Research Expectations," *Research/ Development,* **13**(4), 59–60, April (1962).

THE MILLER CHEMICAL COMPANY CASE

MILLER CHEMICAL CO. (A)

In January 1954, a memorandum was circulated to 25 people in half a dozen operating divisions of the Miller Chemical Company. It came from Dr. Frank Brown,[*] the Vice President for Research and Development. He invited the recipients to attend a meeting to discuss company-wide interest in B-compounds, which were currently the subject of study by the Central Research Laboratories as well as several operating divisions.

Basic research on B-compounds, until recently a laboratory curiosity, had been going on in the Organic Chemicals section of Central Research for about two years, supported by general research funds as well as a small government contract. Recent work in the universities and in two government laboratories suggested some important potential applications for B-compounds in food, textiles, plastics, and other industries.

At the meeting Dr. Brown and members of the Organic section described their current work and recent results from the university and government laboratories working in the field. Members of the divisional laboratories of the Textiles, Agricultural Products, Plastics, Food Products, and Organic Chemicals Divisions also reported on their interests in B-compounds, and several experimental applications to company products were demonstrated.

[*] All names disguised.
Copyright © 1960 by the President and Fellows of Harvard College.
This case was prepared by Professor Albert H. Rubenstein as a basis for class discussion rather than to illustrate either effective or ineffective handling of an administrative situation. Reprinted by permission of the Harvard Business School.

It was agreed that an informal series of seminars be conducted during the coming months to keep all interested parties acquainted with progress in the field and that a guide to the literature of the field be prepared, covering the history of B-compounds, their properties, and potential applications. This report was ready during the late spring and was made available throughout the company. During the rest of 1954 and the first part of 1955 information exchange continued on a sporadic basis, with occasional pleas for better coordination from various groups working in the field.

In July of 1955 Dr. Brown recommended the establishment of a coordinating group on a high technical level to meet monthly, and to study the future potential of B-compounds both for use in current company products and as a saleable product line in its own right. A committee was formed, including administrative and technical people from Central Research, Textiles, Organic Chemicals, Food Products, Agricultural Products, and Plastics. Mr. Paul Wilson, a member of Dr. Brown's staff, was appointed secretary of the committee. He was charged with keeping in close touch with company-wide activity on B-compounds and with initiating some preliminary market studies to ascertain the external commercial possibilities of either the compounds themselves or materials incorporating them.

The most active of the divisional laboratories in this area was the Plastics Division, which had a six-man group working on applications of B-compounds to divisional products. Several successful applications had been made by the middle of 1955 and the group was beginning to produce a small excess of several basic B-compounds for which they had no immediate use within the division. In August of 1955 Dr. Harold Hoffman, leader of the B-compound group in the Plastics Division laboratory, commented to the manager of the Plastics Laboratory that Miller Chemical was dropping the ball in this field.

He mentioned that a dozen other chemical companies were interested in B-compounds and that three of them were already setting up separate operating units to exploit their commercial possibilities. Until six months ago, he indicated, most of these companies had been interested in B-compounds only as components of their existing products, and they had been buying them in small quantities from two chemical specialty houses. With the increased demand for these materials and the broadening of their fields of application, the major chemical companies had become dissatisfied with the delivery and quality of the materials they were purchasing and had decided to begin making their own. As in the case of Miller's Plastics Division, they found that even with minimum sized production equipment there was an excess of capacity for which they had no immediate internal use. As a result of the widespread publicity B-compounds were receiving in the technical literature, many other companies without the capability or the interest in producing them were interested in obtaining sample quantities for trial applications to their products.

By December of 1955 approximately 20 companies, including the original two specialty houses, were offering small quantities of one or more B-compounds for sale. A memo from Miller's patent department to the Director of the Central Research Laboratory described the situation as essentially wide open, since there were few basic patents in the field and there did not appear to be much chance of any one company monopolizing the basic formulations. There was a high degree of art involved at this stage of the process where very exacting control conditions were essential for achieving adequate purity and yield. As a class, B-compounds had several interesting properties, but they varied greatly with the purity attained. The results of a given production run could be classified according to the degree to which the ideal properties were attained, and then could be assigned a place in a hierarchy of potential uses where the combination of critical properties varied. The yield of various grades not only affected total cost of the product, but also influenced the rate at which the highest grades were made available for further applications research, since the production equipment being used throughout the industry at this time was small in size and there were less than 50 people who had sufficient knowledge of the initial processing to produce usable material. Competition for the few qualified people was sharpening, and Miller had already lost two such people—one from the organic section of central research and the man who had initiated work on B-compounds in the Plastics Division Laboratory (Dr. Hoffman's boss).

During this period since the first coordinating meeting in January 1954, Central Research had continued its basic research on B-compounds and had extended its interest to supplying sample quantities to several product divisions (other than Plastics) which were experimenting with applications. The most active of these other product divisions was Food Products, which by the end of 1955 had a substantial list of applications and whose demand for various B-compounds far exceeded the sample scale production in Central Research. The excess capacity in Plastics Division did not include any of the specific compounds needed by Food Products. As a consequence, they were buying semicommercial batches from several of the score of other companies now producing B-compounds, and were also building up their own production capability for certain specific compounds.

In January of 1956 Dr. Brown sent a memorandum to William Marshall, President of the Plastics Division, requesting further information on plans by the Plastics Division Laboratory to substantially increase its efforts in the development and applications of B-compounds. These plans included a new facility, larger scale production equipment, and the hiring of a number of organic chemists to concentrate on B-compounds. He pointed out that there was no desire on the part of corporate management to limit the interest of any of the operating divisions in this field, but that better coordination and integration of the efforts might help Miller Chemical capitalize on the market potential more effecitvely.

During this period, Paul Wilson, assigned by Dr. Brown to study the whole B-compound situation both within and outside the company, had been preparing a series of reports on the situation as he saw it. The first of these was on the internal company situation. Its major conclusions were:

1. In order to recover the company's investment in B-compound research and development over the past few years, the conflict between Central Research and the operating divisions involved in B-compound work should be resolved.

2. The quickest way of recovering this investment was to promote sales of B-compounds outside the company—i.e., to enter into competition with the score of companies already offering them for sale. In order to facilitate this, one of the operating divisions already active in the field should invest in the people and facilities required and be prepared to launch a substantial commercial effort.

3. To back up this commercial effort, the Organic Chemical section of Central Research (which by now had formally established a "B-compound Research Group") would continue to carry on research and limit its production to pilot runs in support of the research. In addition, this group should help the commercial effort in the operating division to maintain its production quality.

4. In order to support a commercial operation, the company might have to broaden its current spectrum of B-compounds to include several which had the most promise for volume production. At present the major effort was devoted to low-volume, high quality compounds for use in current company products.

5. The B-compound activity in Food Products Laboratory was not now capable of turning out high quality compounds due to lack of technical background and proper equipment. The group there, however, was determined to continue in this field and to begin marketing on a commercial scale. The group in Central Research wanted to keep its facilities and personnel, but had reservations about becoming involved in external marketing operations.

6. Whichever way the situation was resolved, a pricing arrangement should be worked out which would handle both interdivisional sales and sales outside the company. This pricing arrangement should help to recover the R and D costs which had by this time amounted to approximately one quarter of a million dollars, exclusive of contract work paid for directly by the operating divisions who wanted specific compounds developed.

In June of 1956 an executive order came from the office of the President, John T. Skelly, defining company-wide responsibility for B-compounds:

> The Central Research Laboratory will have primary responsibility for basic research and development of B-compounds for use within and outside of the company. Food Products, Plastics, and the several other interested Operating Divisions will continue to develop and produce certain specialized B-compounds for application to their own products. The Central Research Laboratory will continue to assist all operations of the company, at their request, in matters relating to B-compounds.

A few days later a member of the Central Research organic group was appointed as official liaison with the B-compound activity in the Food Products Division. Several weeks later a report was issued on interdivisional pricing of B-compounds.

In a second report during the summer of 1956 Paul Wilson gave an optimistic estimate of the commercial possibilities for B-compounds and made the following recommendations:

1. Miller Chemical should not attempt to compete in the high volume market for B-compounds. This was a price business where Miller's competence in producing high quality compounds would not be adequately utilized.

2. Research on high-quality compounds should be intensified and exploited commercially as soon as possible. There was no long-term advantage in withholding special compounds from the market in the hope of maintaining an exclusive position. Commercial exploitation of these high quality products should be carried out through the Food Products Division, which was already active commercially. In order to be able to compete effectively, Miller's position in the basic compounding should be further strengthened, with accelerated work on basic materials for B-compounds.

3. The development and production resources throughout the company should be mobilized for this effort, under the direction of one competent individual with a working knowledge of the basic materials, processing, and applications. He should be assigned a group and a budget and allowed to operate as independently as possible, so as to develop fully the commercial possibilities. The group should not become a captive development group in an operating division.

He ended by summarizing the history of B-compound activity in the Central Research Laboratory and in the several operating divisions. Several weeks later the executive order of June 1956 was amended "in recognition of the critical importance of B-compounds to the company" to give the Food Products Division primary responsibility for exploiting the commercial possibilities of B-compounds. This included the marketing and coordination of all standard B-compounds produced for use both within and outside the company. It included responsibility for initial sales contacts, price quotations and delivery dates, shipping and billing. Transfer prices were established between Central Research and Food Products for high-quality and specialty materials that were produced in small batches in Central Research and marketed by Food Products.

Just before the end of 1956 several informal conferences were held on the future commercial potential of B-compounds. At one of these informal meetings Paul Wilson asked Dr. Hoffman if he would be interested in taking on the job described in his second report. Hoffman replied, "If you are looking for someone to run with the ball, I'd like to be considered. " After consultation with Dr. Brown, the job was offered to Hoffman. When he was first approached, Hoffman was cautious. He knew that Miller Chemical had experienced several previous failures in setting up new groups around a new product or new technology. Some of these projects had suffered from lack of interest on the part of the operating divisions who were the only ones with the production and marketing ability to make them successful. Others had suffered from *too much* interest on the part of several divisions and had been torn apart by intracompany competition.

He was asked to state his conditions for undertaking the venture. Initially the intention was to place him under a department head in the Central Research Laboratory for protection against being used as a service group in one of the operating divisions. He agreed that Central Research was a good location for his group, but he balked at being assigned to one of the section heads in the laboratory.

His idea was to focus all of the relevant talent in the company on the building of a product line which would eventually constitute an operating division of its own. He wanted the initial protection that Central Research could offer, but wanted to report to someone who was interested in his activity as a business venture and not merely as another research project. In addition, if he undertook the job, Hoffman wanted the opportunity to be identified with a business venture in the company and to benefit personally from any commercial and financial success that resulted from his efforts. He had enjoyed his work in the Plastics Division Laboratory, but now felt that it was too confining and wanted a broader scope for his efforts. He wanted to try his hand at managing an activity that had economic as well as technical implications.

During the last few weeks of 1956, Dr. Hoffman was appointed to head up the newly-formed "Special B-compounds Group" (SBC Group), within the Central Research Laboratory. He reported, temporarily, to the Vice President of Research, Dr. Brown. In March 1957 Dr. Brown issued a memorandum to about 50 interested people within the corporation ". . . to acquaint you with the activities of the new SBC Group and ask your cooperation in assisting it." He explained that the group had been formed:

> . . . primarily to take advantage of Miller Chemical Company's skills, experience, and research in B-compounds and their applications by developing materials for sale commercially as well as for use in Miller's own product lines. The market is growing, and as an integrated developer and producer of B-compounds and their derivative materials, Miller Chemical would have many inherent advantages in obtaining a substantial share of the market.

> At present the group is under the general supervision of Central Research. Dr. Hoffman, the group manager, intends to use the experience and knowledge of existing marketing personnel in the company and to take advantage of any applicable business arrangements to sell his products. These specific arrangements are to be worked out. However, the basic marketing policies (including pricing, product policy, and advertising) are to be the responsibility of his group for the foreseeable future. The manufacturing facilities of other divisions and outside vendors will be used for production.

> It would be helpful if you would refer any information or inquiries on B-compounds to this new group. It is important to note that the SBC Group is set up to develop a product line for commercial sale as well as to continue to help serve Miller's internal needs. It is the intention that this new product activity will be reviewed by the New Products Committee to determine its progress and to resolve its future problems relative to sales, manufacturing, product lines, and organizational structure.

In another memo, issued the same week, it was announced that sales responsibility for B-compounds by the Food Products Division would cease immediately. The background for this major change in the method of commercially exploiting B-compounds had been presented to various interested people in the company during the preceding several months, following the appointment of Dr. Hoffman to head the new group. Two of the major arguments in support of the move were 1) that the previous scattered effort had suffered from the lack of strong central direction or a duly authorized person or department to exploit aggressively the potential of this new product area; and 2) that the ultimate profitability was uncertain if it was confined to selling basic B-compounds alone rather than an integrated product line including derivatives and materials incorporating B-compounds. Without this integrated line, it would be hard to justify an effective marketing effort.

In support of the SBC Group, it was also decided to continue basic research on B-compounds in the Organic Chemicals section of Central Research.

In March Dr. Hoffman proposed his annual budget and outlined his product policy. This included the following provisions: 1) only part of the total development effort in B-compounds would be initiated within the group; and 2) some of these projects would be initiated and supported by other divisions within Miller or by outside customers (Government and industrial).

He indicated that the rate of development would be limited by the available manpower and by space for laboratory work and product testing. He expected the group to expand through 1958 due to sales of current products and new developments. Space requirements should increase by 75% by the end of 1958 and manpower requirements should increase by 50%.

His goal was to be self-supporting on a current basis before the end of 1957, with break-even in 1958. This would require a combination of outside support for development projects and product sales of $50,000 per month. For the full year's operations, he anticipated that total costs, including amortization of a proportion of the earlier R and D costs, would be close to $200,000.

In August of 1957 the SBC Group was ready to do some active marketing and the problem of where to get people and how to administer the marketing effort arose. It was agreed that sales should, at least in the early stages, be handled by applications engineers (primarily B.S. Chemical Engineers with previous marketing experience). They should work closely with the development engineers in Dr. Hoffman's group. At the present level of operations, the SBC Group could not afford to support an applications group, so that an arrangement would have to be worked out whereby they could use the services of an existing group on a part time basis.

A temporary arrangement was arrived at with the Marketing Department of the Food Products Division to do a test marketing for six months. Ten per cent of the time of three men from three different sales offices of the Food Products Division would be paid for by the SBC Group, on a direct cost basis, excluding overhead. During this "loan" arrangement, SBC Group would maintain full responsibility and authority for product policy, advertising, promotion, pricing, design, development, and production.

A few weeks later this arrangement was found to be inadequate in both amount and direction. It was suggested that the time allocation by the three applications people be doubled and that thought be given to general nontechnical sales contact on a regional basis, in addition to the specific contacts by the applications people.

A summary of 1957 operations of the SBC Group showed total expenses of $200,000, revenue of $125,000, and a net loss of $75,000. During the last two months of the year, however, the group made a small operating profit after taxes.

In June of 1958 the SBC Group reported on their progress to the New Products Committee and compared their accomplishment with the plans that had been submitted in November 1957. The highlights of the report were:

Sales: The actual sales for the first quarter of 1958 were about 2/3 of those forecast. Sales in the first half were to be almost exactly equal to the forecast. The build-up in orders had been faster than expected, with shipments just beginning (June 1958) to catch up with the backlog. Interdivisional sales had been substantially higher than expected and outside sales substantially lower than expected. The amount of inside service and consultation had exceeded expectations and several of the development projects were taking longer than anticipated.

During the previous six months the Group estimated that it had lost approximately 1/2 million in sales both inside and outside the company, due to being high bidder. Some of the possible reasons for this might be:

1. Lack of technical experience, leading to non-competitive prices.
2. Lack of trained production people. This cost of training had to be absorbed in the pricing.
3. Lack of information on competitors' activity.

All of these factors were receiving attention and the first two were being rapidly overcome. The third was being attacked through the development of a small sales force inside the SBC Group. This did not imply failure of the cooperative marketing arrangement with the Food Products Division, but this arrangement had left many problems unsolved. In addition, continued use of these "borrowed" people was not compatible with the plans to make SBC an integrated product line within one of the major product divisions.

Personnel: Due to faster increases in orders than anticipated, the build-up of personnel had been more rapid than anticipated and by the end of 1958 was likely to be about 30% above the forecast.

Profit and Loss: Due to the level of costs, the need to continue high development expenditures, and the move to new facilities, the forecast date of Break-Even was revised from the last quarter of 1958 to the first quarter of 1959.

Space: Despite a recent move and increase in space, the group was still crowded and would reach a point of desperate over crowding before the end of 1958. With the development of the group had come the addition of people in quality control, purchasing, and administrative work to the point where some production space had been taken over to make room for them to sit. Another 5,000 sq. ft. would keep them for the next 1-2 years.

A few weeks after the distribution of the SBC Group progress report, a meeting was called to discuss the feasibility of integrating the Group into the Food Products

Division. The proposal (by the divisional management of Food Products) was that the Group could be set up as a product department in parallel with the four existing departments within the divisional research laboratory. It would be responsible for product development, pilot production, and maybe some larger scale production. Sales of B-compounds would come under the divisional sales manager. Product policy would be established jointly by the manager of the SBC Department (presumably Dr. Hoffman), the head of the Food Division Laboratory, and the divisional sales manager. Financial accountability would be established through making SBC a cost center. In the transition period, the sales function would be continued by the SBC Group people now assigned to sales, but eventually they would be integrated into divisional sales management.

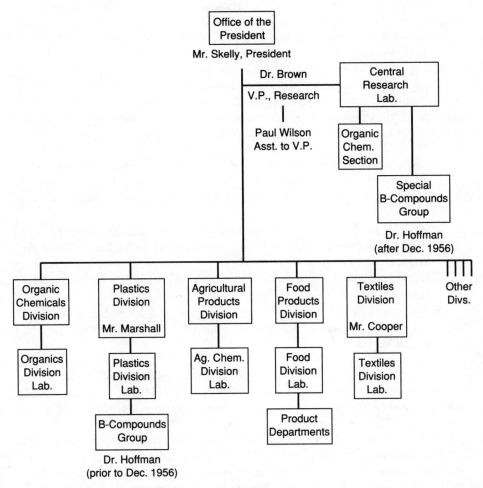

Figure C-1.

Dr. Hoffman reported on this meeting a few days later, indicating that the people attending had exhibited a willingness to be flexible and to find a good, workable solution. He had expressed surprise, however, when he learned that many of the people now in Food Products who would be directly concerned with such an integration move were not currently aware that such a move was under consideration. The meeting had ended with agreement that a lot more discussion was necessary with some of these people before such a plan could be conisdered formally.

At about this time, approximate industry-wide sales estimates were available, showing that Miller had already captured more than 6% of the market for B-compounds and that the planned rate of expansion, if it could be achieved, would lead to a market share of almost 8% by 1962, based on an estimated increase in the total market of more than 600% by 1962.

While this involved a tremendous increase in output for SBC Group in this period, Dr. Hoffman felt that he could gain an even greater market share in that same period if his marketing effort could be materially strengthened. In order to do this, he would need a lot of help from one of the existing product divisions, regardless of the ultimate decision on integration.

A few weeks later, in October 1958, Miller's president, Mr. Skelly, received a memorandum from George Cooper, the president of the Textiles Division, in reply to Skelly's request that he consider the feasibility of the SBC Group being integrated into the Textiles Division. Cooper made these points in his memorandum:

The SBC business, as Miller appeared to be playing it, would remain a relatively low-volume, custom-compounding business with a very large development content in the cost of production. It would also be a very competitive business.

At the moment, sales were approximately 70% internal and 30% external, although they were expected to shift the other way in the future. Currently, losses were running about $14,000 per month. In order, to absorb operating costs, development expenses, and to provide at least a 5% operating margin, approximately $100,000 per month of sales would be required.

With the present facilities this was not possible. An additional allocation of space, operating funds, and capital equipment funds would have to be made available.

In summary, Mr. Cooper recommended that "Although there is doubt that the whole project will ultimately be of benefit to the company as a commercial business, it should be given facilities and funds to get up to a volume of $100,000 per month." This would entail their own target date of May 1959, for this volume, with a 5% operating margin. If they failed to attain this position by then, there was a strong indication that the activity should be disbanded as a commercial venture. If they did attain this position, then a major decision about the future should be made. At that time, the Textiles Division would be interested in integrating B-compounds as a product line of the division.

By this time the issue of the future of the SBC Group was taking up a "dispro-portionate" amount of corporate executive time with respect to the amount of sales the Group was generating relative to total company sales. As Mr. Skelly remarked, "On this basis, I should be spending only a few hours per year worrying about this new venture." Finally, on November 3, a high-level meeting was held including Mr. Skelly, Drs. Brown and Hoffman, Mr. Wilson, and several division managers. The group had met to discuss the reporting relationship of the SBC Group in the event that it was transferred to or integrated into one of the operating divisions. An important factor to be considered was Dr. Hoffman's reservations about the realignment of his reporting responsibility to an operating division at this stage in the development of his venture.

These decisions were made at the meeting:

1. The SBC Group would remain within the Central Research Laboratory until at least June 1959.
2. Dr. Hoffman would continue to develop the Group according to his earlier plans.
3. Additional space would be given to him as soon as possible.
4. Mr. Cooper, at Mr. Skelly's request, would meet with Dr. Hoffman from time to time to discuss problems of the SBC Group and to help guide it toward the status of a profitable business.

INDEX